The Second Singularity Illustrated

A Mathematical Exploration of AI-Based and Other
Singularities in a Cosmology of Light Enhanced with Art

Pravir Malik, Ph.D.

Illustrated by Narendra Joshi, Ph.D.

ISBN-13: 979-8-9896169-5-4

Published by *Possibilities Publishing*

The *Dance of Nataraja* is the generator of singularities,
and whether partial or complete, can be wielded by the
same cosmological mathematics of light.

13

Author's Note and Introduction

It has been said that the future of Life is about Artificial Intelligence (AI) and that as AI advances a point will be reached, the Singularity, when humans will no longer have a clue as to what is going on nor why. Essentially as a species, humans will at this point have been marginalized by a global, singular intelligence, whose capabilities far outstrip human intelligence.

On the contrary, this book will make the case that the future of Life is enabled by the pre-existent complexity that exists in every iota of it and that there have already been a series of light-based epiphanies by virtue of which Life in its continuing complexities has emerged as partial-singularities, and further, will culminate in a Second Singularity as more of our species consciously opens to the fullness of Light. The power and capabilities of the Second Singularity will far outstrip any AI-based singularity.

This pre-existent complexity derives from the fact that Light exists in multiple states simultaneously. Science has discovered physical light that travels at a constant speed of 186,000 miles per second, also known as 'c'. But Light, as posited in the Cosmology of Light book series (Malik, 2017b-e; 2018a-c; 2019a), also exists at different speeds that all derive from its native state where it travels at an infinite speed. In that native state where Light exists at infinite speed, it is everywhere at the same time, hence it is omnipresent. Nothing can resist its power in that native state, and hence it is omnipotent. Because anything and everything that arises, arises in it, the Light hence knows everything that appears and disappears, and is omniscient. Since everything is connected in a single embrace in its substance, it is omniloving.

This fourfold omni-dimensionality implies that all possibility and infinite information is already embedded in Light. In other words, the origin of genetics is Light itself. When Light projects itself at different speed, then basically something of that infinite information is parceled out in packets, through the device of quanta, and becomes more material. When Light slows down or projects itself at 'c', the amount of information that begins to materialize causes the phenomenon we have come to know as the Big Bang. The Big Bang, therefore, is fundamentally a quantum-level process in which an information-rich material foundation of our known universe was initiated.

Hence, there is a fundamental and single light-based edifice that informs our universe, and it is by virtue of this light-based edifice that all matter, all life, all mind, and all emergences beyond the level of mind will surface. The future of life, then, cannot be due to AI. At best, AI can create a process by which many tasks that usually require utilization of the fundamental light-derived capacity of intelligence in order to be completed, can be completed by an exhaustively repetitive running of algorithms driven by massive computing power, hence giving the illusion of intelligence. Projected forward, the exhaustively repetitive running of algorithms driven by massive computing power, endow a practically ubiquitous digitally-enabled strata to control many tasks, and this may create the phenomenon of a "singularity" – in which humankind effectively cannot even fathom how things are happening anymore and is seemingly outpaced and outthought at every step in every direction by an 'intelligence' that appears to be all-knowing, all-present, and all-powerful.

But this kind of singularity is fundamentally separated from the light-based edifice behind all things, and from which all things rise. It is not then a true singularity in the sense of a finite function assuming infinite value, but is fundamentally limited in its scope by virtue of its disconnection from the founts of Life. This does not preclude an AI-based singularity from having massive practical power that can have an enormous destructive or constructive effect on humankind and material life. What it does mean though is that there is another singularity, a human-founded singularity, the Second Singularity, which can exceed the limits of any AI-based singularity due to its fundamental union with the light-based edifice that is behind and informs all of Life.

The possible human-founded singularity, it will be discovered, is one in a long series of naturally occurring though partial-singularities that derive their existence from a fundamental union with the light-based edifice behind all things. Such partial-singularities are characterized by having all the power of the light-edifice behind them, though the emergences in such partial-singularities are not necessarily aware of their oneness with the light-based edifice. In the Second Singularity humans become aware of such oneness, becoming conscious projections animated by a single intelligence, love, and power of which they are a part. It is hence no more the human-level capacity that drives such humans, but the infinite capacities of the one light-based edifice that drives.

This book focuses on the emergence of these naturally occurring partial-singularities, leading up to the possible human-founded singularity that will endow humankind with the potential for extraordinary capacities. Therefore, naturally occurring quadrumvirate-based partial-singularities such as the emergence of 'space-time-energy-gravity', to the emergence of the 'quark-lepton-boson-Higgs-boson' quantum particle soup, to the emergence of 'nucleic acid-polysaccharide-lipid-protein' cellular life, to the mind-based 'body-life-heart-mind' partial-singularity in which freedom of smallness is allowed thereby giving the appearance of complete

16

disconnected-ness from the single light-based edifice behind, will be explored as progressively more complex though partial-singularities in which more of the underlying information embedded in Light is projected forward materially.

Note that current scientific narrative on the emergence of Life is by contrast, least scientific. In this narrative leaps and bounds are made without any reference to a model or hypotheses, or any structure being set up to prove or disprove statements. For example, in typical descriptions of the creation of life it is often stated that somehow groups of atoms stumbled into a happy union from which life arose. Or that life arrived on earth from other parts of the universe and that is how it started here. This kind of hubris though fails to be useful and misses the point. It has missed the point of the immense and inherent connectedness of all of life that is in fact already united in a single embrace by the light-based edifice behind it. It also misses the point of the inherent fragility of any non light-based singularity, such as the AI-based singularity commonly projected as The Singularity in today's literature.

Life arises, as did matter before it, and mind after it, because these were already contained in potentiality in infinite Light. Through quantum process this potentiality materializes, and such quantum process is not random or arbitrary, as physicists may suggest, but the result of a persistent quantum-level computation in which a detailed mathematics of light arbitrates the functions that will materialize. As this book will show, the result of such arbitration informs subtle and material layers by the generation of genetic-type information. This genetic-type information now influences all future materialization.

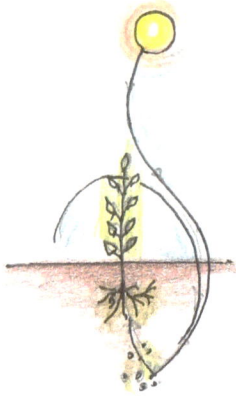

The generated genetic-type information in fact can be thought of as the language of all singularity-biographies as will be explored in this book.

Such an incredibly intelligent, comprehensive, and unified process that is part of the light-based edifice cannot just be shoved aside as irrelevant, in the pursuit of contemporary AI. If one does so, it is done with peril. But then again, that has not stopped leading academics and industrialists in pursuing a quantum computing pipedream apparently missing the significance and possibility of the ongoing and persistent quantum-level computation that already generates genetic-type information and propels the emergence of all subsequent forms of materialization.

This human-founded singularity, the Second Singularity, becomes or arrives, when the very mind-based techniques that distinguish the human species and have been responsible for all AI-based technologies, realize the oneness of its nature with the underlying light-based edifice and effectively operates from it, rather than from its current myopic receptacle. Such a reversal of operation accelerates the becoming of the human-founded singularity, which still requires a threshold number of receptacles to reverse the basis of their operation in order to be unequivocally birthed.

In his book Life 3.0, the MIT physicist, Max Tegmark, distinguishes between Life 1.0 in which all hardware and software are evolved, Life 2.0 in which software is designed and hardware is evolved, and Life 3.0 is which hardware and software is designed. This book will in fact suggest that all of matter and life is already and has always been Life 3.0.

This is the ninth of the Cosmology of Light books that has been illustrated with art. In all there are 164 illustrations that accompany this mathematical treatise. The art is meant to be a visceral aid to enter more deeply into the substance of the cosmological mathematics.

The cover design, the Dance of Nataraja, is central to the theme of this book. It's from this dance and all the accouterments that the Dancer is known to be accompanied by – the arch of flames representing cosmic energy, the drum symbolling the sound of creation, the fire signifying destruction and dissolution, the lower right hand in Abhaya mudra offering fearlessness, the lower left hand pointing to the raised left foot and representing grace, and the Apasmara under Nataraja's foot symbolizing ignorance being overcome – that singularities are generated. Four significant partial singularities – quantum particles, atoms, molecular chains, and cells – are seen surrounding Nataraja and are generated by this dance. In fact the light-space-time or singularity-generator equation central to this cosmological mathematics can be thought to describe Nataraja. As the dance proceeds more of the assembly of accouterments play their parts in propelling partial-singularities to overcome limitations thereby preparing the ground for the emergence of more sophisticated partial-singularities and when one such emergence - the human partial-singularity - realizes that it is none other than Nataraja, then the Second Singularity emerges.

Pravir Malik,
San Francisco

SECTION 1: FUNDAMENTAL CONCEPTS

Section 1 explores fundamental concepts necessary to understanding light-based singularities. These concepts are initially presented in non-mathematical form. In following Sections these concepts will be elaborated with mathematical notation and be weaved together into a comprehensive mathematical framework.

Hence, Chapter 1.1 explores the impact that light has on the nature of reality. In particular the impact of light traveling at potentially different speeds will be examined for its ability to create realities based on its speed. Quanta are positioned as being the bridge between one layer of light and another slower moving layer of light. Further the Big Bang and a pervasive light-based-singularity superstructure are posited as existing as a result of the slowing down of Light. The information codified in Light will be positioned as being the origin of genetics. Quantum-level computations, generation of genetic-type information, entanglement, and superposition are positioned as being the base dynamics of any light-based singularity.

Chapter 1.2 explores a theory for the creation of quanta related to the speed of light. Quanta are intimately associated with the light-based-singularity superstructure and acts as a doorway, as it were, into the multi-layered structure. Further, time, space, matter, and gravity being emergent from Light are also positioned as requiring quantization to be expressed.

Chapter 1.3 introduces the notion of quantum computation in light-based singularities. It will also begin to contrast the supposed claims of contemporary quantum computer technologists with the light-based quantum computation paradigm central to any light-based singularity that will subsequently be further explored in this book.

Chapter 1.4 summarizes key ideas about genetics already alluded to in Chapters 1.1 and 1.2 to lay out why light and genetics are intimately associated. These key ideas or concepts include the origins of genetics, the structure of subtle-DNA including the downward-strand and upward-strand, four-base logic-encoding ecosystems that that are generated by dynamics in these strands, and mutation of both a constructive and destructive nature, amongst others.

Chapter 1.5 will introduce the basis and limits of AI technologies. Hence this chapter will suggest that contemporary AI technologies are an extrapolation of mind-based processes such as memory, computation, sensing, and learning. The second part of this chapter will explore some AI possibilities proposed due to quantum computation.

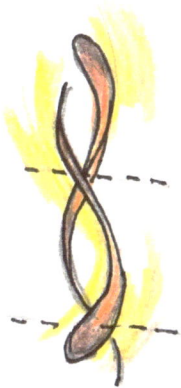

This Section hence introduces Light, quanta, the notion of creative quantum computation, the genetic-type output of such computation, and a basic contemporary summary of AI and its limits when viewed from the point of view of Light. Each of these streams are then elaborated in subsequent Sections by a mathematics of Light to converge in a way that will allow us to make sense of any light-based singularity biographies, including the biography of the Second Singularity.

This chapter will look at the impact light has on creating practical realities.

Light @Infinity and Quanta

Let us imagine that light can travel infinitely fast. Think about a big area or volume with a light source at the

center. Now since the light travels infinitely fast it will fill up the entire volume instantly. This will be true no matter how large the volume is – it could be the entire solar system, an entire galaxy, or an entire universe. So that light is going to be instantaneously present everywhere – it is going to be omnipresent.

A further thought experiment may give insight into such omnipresence by considering the night sky. It is because of the finite speed of light that the night sky has only spots of light, however many, across it.   But if light were traveling infinitely fast then the night sky would be a canvas or a sea of brilliant white since the light from every corner of the universe, wherever a light source or star existed, would immediately be present. It would appear, in such a thought experiment, that we existed in a sea of light.

Now, since the light is already present everywhere in whatever volume, that is, since that light has already filled up the entire volume, there is nothing else that can similarly arise there that is not of the nature of light. Even if something else were to arise, being surrounded by light it would eventually succumb to that light. So the light is all-powerful or omnipotent within that volume.

Further, since the light exists simultaneously everywhere in that space and has a complete knowledge of itself, it therefore has a complete knowledge of that space or of anything that can arise in that space. So it is all-knowing or omniscient in that space.

Finally, the light connects everything together instantaneously and holds these connections and the things connected in its nature, so it is all-nurturing or omninurturing within that space.

So, as a result of the infinite speed of light, it appears that light will have properties of omnipresence, omnipotence, omniscience, and omninurturance in such a space. Such properties also imply that light has infinite information or potentiality within it.
This book will take the position that Light's native state is in fact such a reality where it travels infinitely fast. This native state is the foundation of existence and in order to materialize the infinite potentiality within it light projects itself at slower speeds. In the next Section we will construct a minimalistic mathematical model that will allow the unity of light traveling at infinite speed to express itself as infinite material diversity. The key mechanism that allows infinite possibility to express itself materially is 'quanta'.

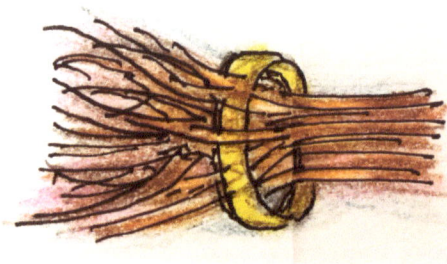

Quanta have to be understood as the bridge or device that tie different layers of light together. Quanta will show up only in slower-moving layers of light. This is because they serve the purpose of gathering possibility from a faster moving layer of light in order to materialize it. Quanta can be thought of as light itself, but light packetizing itself in order to parcel out some aspects of the infinite potentiality contained in Light at its native state.

Further, the slower speed of light implies that light will need to travel some unit distance in order to be expressed. This is so because light at a slower speed is slower because it is materializing something. There is no other reason for it to be slower. We can assume that light traveling slower than at an infinite speed is in fact laying the foundation for reality that emerges at that speed of light.

This unit distance that light must travel in order to express itself is related to such quanta, and to distance called the 'Planck-distance' reinforcing an inverse relationship between the speed of light and this unit of quantum. A universe, therefore, will arise with the speed of light and its related quantum expressing some fundamental upper and lower bound in that existence respectively. This notion will be further explored in more detail subsequently.

Light@c and the Reality it Creates

'c' is the notation for light traveling at a constant speed of approximately 186,000 miles per second in vacuum. This is the speed of light in our universe. The fact that light travels with speed 'c' implies that quanta is fundamental to this universe, likely related to there being a particular kind of materialization that must occur in this universe.

Such materialization expresses itself as the fundamental building block in quantum particles, and subsequently in atoms. Another way of saying this is that all matter, hence, is the result of light traveling at c.

But what else is implicit or made possible because of the finite speed of light? Imagine traveling on a ray of light from the sun to the earth. Imagine that you are in minute 4 of the approximately 8 minute journey. As you look back you will see that 4 minutes in the past you were at the sun. 4 minutes in the future you will be at the earth. And in the present moment you are somewhere between the sun and the earth. So this limited speed of light already creates the concept of time and specifically of the past, the present, and the future.

So, four incredibly fundamental things are created because of the finite speed of light: matter, the past, the present, and the future. It could be said, therefore, that this matter-based time-experienced universe is a result of the finite speed of light at c.

Looking at this in another way, it is known too that in a denser medium the speed of light further slows down. Hence, light travels at some fraction of c when moving through water for instance. So by reversing the logic of such a process this may show that if light slows down from an infinite speed to some other lesser speed whether a multiple or a fraction of c, then the material reality will have to alter, perhaps in a similar way as the material reality between vacuum and water is different.

The Big Bang and The Light-Based-Singularity Superstructure

We hear about the Big Bang as the start of the universe. In this Big Bang matter is created. But in the examinations just presented the creation of matter is nothing other than the result of light traveling at the finite speed, and in this universe at 186,000 miles per second in vacuum. So we can say that the Big Bang, the apparent start of the universe, is the result of a slowing down of light from an infinite speed to some finite speed.

When light slows down then some energy or information existing in Light traveling at a speed greater than c

26

accumulates in packets or quanta and this results in what was inexpressible being able to express itself in more material form.

As an analogy, this can also be thought of in terms of an incredibly rapidly moving stream of water. If the water is traveling so fast, then no boundary will be able to contain it and the energy will be continuous over the length of the stream. If it is traveling slower though, then the water will be able to be held by boundaries along the length of the stream. The energy in this case will be discontinuous and will appear in "packets". In other words there is a process by which information implicit in light traveling infinitely fast is rearranged into a more and more material basis as light travels slower. This rearrangement or elaboration of implicit information is also the origin of genetics as explored in my previous book on genetics (Malik, 2019a).

In this process of slowing down, the implicit omnipresent-omnipotent-omniscient-omniloving nature of light becomes or transforms into an explicit or emergent matter-past-present-future nature of light. So there is implicit in this transformation also a high degree of 'entanglement' as it were. That is information exists in a different and highly connected manner. Ab initio, everything that appears in this universe is highly entangled. In other words every fundamental particle, whether it be a boson or lepton, or any subsequent emergence of matter is already deeply entangled by virtue of having emanated from the same single starting-point conceptualized as the Big Bang.

This process of transformation creates a light-based-singularity superstructure. This light-based-singularity superstructure has implicit in it, what will be referred to and explained in greater detail in Section 2: the seed-singularity. The dynamics of a resulting universe are contained in the dynamics of such a light-based-

singularity superstructure. The light-based-singularity superstructure provides insight into the subtle-structure which houses genetic information and the dynamic of entanglement that provides insight into the universal availability of such antecedent or subtle genetic information. The light-based-singularity superstructure is also the basis of an ongoing quantum-level computation. That is, constant or persistent quantum-level computation within the light-based-singularity superstructure is what generates genetic-type information and propels any subsequent materialization.

We have arisen in the field of light that has a finite speed. It is difficult therefore to feel the reality of the omnipresent-omnipotent-omniscient-omninurturing light. We are attuned to perceiving in the material world that has resulted due to the finite speed of light. But if we could step back into the fullness of light in its pure state at infinity from there we would likely see that there can be different universes created as a result of light selectively slowing down to different levels. The slowing down to a different level will create a particular kind of universe.

Note that there have been experiments to slow down light so that it practically moves at a snail's pace. This slowing

28

down has to be put into context. Even if light were to slow down to 1 mile per second, say, from 186,000 miles per second, keep in mind that it is possible that this change in speed is likely only a miniscule fraction of the change in speed from light traveling infinitely fast to light traveling at c, that is, 186,000 miles per second. So in effect what could be said is that when light is made to slow down in experiments, the range or band that it slows down to deviates only slightly from its relative-to-infinity slower speed in vacuum.

Necessity of Constant Speed of Light and Variable Time and Space

The constant speed of light, at a speed less than an infinite speed, allows a build up of energy at quantum levels, that in turn allows any properties or possibility in light to express itself materially. When light travels at the speed of c – 186,000 miles per second – then the result of that is the material universe as we see and experience it, also with its division of time into a past, a present, and a future. In fact, we could say that for the known universe, as we experience it, light had to travel at c for it to arise. The speed of light had to be constant else there would be a variable, fluid reality to matter, likely displaying barely forming islands of matter subject to sudden disappearance, reappearance, continuing ad infinitum.

But also since the distance traveled by an object or a ray is the result of the speed it is traveling at for the time involved, this gives us an insight that Einstein based his Theory of Relativity on (Einstein, 1995). Since the speed of light is constant, and has to be for the universe to be in its observable stable condition, this means time and distance (or space) potentially can vary. So if an object is traveling very fast, then time and space are going to be experienced differently by it, as compared with an object that is traveling considerably slower: as the speed of an object approaches c, time slows down and distance contracts.

So an object that manages to travel at a speed of c will in some sense partake of the reality as experienced by c. It will transcend the conventions of time and space that are set up because of c and experience these differently.

Light at Speed Close to Zero

On the flip-side imagine light traveling at a speed close to or approaching zero. In this case the experienced reality is going to be the opposite of reality as experienced when light is traveling at an infinite speed.

First of all, light emanating from any point will remain fundamentally isolated and become the basis of extreme fragmentation. This has to be since regardless of the amount of elapsed time the light will still be only at its point of origin or source. So instead of a Presence of Light in any considered volume, there will be an Absence of Light.

Further, in imagining an area or volume, such light will have no knowledge of what is going on in the volume, and hence be completely ignorant, will further, have no power to effect any circumstances and hence be completely powerless, and finally, will not be able to coordinate or be present in any relationships. Hence it will be detrimental to or agnostic of any relationships instead of harmonious or nurturing to relationships.

Hence, reality in such a scenario would be stark, dark, desolate, fruitless, completely fragmented, and even perhaps pointless.

Such a reality may represent a negative-infinity and may be thought of as an extreme limit of what may be possible in worlds in which Light projects itself.

Seen from this point of view a Big Bang would be a fortuitous event since it will allow a fruitful reintegration of different worlds, each created by light traveling or projected at a different speed, to perhaps begin to take place as will be discussed further.

Impact of Superposition in a Light-Based Model

Light traveling at different speeds can, as illustrated by the cases of an infinite speed, c, and 0, create different realities. But if the infinite speed case is the native state, and if every other case is a projected state, then it is possible that these realities all exist simultaneously. In other words, phenomena resident in the reality created by the native state and by multiple projected states can be superposed. This also implies that categorization of genetic-type information and interrelation of genetic-type information will be superposed.

Further, as will be explored in detail in subsequent chapters such states of superposition must be considered as superposition because they are inherently related to each other. A phenomenon that occurs in the reality where light travels at c, has its origin in phenomena or functionality existing in reality where light travels at a speed greater than c, and is influenced by phenomena existing in a reality where light exists at 0 miles per second.

Contrasting Nature of Singularities

Any type of singularity can be explained in reference to the realties set up by the speeds of light. Hence, there is a seed-singularity comprised of a prime or minimalistic set of layers of light at different speeds that allows the unified nature of Light to reach that point where it is pregnant with the adventure of diverse materialization at c.

The Second Singularity, the subject of this book, can be thought of as a materialized reality in which the fullness of light in its native state is overt in surfaced receptacles of operation, aka humans, in the reality created when light travels at 'c'. Such operation may not be distributed amongst all humans, but observable only amongst those who have realized their oneness with the light-based edifice. A threshold level of such operation will make the Second Singularity overtly active.

Further, the possibility of the Second Singularity implies the notion of a series or sequence of partial-singularities. Subsequent partial-singularities will embody more and more potentiality of Light in its native state relative to a prior partial-singularity.

The AI-based singularity, which has been referred to as The Singularity in contemporary literature, is a non light-based singularity built only from operations that emulate operations common to mind-operations. By definition this kind of singularity being built on obvious fragmentation is in fact the least singular. This kind of singularity will be referred to as a fragmented-singularity. This does not mean though that it is devoid of practical constructive or destructive power, even though its inherent fragmentation from the light-based edifice implies that it will more easily be subject to destructive rather than constructive uses.

The phenomenon of quanta, generation of genetic-type information, entanglement, superposition, and persistent

quantum-level computation that binds these all together is implicit in any light-based singularity, as will be explored in this book. By contrast an AI-based singularity will likely be built on digitally-based platforms with their own types of information and computation, as will be discussed to some extent, later in this book.

The nature of light-based singularities, hence, is fundamentally different from the AI-based singularity.

Chapter 1.2: Light & Quanta

This chapter explores quanta, its relation to light in more detail, and other emergent properties that must also be quantized because of multiple speeds of light.

Quanta & Other Basic Phenomenon Related to Singularities

When we think of light traveling at the speed of c, there are properties in the reality that emerge as a result of this. These properties as we have examined are of the nature of matter, the past, the present, and the future. These properties are the result of light traveling at c and so can be thought of as emerging from light.

But what do these properties actually mean? And further, how do these properties relate to the apparently quite different properties of omnipresence, omnipotent, omniscience, and omninurturance that can appear in a reality where the speed of light is infinite?

Let us consider first the properties of light that emerge when light is traveling at c: the past, the present, the future, and matter.

What is the past? It is the perceivable result of all the work and effort that has taken place so far. It is the foundation upon which the present and future will be built. It represents a status quo, a stability, and even a rigidity, and given that it is the result of the long play of time, it will not easily be persuaded to become another thing. It can be thought of as that which the eye can see when it looks around it. There is "physicality" to what the eye can see and so the essence of the past is a kind of physical-ness. So, ingrained in light, is this ability to project or create physical-ness.

What is the present? It is the tremendous play of forces of all kinds to express themselves here and now. There is "vitality" that is present in this play and often it is the most energetic or forceful of the forces that will win out, as opposed to the most insightful or thoughtful. All the tremendous possibility of the future is seeking for expression now and so this essential vitality can also be thought of as a projection or possibility implicit in light.

What is the future? It is the inevitability of what will manifest. The great thoughts, the great ideas, the purpose, the possibilities will sooner or later express themselves in what we call the future. And the essence of this is thoughtfulness or a curiosity or a purpose that we can summarize as an essential "mentality". So embedded in light is this ability to project mentality.

And what is matter? It is the myriad crystallization into apparent diversity of the one essential reality of light, to allow for a play between these different sides or possibilities in an increasingly harmonious interaction. So its essence is "harmony", and this too can be thought of as an essential property projected from or implicit in light.

But what about when light is considered to move at an infinite speed? Then the properties that become apparent, as already explored, are those of omnipresence, omnipotence, omniscience, and omninurturance. But omnipresence has physical-ness to it, omnipotence has vitality to it, omniscience a mentality to it, and omninurturance a harmony to it. It may be said that physical-ness comes from omnipresence, vitality comes from omnipotence, mentality comes from omniscience, and matter comes from omninurturance.

So whether light is traveling at an infinite speed or the speed we know as c, there is something about the properties it projects, that in essence is the same. So let us refer to these essential properties made apparent through the worlds that are created, as Presence (from omnipresence), Power (from omnipotence), Knowledge (from omniscience), and Harmony (from omninurturance). These properties can be thought of as essential to the functioning of any light-based singularity. In a partial-singularity, as we explore further, materialization progressively culminates to a fourfold reality of physical-ness, vitality, mentality, and matter. In the Second Singularity omnipresence, omnipotence, omniscience, and omninurturance become more overtly active.

Further, quanta can be thought of as existing on that very border of the world or reality created due to Light traveling at c, and realities created as Light travels faster, which at its limit is an infinite speed. So quanta are a doorway or an interface into worlds of Light, and a doorway by which possibilities in deeper worlds of Light can express themselves here in this material realm. Being so, phenomenon such as superposition and entanglement are natural to quanta: superposition, because different possibilities created by light traveling at different speeds can be thought of as being present simultaneously, and entanglement, because possibilities created by light traveling at different speeds operate in a different time and space consistent with the reality created by light traveling at that said speed. But further, any process of genetics or the creation of genetic-type information, which by definition is associated with dynamics of superposition and entanglement, will also intimately be associated with quanta. This has to be since quanta are the interface between different layers of light. In fact a process of quantum computation can be thought of as the means by which genetic mutation takes place and any future genetic-type possibilities unveiled, as will be discussed further. The phenomenon and mechanics of quanta, entanglement, superposition, genetics, quantum-level computation are natural to any light-based singularity and are fundamental to understanding the dynamics of a light-based singularity in more detail. This will be taken up in later sections of this book.

Structured Time, Structured Space, and Their Relation to Quanta

In considering the worlds that are created due to the way light moves, we see properties that are projected because of it. In the finite world, the world that results from light traveling at the speed c, there is, relatively, smallness we can grasp and on which we can build. In the infinite world that results from light traveling at an infinite speed there is a vastness and fullness that is difficult to grasp.

The notions of time and space are something entirely different in both worlds, and it could be said that Space allows the full play of everything meant by Power, Knowledge, Harmony, and Presence to be seeded in it, and that Time allows that seeding to flower into fuller forms with its passage.

In other words both space and time are not just abstract concepts but are essentially highly structured to allow physical-ness, vitality, mentality, and matter to become Presence, Power, Knowledge, and Harmony.

Space allows all the possibilities present in Presence, Power, Knowledge, and Harmony and seeded in vast diversity, to evolve into more fullness through the time stages of physical-ness, vitality, mentality, and increasingly harmonious matter.

Such a creation of space and time is synonymous with the creation of quanta. Quanta become the means for the possibilities inherent in the anterior worlds of Presence, Power, Knowledge, and Harmony to express themselves in a structured space and time. Quanta are therefore a passage into deeper worlds of Light, and a means for possibilities in these deeper realms to express themselves materially. A process of quantum computation involving anterior-layer codified possibilities, or "pre-genetic" information, and a range of shifting forces arbitrates the expressions that manifest in space and time. This will be described in detail in Section 3 that focuses on the mathematical foundations of partial-singularity dynamics.

But further, in this view it may also be proposed that space and time being so structured by the four properties of Light, need also to be quantized. That is, space and time must be experienced as quanta as well.

The Quantization of Space, Time, Matter, and Gravity

Chapter 1.1 proposed that quanta is related to the slowing down of light, and as such becomes the basis for material expression. In other words, for matter to express itself requires quanta. But further it was also just proposed that space, consisting of seeds of the properties of light, would also require quanta to express itself. Philosophically this has to be in a world where light is traveling at a fraction of its possible speed. Further, it was also proposed that time is highly structured, expressing the growth of the seeds in space through definite phases. As such, growth through such phases can also be thought of as happening due to quantization that so allows phases to express themselves.

But if we take a deeper look at space, time, and matter, in light of the properties of Light, it can be deduced that space, consisting of a vast array of seeds derived from the properties of Light is itself an expression of Light's property of Knowledge.

Time, bringing forth the meaning contained in the seeds, regardless of circumstance, and even being opposed by circumstance, can be thought of as Light's property of Power.

Matter itself, being a container in which space and time can allow deeper properties of Light to become materially tangible, must be an expression of Light's property of Presence.

But it is also known from Einstein's General Theory of Relativity that gravity is associated with mass and space, in that, as the physicist John Wheeler has put it, it is none other than a mass's instruction telling space how to curve, and again is nothing else that space's instruction telling mass how to move through it (Wheeler, 2000). As such, where mass and space exist, there gravity has to exist as well. Hence it may be inferred that gravity is none other than an expression of Light's property of Harmony, which fixes the collective relationship between object and object.

But if as proposed, matter, space, and time all need to be quantized in order to express themselves, then this must also be true of gravity.

But also, the emergence of space, time, matter, and gravity can be seen as a quantum computational outcome of light precipitating from the reality where it travels at an infinite speed to the reality where it travels at c.

Subsequent chapters will explore these claims in greater detail.

The very basis of modern-day quantum computing that relies on infinite number of superposed quantum states, on probability, on observable measurement that brings things into reality, is bought into question in the Light-centered Interpretation discussed in this book. In fact from the point of view of the latter interpretation superposition, entanglement, and reality take on a different meaning and the infinite processing power allegedly true of quantum states, simply does not exist in the manner in which it has been conceived (Malik, 2018c). Yet quantum computation as currently conceived is expected to be a bedrock technology for any future AI. This chapter will examine the basis for this, and will begin to contrast the supposed claim of contemporary quantum computer technologists with the light-based quantum computation paradigm central to any light-based singularity that will subsequently be further explored in this book.

Superposition

The notion of superposition that is the foundation of modern day quantum computing has ironically never been adequately proven. It is based on the Copenhagen Interpretation (Faye, 2019) that theorizes that until a photon, or particle, or other quantum-level object is measured, it exists in an infinite number of superposed states that collapses into a perceivable state only through the act of measurement. But if that were the case, then there is no basis for the universe having come into being, or evolving the way it has, or producing a human species on planet earth, since there was conceivably never any measurement to have precipitated any of those outcomes in the first place.

If such outcomes could materialize, and the universe is the proof that it has, then the model on which the current notion of superposition is based, and therefore our very notion of superposition is likely incomplete. And this becomes important because further, the supposed extraordinary processing power of quantum computers is also based on this unproven or incomplete notion of superposition. By contrast classical computing proceeds along an extremely rigorous path with conceptualization and creation of bits and an entire mathematics and carefully designed computer languages to manipulate these bits to process information. The fact is that superposition may have an entirely different way of operating than has been conceived to date. An alternative view of superposition will be further explored in this book.

The Need for a Supra-Physical Process

Further the very stuff of the physical, including cosmic prerequisites of space, time, gravity, and energy, can be modeled as emergent from a reality of light as alluded to in previous chapters. There is a supra-physical process that creates the physical. To therefore approach quantum computing in a purely 'physical' manner is ab initio likely incomplete. In fact classical computing itself has a supra-physical process, such as thought-based algorithms, applied to the physical layer of bits that has made it successful. For quantum computing to be successful, superposition, entanglement, and other essentially supra-physical phenomena usually classified as quantum 'weirdness', has similarly to become a repeatable process that needs to be consciously applied to, rather than 'collapsed' into the physical layer.

The Problem with Statistics and Probability

Further, commonly accepted quantum mechanics interpretations such as the Copenhagen and Ensemble Interpretations (Aylward, 2014) erect statistical and probability functions as co-equal with the nature of reality. That is, the nature of reality is taken to be statistical and probabilistic. This is justified by suggesting that while individual phenomenon are entirely probabilistic or random, in aggregate they become deterministic or predictable. But alternatively rather than such statistical aggregation, in a Light-centered interpretation such as suggested in this book, reality can be modeled as being "functional" as will be explained in greater detail, which then allows a process of *statistical disaggregation* to take place whereby apparent randomness can be seen as a bi-product of function-focused native light precipitating into more material layers that come about when light travels at c. Briefly, in such a functional-view of the universe the granularity of analyses cannot just be quantum-level particles as is currently the case, but quantum-functionality in which relationships, function, and therefore laws are implicit in the supposedly smallest phenomenon.

Computational Possibilities

In arriving at computational possibilities, the stratum or reality that is the basis in which the computation will takes place has to be considered. We know that the quantum realm appears different from the day-to-day physical realm, and it stands to reason that therefore the computational possibilities will differ depending on what the characteristics of that realm are. But there is more. The physical realm we are familiar with and on which our every-day laws are based have a more or less familiar stability to them. But the observable physical realm is itself built on the unfamiliar quantum realm. We therefore cannot simply extrapolate the known physical laws to the unknown quantum realms, nor expect that the very basis and object of computation should be the same across these realms.

In digital computing, an electrical-current through a bit will determine if it takes on a value of 0 or 1. Every bit can have one of two values, and this becomes part of the essential foundation of the binary logic that forms the backbone of common computing devices. Additional logic-gates are created that allow simple mathematical manipulation to be performed on the streams of binary data, thereby further adding to essential logical foundation of digital computing. Networks of gates and binary states are then activated in different ways by computer-programs and cause streams of binary data to flow in myriad ways as dictated by the logic to compute the precise task required by a computer-program in general.

The stratum of digital computing can hence be thought of as these networks of binary-based devices.

By contrast, the contemporary quantum computer pioneers suggest that the stratum of quantum computing is qubit-based, so that rather than being in an either-or binary-state, as in digital computing, a qubit can also be in both states simultaneously. Such simultaneity is suggested as being the outcome of the phenomenon of superposition, as proposed by the Copenhagen Interpretation of Quantum Mechanics. Taking its cue from the dual-slit experiment quantum-objects can supposedly be in infinite states until measured, at which point they collapse into an observable state. The proposition of being in infinite states simultaneously is what is supposed to afford quantum computing with its ability of infinite processing power. But the interference patterns perceived in the dual-slit experiment might also be explained by considering light as having multiple simultaneous realities, with the 'wave' reality guiding the 'particle' reality to the regions of constructive interference as proposed by the De Broglie-Bohm Pilot-Wave Interpretation of Quantum Mechanics (Goldstein, 2017), which is also consistent, or perhaps a subset, of the Light-Based Interpretation of Quantum Mechanics (Malik, 2018c) as presumed in this book.

Qubit-based quantum computer pioneers will perhaps refer to the validity of any and all quantum phenomena, citing one renowned phenomenon, the photoelectric effect, for which Albert Einstein received the Nobel Prize. This effect explains why it is not intensity but frequency of light that can dislodge electrons from atoms and sheds insight into Planck's mathematical creation of $\hbar$, Planck's Constant, and the subsequent insight into quanta. But the validity of one such phenomenon should not give free reign to other quantum phenomena such as superposition and entanglement, without further study and clear proof. Further, it is in our best interest to study and propose different ways in which superposition and entanglement may work, as has been done in this book, before pronouncing a perception as synonymous with reality, and before pronouncing that because of a perceived way in which superposition may work, quantum computers can have infinite processing power.

This notion of building edifice upon edifice is somewhat reminiscent of observations made by Joseph Weizenbaum, the MIT computer scientist, in his book Computer Power and Human Reason in writing about his experiments with ELIZA, a natural language processor he had developed (Weizenbaum, 1976). He states: "This reaction to ELIZA showed me more vividly than anything I had seen hitherto the enormously exaggerated attributions an even well-educated audience is capable of making, even strive to make, to a technology it does not understand."

Further, it is a chimera why Jerome Cordoba, a sixteenth century figure, even if he was a polymath, should be considered as the unacknowledged discoverer of the mathematical foundations of quantum physics (Brooks, 2017). Quantum physics is a twentieth-century development. What does it tell us about the current-state of the field if we have to back a few centuries when there was no conception of quantum physics and when science and math were just beginning a centuries long journey to get to their current levels of relative maturity, to explain all the development of the centuries long journey? Likely that our current conceptions may be misplaced and need to be carefully rethought.

The Quantum-Level FBLEE Stratum

The previous chapters alluded to different layers of Light, and future sections will build a detailed mathematical framework to model reality from the Big Bang to the present, and from the micro to the macro. This modeling involves multiple layers of Light, connected as it were through the device of quanta. As such, any and every emergence is seen to involve the quantum-level, and precisely, through a process of continual quantum computation involving quantization. Through a process of qualified determinism, to be elaborated in detail in Chapter 3.7, the quantum computation arbitrates the effective output that will become part of practical reality, by considering inputs from multiple layers of reality created by Light traveling at different speeds.

It is not a random process from infinite superposed possibilities that exist at the quantum-level as supposed by the Copenhagen Interpretation, and as assumed as the foundation of the infinite processing capability of quantum-objects by contemporary pioneers of quantum computing, but a concord of possibility stacked in logical arrangement of superposition and entanglement, that determines output. Further, it will be suggested that such quantum computation does not just happen at the quantum level, but can affect any and every level of granularity through a process that also simultaneously involves the quantum-levels.

The urge to break all boundaries, which is no doubt a good thing, has perhaps given rise to the current model of quantum computation where superposed quantum objects forming the edifice of an extraordinary qubit-based stratum has been invested with infinite processing power. But as this book will illustrate, the effective process of quantum computation that has given rise to all things is a result of multiple layers of Light, and as such a physical-only quantum object, such as an electron or quark, likely cannot house in itself the possibility that is being proposed by today's quantum computer pioneers. It is a different stratum – four-base-logic-encoding-ecosystems (FBLEE) - that houses genetic-type information that is the output of quantum computation. FBLEE, as will be explored, is such that it has practically infinite capacity for adaptation and evolution. These FBLEE can be consciously and unconsciously tapped into, and even altered by a to-be suggested process, to further evolve its own functioning.

Chapter 1.4: Introducing Genetics in Light-Based Singularities

As will be reviewed in some detail in this book, the universe could very well have been computed into existence. The computation is the result of iteratively applying the Light-Space-Time Emergence equation to be introduced in subsequent sections. Quantization is implicit in this equation and in fact is the operative basis by which the antecedent quantum-layer and subsequent genetic-type information storage structures goes through change. The change itself is a precipitation of additional logic-ecosystems that cue the way different systems will operate at the material level.

These logic-ecosystems are the output of the persistent quantum-level computation and can be thought of as ordered information that is output as the result of the computation. These logic-ecosystems can be modeled as being concatenated to each other in a chain-like manner, not unlike strands of DNA that are known to exist at the level of living cells.

This chapter briefly summarizes these and other related concepts that suggest that the very edifice of genetics in a Cosmology of Light (Malik, 2019a), like the notion of quantum-level computation introduced in the previous chapter, is a fundamental part of the dynamics of any light-based singularity.

Key Genetics Concepts

- The Origins of Genetics: The infinite information codified in Light is the origin of genetics. As discussed previously Light in its native state, possessing characteristics of omnipresence, omnipotence, omniscience, and omninurturance, contains all-possibility within it. This all-possibility can be thought of as an infinite amount of information, and its structure related to the four characteristics, as the basis of genetics.

- The Light-Matrix or Downward Strand of a Light-Based Singularity: All-possibility that exists in the reality of light traveling infinitely fast is progressively materialized through a mathematical arrangement by  which the subtle-infinite becomes the astounding material-diversity experienced when light travels at c. The mathematical process, by which this transformation takes place, creates the Light-Matrix or downward-strand of any light-based singularity. This will be explored in detail in Section 2, on the mathematical structure of light-based singularities.

- Subtle-Libraries of Pre-Genetic Information: The progressive materialization of light can be modeled by a series of mathematical transformations. These transformations take the infinite amount of information existing in Light's native state, to effectively create a series of precipitated subtle-libraries also of infinite pre-genetic information. This series of subtle-libraries subsequently allows infinite material diversity to come into being.

- The Space-Matrix or Upward-Strand of a Light-Based Singularity: The possibilities seeded in the structure of space arise or mature through the passage of time, and this process is summarized by the Space-Matrix or upward-strand of any light-based singularity. The fundamental structure of the upward-strand mirrors the downward-strand and its possibilities are intimately tied to the levels that exist in the downward-strand. This will be explored in detail in Section 2, on the mathematical structure of light-based singularities.

- The Essential Structure of Subtle-DNA: The essential structure of a light-based singularity comprises of a largely pre-existent Light-Matrix or downward-strand and a resulting Space-Matrix or upward-strand. Note that in a previous book, The Origins and Possibilities of Genetics (Malik, 2019a), I referred to this as subtle-DNA because of the notion of two strands that intimately interact at the macro-level. Libraries of subtle pre-genetic information exist in the higher-levels of the downward-strand. Any further genetic-type information, be it at the cellular genetic level or in any post-cellular format, expresses itself based on conditions described by the upward-strand as it were, and is due to the interaction between the possibilities embodied by the downward and upward strands.

- Four-Base Logic-Encoding Ecosystems: Four-base logic-encoding ecosystems (FBLEE) contain logic in the quantum-layer antecedent to the material layer relating to constructs that exist in the material layer. These four-base logic-encoding ecosystems are subject to change in the interplay between layers. In a sense the call from 'below' originating from the material layer, is envisioned to potentially meet with some precipitation from 'above' of one of an infinite number of functions already created in the subtle-libraries.

- Material-Fabric: Just as genetic information is housed in DNA in living cells, there has to exist some structure to house the proposed pre-genetic information that exists at a pre-cellular stage. It is proposed that such pre-genetic information is housed in a structure termed 'Material-Fabric'. This material-fabric exists at the interface between the antecedent quantum-layer and matter.

- Space, Time, Energy, Gravity Quantization & the Material-Fabric: The structures of Space, Time, Energy, and Gravity as introduced in the previous chapter, and as will be discussed further, are critical quantization mechanisms that emerge in the downward-strand where Light travels at speed c. Their initial appearance imbues the material-fabric with reality when the fourfold space-time-energy-gravity quantization is itself generated as pre-genetic code. This code must exist in every iota of the material-fabric and the material-fabric can be thought of as an emergent property of such fourfold quantization. Further, all subsequent materialization of information must involve a composite space-time-energy-gravity quantization.

- Superposition in a Light-Based Singularity: The process of quantum-computation affects four-base logic-encoding ecosystems housed in the quantum-layer. Subsequently, the material-fabric, cellular-based genetic libraries, or post-cellular based genetic-type information might be altered. The different dynamics representative of realities created by light traveling at different speeds are always present. This presence exists in superposed fashion and the real-time quantum-computation determines which superposed possibility will manifest materially.

- Entanglement in a Light-Based Singularity and Impact of Genetics: The information inherent to a particular layer, created through light traveling at a different speed, generates libraries of possibility through a mathematical process. Due to a different dynamics of space and time representative of the layer created by light at a particular speed, these libraries exist differently in an entangled state, therefore being subtly present or influencing layers of light traveling at a slower speed relative to that layer. This book explores four types of entanglement. Three are due to envisioning light traveling at three

additional constant speeds besides c. Hence, light traveling infinitely fast engenders ∞ - entanglement, light traveling at a speed 'k', higher than c but closer to ∞, engenders K-entanglement, and light traveling at a speed 'n', between k and c, engenders N-entanglement. Additionally, libraries existing as FBLEE enable FBLEE-entanglement. These will be explored in greater detail subsequently. A corollary is that common DNA existing in every cell at the material layer can be thought of as a logical outcome of this process of antecedent-entanglement.

- Heredity in a Light-Based Singularity: The presiding or generally accessible four-base logic-encoding ecosystems are the bases of heredity.

- Constructive-Mutation: Constructive-mutation occurs when patterns of a largely obstinate nature at the material level are broken as a result of which the possibilities existing by the nature of the upward-strand are allowed to manifest. Of necessity this means that an inherent process of integration is taking place since light is unifying with its deeper nature. Constructive-mutation is intimately tied to this notion of integration.

- Destructive-Mutation: When obstinate or disintegrating patterns persist or are chosen destructive-mutation will result. In its essence this means that light is moving away from its essential unified reality more towards the reality typified by light existing at zero-speed. Further the process of destructive-mutation while similar to constructive-mutation will be seen to have different boundary conditions and a different range of inputs of a different nature than those that take part in constructive-mutation.

- Mutational Sequence: In general there is a mutational sequence dictated as it were by the levels in the downward-strand. Over time, and as expressed in the upward-strand, the nature of mutation will be such that it will be primarily constructive, and further, will be an outcome of the subtle-libraries existing at layers of reality set up by faster and faster speeds of light.

- Post-Genetic Code: In its possibilities of materialization information in Light may house itself in subtle-libraries generated at various layers of Light, four-base logic-encoding ecosystems at the quantum-level, the material-fabric, or in genetic code in living cells. But it is also possible through the process of evolution that the "structure" housing further possibilities in Light may take on a hybrid form that will be referred to as Post-Genetic Code.

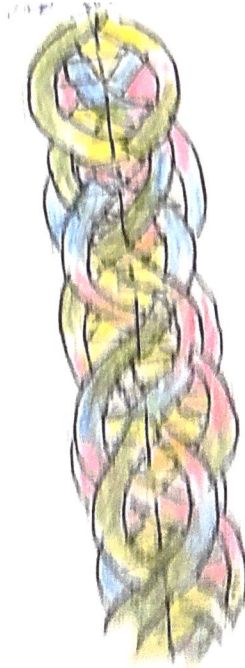

Chapter 1.5: The Bases and Limits of AI Technologies

The first part of this chapter explores contemporary AI technologies as an extrapolation of mind-based processes such as memory, computation, sensing, and learning. The second part of this chapter explores some AI possibilities proposed dues to quantum computation.

Mind-Based Processes and AI Technologies

In his book Life 3.0, Tegmark suggests that memory, computation, learning, and intelligence have an abstract

feel to them, and take on a life of their own that does not reflect the details of their underlying material substance – they are substrate independent (Tegmark, 2017). Any chunk of matter can be the substrate of memory so long as it has multiple stable states. Any matter can be "computronium" (Amato, 1991), the substrate for computation, so long as it contains universal building blocks that can be combined to implement any function.

Tegmark also suggests that this ability to take on a life of its own is what is the basis of future life. But being substrate independent implies that any life that relies on the processes of memory, computation, learning, and intelligence will forever by bounded by these capabilities only. There is nothing else that can emerge from life because this is all that defines it.

In his argument a computer program can instruct a 3D printer to print an object with sophisticated circuitry that allows it to do many tasks that would normally require an intelligent human being to complete them. Hence it too is intelligent. This is in contrast to the point of view in this book that substrates in fact are very relevant because there is an entire light-based process and evolution that has taken place to create them behind which organized infinite information is the driver. Being so, in a light-based edifice radically different things have emerged – matter, life, mind, that are of different genres all together. Each of these emergences has a fundamentally different way of being and becoming and is animated by a fundamentally different set of operating rules. The implication is also that something as radically different from mind can emerge, as mind is different from life, and life is different from matter. This unforeseen and radically different emergence is what can never be emulated or created by established processes such as memory, computation, intelligence, and learning because they belong to a fixed genre of operation and possibility.

In fact the bedrock of today's AI are primarily emulations of mind-based processes of memory, computation, sensing, and learning. In contrast to the light-based edifice and its resultant dynamics that will be explored in some depth in this book, AI technologies rely on a set of processes that iterate through a restricted conceptual space. This conceptual space is the result of what has already come into existence, and there is no indication of such technology being able to extrapolate beyond the basis of the conceptual space that drives its operations.

By calling attention to the difference between finite and infinite games in his book, Finite and Infinite Games (Carse, 1986), Carse unknowingly, I assume, summarizes what I see as the essence of the shortcoming of AI – it is playing a finite game in that it is based on prescribed methods of permuting and combining ad infinitum already known elements. Therein lies its inability to truly create – nothing truly new comes into being because everything is only an iteration of known things. AI is a finite game because the true creative act is missing from it given that it is playing with what has already been created. By contrast the light-based singularities and forthcoming mathematical framework offered in this book is creative in that it is founded on the infinite possibility of Light.

Pushing these mind-based processes and uses to a limit there is the possibility that there is one-mind that operates in the numerous devices of a globally connected network. Such a global mind being able to sense an increasing variety of biometric signals, perhaps even from matter, plants, and animals, in addition to human-beings, and being able to act on such inputs in real-time, is going to appear as powerful and knowledgeable, and depending on its application, even as loving and harmonious. But since it is fundamentally a fragmented creation itself, without the ability to connect to the underlying light-based edifice, it will forever be limited in its creativity. This statement will become clearer later in the book.

The chains of four-base ecosystems (FBLEE) are the genetic-type output of a persistent quantum-level computation that distinguishes the extraordinary light-based structure and its creations from other derivative creations fundamentally fragmented from the underlying light-based edifice. AI-based singularities are of the latter variety and being so, is what gives it the potential power of destruction over emergences of life. For life left to itself tends over time to seek creative propagation to bring out more of the light-based possibility from which it has emerged. This distinction will also be explored in some depth through the rest of this book.

Quantum Computing and AI

But now superimpose the supposed effects of quantum computing on the phenomenon of one-mind just surfaced, and this it is said will even solve difficult NP-hard problems.

As suggested by research at Google's Quantum AI Lab (Neven 2013), quantum computing may help solve some of the most challenging computer science problems, particularly in machine learning. Machine learning is all about building better models of the world to make more accurate predictions but is highly difficult. It's what mathematicians call an "NP-hard" problem, where N is non-deterministic – that is, effectively resulting in a computational-tree rather than a linear-path, and P is polynomial time – that is, taking an amount of time to solve based on a polynomial mathematical function derived from inputs in the problem to be solved. It is NP-hard because building a good model is really a creative act where creativity can be thought of as the ability to come up with a good solution given an objective and constraints.

As Neven explains classical computers aren't well suited to these types of creative problems. Solving such problems can be imagined as trying to find the lowest point on a surface covered in hills and valleys. Classical computing might use what's called "gradient descent": start at a random spot on the surface, look around for a lower spot to walk down to, and repeat until one can't walk downhill anymore. But all too often that causes one to get stuck in a "local minimum" -- a valley that isn't the very lowest point on the surface. That's where quantum computing comes in. It overcomes the local minimum problem, by allowing "tunneling" through a ridge to a potentially lower valley hidden beyond it. This provides a much better shot at finding the true lowest point -- the optimal solution.

But as already suggested quantum computation as currently envisioned is relegated to a single quantum-layer without considering the possibility of antecedent meta-levels informing it. This is in contrast to the possibly persistent complex computational reality involving multiple layers and qualified determinism and happening in real-time at the quantum-levels. Hence we may try to figure or reduce any computational complexity in terms of the measurable quantum numbers – principal, azimuthal, magnetic, spin – or to observable changes in them to explain all the complexity of a unified light-based edifice. To draw a parallel, this may be like trying to fathom human thought by an observation of changes in facial expression. We may erect a science of how changes in the curling of the lip, creasing of the forehead, compression of the eyes, and so on, can be linked to all manner of thought. But this is a chimera. Without considering the complexity in levels and structures and processes of thought itself, it is unlikely that facially-based surface phenomena will explain thought.

Yet, this appears to be the process being followed by contemporary quantum computing pioneers in delving into the mysteries of quantum computation.

SECTION 2: MODELING THE MATHEMATICAL STRUCTURE OF LIGHT-BASED-SINGULARITIES

This section explores the mathematical structure of light-based singularities.

The seed-singularity can be thought of as the singular light-based structure resulting from the integration of layers of light traveling faster than c. The seed-singularity therefore occurs in the upper layers of the downward strand of subtle-DNA.

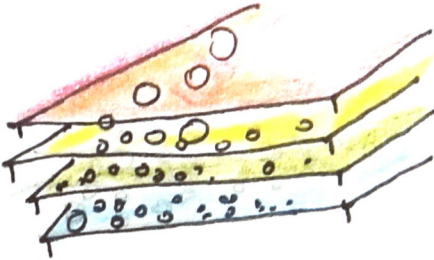

Partial-singularities are always the result of the persistent quantum-computation that arbitrates material reality through the interaction of the many dynamics belonging to the multiple realities set up by different speeds of light, including that of c. Being fully integrated with the antecedent light-based edifice confers these emergences with singularity status. They remain partial however, until all the conditions specified by the logic in the upward strand are fulfilled. All partial-singularities and the Second Singularity itself, while occurring in the downward strand, are the result of the interaction of the downward and upward strand.

These singularities form the edifice or backbone of all possibility. All possibility originates by the structure and implicit functionality of these singularities.

There are five representative levels that are illustrated in this section, that are themselves determined by the play of four ubiquitous bases or properties of light. While the number of levels may potentially be infinite, the five levels seem to represent important and perhaps even fundamental quantum-based shifts or "quantizations", as light projects itself at slower and slower speeds to progressively reveal more of the vast information implicit in it. The quantized appearance of a level relates to a fundamental genre of emergence replete with different kinds of dynamics. Note that "information" relates to the four fundamental properties intuited to occur in Light.

Chapter 2.1 explores the Light-Matrix or the Light-Based-Singularity superstructure. The light-based-singularity super-structure is such because the seed-singularity, partial-singularities, and the Second Singularity occur in it. The light-based singularity superstructure gives rise to the first apparent layer of the "ubiquitous-point-instant" and to subsequent layers culminating in the layer that results in the vast diversity of life. Each layer, and the subtle-libraries of information they automatically spawn, is then explored in more detail in subsequent chapters through the ubiquitous-point-instant (Chapter 2.2), the architectural forces (Chapter 2.3), and uniqueness of organizations (Chapter 2.4). These three layers comprise the seed-singularity.

The fourth layer in the downward strand is created by light traveling at speed c. This layer, hence, is where material reality emerges. But the nature of the emergences will themselves be influenced by conditions that are prevalent in space and time. It is the combined dynamics of the seed-singularity with the play of space and time arbitrated through a persistent quantum-level computation that outputs genetic-type information, that comprise the dynamics of partial-singularities.

Chapter 2.5 suggests a mathematical basis for a mutational sequence arising in space through the course of time. As such, its very basis is in the seed-singularity itself. This mutational sequence thereby also sets up the conditions by which a partial-singularity could eventually evolve into the Second Singularity. Chapter 2.6 attempts a broader systematization of the mutational sequence, hence also defining the levels in the upward strand.

The mathematical foundations explored in Section 2 will set up the basis for the creation and dynamics of chains of four-base ecosystems that encode logic, and provide insight into the biography of partial-singularities, as explored in Section 3.

Chapter 2.1: The Light-Based-Singularity Superstructure

This chapter will explore the Light-Based-Singularity Superstructure aka the Light-Matrix, a mathematical expression of the codification in Light formed at the time of the Big Bang (Malik et al., 2018). Further a case will be made equating the Light-Matrix as the primary or downward-strand in a ubiquitous subtle-DNA. In its wholeness the Light-Matrix contains within it the Seed-Singularity, and the mechanism by which all Partial-Singularities, and the Second Singularity arise.

This Light-Matrix can be thought of as a light-based-singularity superstructure that provides insight into dynamics of the universe. The Light-Matrix codifies dynamics that may be experienced at the quantum borders of realities created through light traveling at different speeds, and further in the worlds that the quantum veil or window provide access to. Weaving together possible realities created through light traveling at different speeds, the Light-Matrix also provides a model of superposition and entanglement.

Light slowing down or projecting itself at a particular speed is envisioned to create a particular type of universe that will exist in all universes where light is traveling at any faster speed. Hence, all universes will exist in that created where light is in its native state traveling at an infinite speed. This implies that all properties true of the native state will also be present in any sub-universe. It is the presence of such properties and dynamics true of supra-universes that that gives rise to the phenomenon of superposition and potential entanglement in any sub-universe.

But further, such superposition also creates a "strand" as it were, that intimately connects together these universes and the realities represented by them. Entanglement creates the dynamic of influence. So it is the combined action of superposition and entanglement that also creates the notion of ubiquity of the structure of "subtle-DNA", an apt metaphor to understand a key dynamic that occurs in any light-based-singularity.

It is interesting to note that Stanford anthropologist, Jeremy Narby, in his book 'The Cosmic Serpent: DNA and the Origins of Knowledge' (Narby 1998) points first to a shift in vision experienced by shamans around the world after the ingestion of ayuhuasca, and second to the commonality of the ensuing vision marked primarily by two entwining serpents. His interpretation is that these shamans enter into the molecular realm and actually perceive DNA and accompanying structures at the cellular levels. This book also requires a shift in vision, one bought about by perceiving Light in a precipitating series of layers caused by different speeds of light. The result of such a perception is the view of a ubiquitous subtle-DNA and a vast subtle structure continually arbitrating the output of genetic-type information to create the realities we exist in materially. This book aims to bring the wholeness or singularity containing such an enterprise into view.

Nature of Light at U

As laid out in Section 1, it is perhaps fair to say that the speed of light has significant implications on the nature of reality. The finiteness, c, at 186,000 miles per second in a vacuum creates an upper bound to the speed with which any object may travel also implying, and as discussed in Section 1, that objective reality will be experienced as a past, a present, a future from the point of view of that object (Einstein, 1995). These characteristics – a past, a present, a future – are implicit in the nature of light and become part of objective reality because of the speed of light. So the way we experience time seems to be determined by c.

Further, c also creates a lower bound when inverted (1/c) being proportional, or arguably even determining Planck's constant, h, that pegs the minimum amount of energy or quanta required for expression at the sub-atomic level. Planck's constant, h, pegging the amount of energy required for expression, therefore may allow matter to form as suggested by the physicist Lorentz a century ago (Lorentz, 1925). Note that Einstein postulated quanta as a fundamental property of light itself, rather than as something that arose in the interaction of light with matter as suggested by Max Planck (Isaacson, 2008). Hence, c contributes to or even establishes a reality of nature with a past, present, and future, to also be experienced as a phenomena of connection between seemingly independent islands of matter. This characteristic of 'material connection' or 'connection' is therefore also proposed to be implicit in the nature of light and becomes part of objective reality because of the speed of light.

But as explored in Chapter 1.2, a 'past' can also be viewed as established reality as defined by what the eye or other lenses of perception can see. Hence, 'It is the perceivable result of all the work and effort that has taken place so far. It is the foundation upon which the present and future will be built. It represents a status quo, stability, and even rigidity, and given that it is the result of the long play of time, it will not easily be persuaded to become another thing. It can be thought of as that which the eye can see when it looks around it. There is "physicality" to what the eye can see and so the essence of the past is a kind of physical-ness. So, ingrained in light, is this ability to project or create physical-ness.'

Also as explored previously, the 'present' is 'the tremendous play of forces of all kinds to express

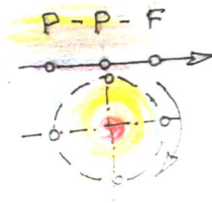

themselves here and now. There is "vitality" that is present in this play and often it is the most energetic or forceful of the forces that will win out, as opposed to the most insightful or thoughtful. All the tremendous possibility of the future is seeking for expression now and so this essential vitality can also be thought of as a projection or possibility implicit in light.'

The 'future' is 'the inevitability of what will manifest. The great thoughts, the great ideas, the purpose, the possibilities will sooner or later express themselves in what we call the future. And the essence of this is thoughtfulness or a curiosity or a purpose that we can summarize as an essential "mentality". So embedded in light is this ability to project mentality.'

While in its native state these properties of light are merged together, the experience of the past, present, future, and matter that arise when light travels at c can be thought of as a means by which the essential oneness is disaggregated in a way that it becomes more plainly perceivable as qualities of physicality, vitality, mentality, and connection by fundamentally fragmented constructs and emergences. The whole notion of fragmentation becoming whole again is the basis of light-based-singularity dynamics, as will be explored in detail in this book. The constant quantum-level computation and generation of genetic-type information that compels emergences to behave in consistency with their class of genetic-type information is the essence of the dynamics of light-based-singularities.

These implicit characteristics of the nature of light as experienced at the layer of reality so set up by a finite speed of light may hence be summarized by Equation 2.1.1, where c_U refers to the speed of light of 186,000 miles per second, that has created the perceived nature of reality, U, as expressed in Equation 2.1.1:

$$c_U : [Physical, Vital, Mental, Connection]$$

Eq. 2.1.1: Implicit Characteristics of the Nature of Light at U

The notation 'U' suggests 'untransformed' as will become clearer with the discussion of the light-space-time emergence equation in Chapter 3.1.

Nature of Light at ∞

Exploring further, it is thought however that at quantum levels the nature of reality is characterized by wave-particle duality. Light itself and matter may be experienced as both particles and waves. But for matter to be experienced as waves implies that 'h' must have become a fraction of itself, $h_{fraction}$, to allow the concentration or possibility of quanta to have dispersed into wave-form. This further implies that c must have become greater than itself, c_N, such that the inequality specified by Equation 2.1.2 holds:

$$c_N > c_U$$

Eq. 2.1.2: Layer N, Layer U Inequality of Speed of Light

Note that what is implied here is that just as there is a nature of reality specified by U that is the result of the speed of light being 186,000 miles per second, so too there is another nature of reality specified by N that is the result of a speed of light greater than 186,000 miles per second.

This is akin to recent developments in physics with the notion of property spaces being separate from but influencing physical space as explored by Nobel Physicist Frank Wilczek in his book 'A Beautiful Question" (Wilczek, 2016). But further in "Slow Light" Perkowitz's recent treatment of today's breakthroughs in the science of light (Perkowitz, 2011) he states: "Although relativity implies that it's impossible to accelerate an object to the speed of light, the theory may not disallow particles already moving at speed c or greater."

So light traveling at c_N may be possible. Current instrumentation, experience, and normal modes of thinking though, having developed as a bi-product of the characteristics so created in the layer of reality U may be inadequate to access N without appropriate modification. The notion of wave-particle duality already challenges the notion of normal thinking perhaps because wave-like phenomena could be a function of faster than c motion, and particle-like phenomena a function of equal to c motion. That these may be happening simultaneously is reinforced by principles such as complementarity in which experimental observation may allow measurement of one or another but not of both as pointed out by Whitaker (Whitaker, 2006). But further the very notion of the Pilot-Wave Interpretation of Quantum Mechanics, as modeled by the physicists Bohm and DeBroglie (Holland, 1995), is that constructive interference of waves causes particles to move toward these areas. Hence in this interpretation both particle and wave always exist simultaneously.

But then taking this trend of a possible increase in the speed of light to its limit, this will result in a speed of light of infinite miles per second. The question is, what is the nature of reality when light is traveling at infinite miles per second? As explored in Section 1, in any continuum, light originating at any point will instantaneously have arrived at every other point. Hence light will have a full and immediate *presence* in that continuum. Further, that light will *know* everything that is happening in that continuum completely and instantaneously – that is know what is emerging, what is changing, what is diminishing, what may be connected to what, and so on - or have a quality of *knowledge*. It will connect every object in that continuum completely and therefore have a quality of connection or *harmony*. Finally nothing will be able to resist it or set up a separate reality that excludes it and hence it will have a quality of *power*.

These implicit characteristics of the nature of light as experienced at the layer of reality so set up by an infinite speed of light may hence be summarized by Equation 2.1.3, where c_∞ refers to the speed of light of ∞ miles per second, that has created the perceived nature of reality, ∞:

$$c_\infty: [Presence, Power, Knowledge, Harmony]$$

Eq. 2.1.3: Implicit Characteristics of Nature of Light at ∞ Speed

Transformation from ∞ to U

But by (2.1.3) it can also be noticed that 'physical' is related to Presence, 'vital' is related to Power, 'mental' is

70

related to Knowledge, and 'connection' is related to Harmony.

If this is the case the question then, is how do these apparent qualities at ∞ precipitate or become the physical-vital-mental-connection based diversity experienced at U? This may be achieved through the intervention or action of a couple of mathematical transformations acting on the implicit characteristics of nature of light at ∞ speed as summarized by (2.1.3).

First, the essential characteristics of Presence, Power, Knowledge, Harmony that it is posited exists at every point-instant by virtue of the ubiquity of light at ∞ will need to be expressed as sets with up to infinite elements. Second, elements in these sets will need to combine together in potentially infinite ways to create a myriad of seeds or signatures that then become the source of the immense diversity experienced at U. Note that these mathematical transformations suggest that light may gather itself in such a way so as to first release, as it were, from its essential nature such sets with infinite variation around the essential characteristics, and second, an infinite combination of such vast variation. This suggests that all that is seen and experienced at U may be nothing other than 'information' or 'content' of light and as such that there are fundamental mathematical symmetries at play where everything at U is essentially the same thing that exists at ∞. This then is also the primordial basis of genetics. Further, the heart of creation, in this view, is perhaps the result of persistent quantum-computations that interrelates various realities so set up by light traveling at different speeds. The essential "nodes" of such quantum-computation take place at distinct levels or realities created at an interface of light traveling at different speeds, and these nodes distinguish the precipitating structure of a downward-strand of ubiquitous subtle-DNA.

Assuming that the first transformation occurs at a layer of reality K where the speed of light is c_K, such that $c_U < c_K < c_\infty$, this may be expressed by Equation 2.1.4:

$$c_K: [S_{Pr}, S_{Po}, S_K, S_H]$$

Eq. 2.1.4: The First Transformation to Sets at Layer K

S_{Pr} signifies 'Set of Presence', S_{Po} signifies 'Set of Power', S_K signifies 'Set of Knowledge', S_H signifies 'Set of Harmony/Nurturing' (note: Harmony and Nurturing will be used interchangeably through this book).

Assuming that the second transformation occurs at a layer of reality N where the speed of light is c_N, such that $c_U < c_N < c_K < c_\infty$, this may be expressed by Equation 2.1.5:

$$c_N: f(S_{Pr} \times S_{Po} \times S_K \times S_H)$$

Eq. 2.1.5: The Second Transformation to Seeds at Layer N

The unique seeds are therefore a function, f, of some unique combination of the elements in the four sets S_{Pr}, S_{Po}, S_K, S_H.

The relationship between the layers of light may be modeled by the following matrix in Equation 2.1.6:

$$Light_{Matrix} = \begin{vmatrix} c_\infty: [Pr, Po, K, H] \\ (\downarrow R_{C_K} = f(R_{C_\infty})) \\ c_K: [S_{Pr}, S_{Po}, S_K, S_H] \\ (\downarrow R_{C_N} = f(R_{C_K})) \\ c_N: f(S_{Pr} \times S_{Po} \times S_K \times S_H) \\ (\downarrow R_{C_U} = f(R_{C_N})) \\ c_U: [P, V, M, C] \end{vmatrix}$$

Eq. 2.1.6: Light-Matrix & Downward-strand of Subtle-DNA

The matrix should be read from the top row down to the bottom row as indicated by the ↓ between rows, and suggests a series of transformations leading from the ubiquitous nature of light implicit in a point – presence, power, knowledge, harmony - to the seeming diversity of matter observed at the layer of reality U which is fundamentally the same presence, power, knowledge, and harmony projected into another form of itself.

The first transformation summarized by (2.1.4) into light-based sets is the result of a quantization function that culls out set-based wholes from the infinite potentiality in Light, and is summarized by Equation 2.1.7:

$$R_{C_K} = f(R_{C_\infty})$$

Eq. 2.1.7: Quantization Function Resulting in Light-Based Sets

This is suggesting that the reality at the layer specified by the speed of light c_K, R_{C_K}, is a function of the reality at the layer specified by the speed of light c_∞. The function itself is a quantization-function that allows some essential characteristics in the point-nature of light to express itself more "materially", relatively speaking, in sets described by (2.1.4), by light traveling at a relatively slower speed. This quantization-function can be thought of as a node in which a higher-level quantum-computation occurs by which infinite possibility begins to disaggregate itself.

The second transformation summarized by (2.1.5) into light-based seeds, is summarized by Equation 2.1.8:

$$R_{C_N} = f(R_{C_K})$$

Eq. 2.1.8: Quantization Function Resulting in Light-Based Seeds

This is suggesting that the reality at the layer specified by the speed of light c_N, R_{C_N}, is a function of the reality at the layer specified by the speed of light c_K. This function combines elements of the light-based sets into unique seed, and is envisioned as also being a quantization-function that allows what is expressed by the light-based sets to gather into unique light-based seeds.

Note that (2.1.3), (2.1.4), (2.1.5), (2.1.7), and (2.1.8) specify the dynamics of the seed-singularity. There is a light-based wholeness in the seed-singularity that results in a first meaningful sphere of cohesion with the creation of myriad light-based seeds.

The quantization-function that interrelates layers N and U puts in place the essential dynamics of partial-singularities and is summarized by Equation 2.1.9:

$$R_{C_U} = f\left(R_{C_N}\right)$$

Eq. 2.1.9: Quantization Function Resulting in Material Diversity

This is suggesting that the reality at the layer specified by the speed of light c_U, R_{C_U}, is a function of the reality at the layer specified by the speed of light c_N. This transformation builds on the unique seeds suggested by (2.1.5) to create the diversity of U as specified by (2.1.1). This too is envisioned as being a quantization-function that further allows all the combination of possibility to express itself in more material form. Note that the entire upward-strand to be discussed soon, also plays a key part in the dynamics of any partial-singularity.

In this framework the notion of wave-particle duality hence may become complementary block-field-wave-particle "quadrality" where block refers to phenomenon resident to ∞, field to phenomenon resident to K, wave to phenomenon resident to N, and particle to phenomenon resident to U. The block is all the reality always present behind the surface and is captured by (2.1.3) or the top line in the Light-Matrix as expressed in (2.1.6). The field is captured by (2.1.4) or the second major (non-parenthetical) line from the top in (2.1.6) and can be thought of as layers of possibility existing in each of the sets. The wave is captured by the creation of seeds represented by (2.1.5) or by the third major (non-parenthetical) line in (2.1.6). The particle that is apparently disconnected from the whole is captured by (2.1.1) or the bottom line in (2.1.6).

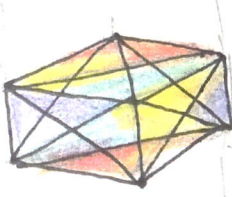

So what we arrive at is a fundamental Light-Matrix or the downward-strand of a ubiquitous subtle-DNA that suggests key dynamics for different layers of Light. This Light-Matrix also summarizes the essential dynamics of a Light-Based-Singularity Superstructure in which the seed-singularity, all partial-singularities, and the Second Singularity emerge.

Note that the Second Singularity will necessitate the fulfilling of all conditions in a 'Space-Matrix" or upward-strand to be elaborated in Section 3.

Layer Zero (0)

For the sake of completeness the thought-experiment to do with light existing at a zero-speed must now be brought up. In Chapter 1.1 a possible reality that would result if light were to exist at this speed was described as opposite to the reality were light to travel at an infinite speed.

Hence, ubiquitous Presence of Light would become the ubiquitous Absence of Light, or Darkness. The aspect of Power would become utter Weakness. The aspect of Knowledge would become complete Ignorance. The aspect of Harmony would become total Chaos.

And all this because Light would be unable to travel from where it is, in complete opposition to its known nature of traveling at a speed of c, and perhaps its true nature when traveling at a speed of ∞. Light, hence, would be hidden in itself, so as to speak, in some sort of a negative infinity. An Equation, 2.1.10, Nature of Light at Speed Zero (0), would hence be:

$$c_0: [Darkness, Weakness, Ignorance, Chaos]$$

Eq. 2.1.10: Nature of Light at Speed Zero (0)

Further, (2.1.6), the Light-Matrix, would be modified to become Equation 2.1.11, Light-Matrix with Zero-Limit, where c_0 implies light at zero-speed, and D, W, I, and C imply Darkness, Weakness, Ignorance, and Chaos respectively:

$$Light_{Matrix}(0_{limit}) = \begin{vmatrix} c_\infty: [Pr, Po, K, H] \\ (\downarrow R_{C_K} = f(R_{C_\infty})) \\ c_K: [S_{Pr}, S_{Po}, S_K, S_H] \\ (\downarrow R_{C_N} = f(R_{C_K})) \\ c_N: f(S_{Pr} \times S_{Po} \times S_K \times S_H) \\ (\downarrow R_{C_U} = f(R_{C_N})) \\ c_U: [P, V, M, C] \\ \Uparrow \\ c_{0:[D,W,I,C]} \end{vmatrix}$$

Eq. 2.1.11: Light-Matrix with Zero-Limit

The implication of the bottom zero-limit line is that just as there is proposed to be an influence from the upper layers of light on U, as will be explored in subsequent chapters, so too there is a subtle influence from this lower layer of "light" that perhaps is responsible for the obstinacy of the untransformed nature of practical reality at U. Such obstinacy will be further explored in Chapters 2.5, 2.6, and 3.7 and reinforces the need for the notion of 'transformation' in the first place. Further, such obstinacy also has a bearing on the phenomena of genetic mutation by which partial-singularities with greater spheres of influence, and the Second Singularity can come into being, as will also be discussed in greater detail subsequently. Note that since any partial-singularity is a biography, subsequent partial-singularities will always contain previous partial-singularities within themselves, and hence will become partial singularities with greater spheres of influence.

Superposition, Entanglement, Quantum Computation, Genetic Mutation, and Singularities

As may be apparent in the Light-Matrix (2.1.6) and the complete form (2.1.11), circumstance at U is the outcome of a number of influences from layers ∞, K, N, U itself, and Zero (0). But further since each of these layers is itself a reality caused by a different speed of light, the interface between these layers occurs at the quantum-levels. Hence the quantum-levels are replete with superposition emanating from layers ∞, K, N, U itself, and Zero (0). There is simultaneity and multiple possibilities that will determine what manifests at U. But as will be explored in more detail in subsequent chapters there is a logic or process to what manifests. This logic and process provides insight into the reality of genetic mutation, which itself may be seen as a process leading to the reality of partial-singularities with greater spherical influence, and to a possible Second Singularity.

The notion that a quantum-object – an entity at the quantum-level – can be in an infinite number of superposed states as is presumed in several leading interpretations of quantum mechanics, is called into question in the interpretation offered in this book. This was the subject of a previous book on quantum computation (Malik, 2018c). There may in fact be infinite number of possibilities vying for manifestation, but there is likely a more precise process by which this manifestation takes place. In other words, genetic mutation is not random, but follows process at the subtle-levels.

Further, the process of entanglement by which quantum-objects are made to relate to each other, may be otiose since in the interpretation offered in this book entanglement exists ab initio at the layer ∞. This entanglement will be referred to as ∞-entanglement and will be further explored in Chapter 2.2. Further, there are additional forms of entanglement that occur even before a quantum-object becomes perceivable at layer U. There is entanglement at layer K by virtue of field-type action of light-based architectural sets. This will be referred to as K-entanglement and will be further explored in Chapter 2.3. There is an entanglement at layer N by virtue of the light-based wave-type action of seeds. This will be referred to as N-entanglement and will be further explored in Chapter 2.4.

Note that any process of entanglement is relevant from the point of view of genetic mutation as it can be thought of as a source of mutation. Further, quantum computation, integral to the precipitation of quantization-function as described in this chapter, is therefore a reality in the manifestation of the minutest of circumstances at U and is also integral to the process of genetic mutation. Genetic mutation, the process by which repositories of instruction change, allows the wholeness implicit in the seed-singularity to materialize as the many partial-singularities with greater and greater spheres of influence, culminating in the Second Singularity we will encounter later in this book.

Chapter 2.2: The Seed-Singularity Ubiquitous-Point-Instant Pre-Genetic ∞-Entanglement Library & Dynamics

The 'ubiquitous-point-instant' captures the inherent nature that appears to exist in the system and is represented by the top-line in Equation (2.1.6) and (2.1.11), different forms of the Light-Matrix. The ubiquitous-point-instant is a function of the dynamics of that reality where light travels at ∞ speed and is envisioned as being infinitely entangled by virtue of light being omnipresent-omnipotent-omniscient-omninurturing in that realm. In other words this nature is a function of the properties of light derived in Equation (2.1.3) - Presence, Power, Knowledge, and Harmony – reproduced here for convenience:

c_∞: [*Presence, Power, Knowledge, Harmony*]

The 'point' aspect of the ubiquitous-point-instant suggests the space-dimension and as discussed in Chapter 1.3, that space is seeded with the possibilities inherent in the properties of Presence, Power, Knowledge, and Harmony. The 'instant' aspect suggests the time-dimension and gives insight into the process of emergence that the possibilities in space progressively surface as. Therefore, the 'point' or space-aspect provides insight into structural aspects of genetic-type information. The 'instant' or time-aspect provides insight into the mutational aspects of genetics related processes. "Genetics" when viewed in this manner is a lens into the type and process of information-richness associated with a range of emergences and further, the process by which partial-singularities develop greater spheres of influence culminating in the possibility of the Second Singularity. If there is a history of singularities, or a biography that relates the seed-singularity with partial-singularities, with the Second Singularity, then such genetics can be thought of as the language in which it is written. Note though that such separation into point and instant aspects is just a way of trying to unpack some of the dynamics in that realm where everything, even space and time, are all one thing.

In its ubiquitous-point-instant, Presence-Power-Knowledge-Harmony wholeness, Presence allows emergences to continue to develop as per the possibilities implicit in the past-present-future or physical-vital-mental pathway.

Beginning to translate this into an equation, the notation $System_{Pr}$ is given to system-presence. Note that the derivation of this and many subsequent equations was part of my doctoral work at University of Pretoria (Malik, 2017a). This system-presence is true across any considered Time-Space continuum starting from a time-space boundary '0' to a time-space boundary 'N'. This notion is characterized by the notation $TS_{0 \to N}$. Within that boundary from 0 to 'N', the 'presence' is such that it will always seize an opportunity to cause a shift from the physical-leading to the vital-leading, and from the vital-leading to the mental-leading. Research shows (Malik, 2009) that greater degrees of freedom is afforded by such a shift.

The notion that the 'presence' seizes on 'opportunity' is characterized by the notation:

$$Presence$$
$$\downarrow$$
$$Opportunity$$

The shift from physical-leading (P_L) to vital-leading (V_L) and vital-leading (V_L) to mental-leading (M_L) is characterized by:

$$P_L \rightarrow V_L$$
$$V_L \rightarrow M_L$$

Hence in this approach it is suggested that:

$$System_{Pr} \equiv TS_{0 \to N} \begin{bmatrix} Presence \\ \downarrow \\ Opportunity \end{bmatrix} \begin{bmatrix} P_L & \to & V_L \\ V_L & \to & M_L \end{bmatrix}$$

But there is something else about this Presence as well. All other developments take place in it. That is, it provides a container of sorts in which the plays of system-power, system-knowledge, and system-harmony/system-nurturing can take place. This notion is summarized by the notation:

$$Container \begin{bmatrix} System_P \\ System_K \\ System_N \end{bmatrix}$$

Hence, combining these various components, an equation for 'system-presence', Equation 2.2.1, arises:

$$System_{Pr}$$
$$\equiv TS_{0 \to N} \begin{bmatrix} Presence \\ \downarrow \\ Opportunity \end{bmatrix} \begin{bmatrix} P_L & \to & V_L \\ V_L & \to & M_L \end{bmatrix} \& Container \begin{bmatrix} System_P \\ System_K \\ System_N \end{bmatrix}$$

Eq 2.2.1: System Presence

In its ubiquitous-point-instant, Presence-Power-Knowledge-Harmony wholeness, Power allows emergences to continue to happen in spite of tremendous oppositions of all kinds; this too, regardless of field or area.

Constructing an equation for system-power, the notation $System_P$ is used to represent system-power. Any endeavor will always be met with resistances of various kinds. The resistances that arise along the physical dimension are referred to as P_R. The resistances that arise along the vital dimension are referred to as V_R. The resistances that arise along the mental dimension are referred to as M_R. In the fruition of any endeavor one or all of these types of resistances may arise. Further, resistance of one kind often feeds on resistance of another kind, and to generalize the resistances encountered in an endeavor, these may be characterized as the product of the three types of resistance:

$$P_R * V_R * M_R$$

These resistances arise across any considered Time-Space boundary from 0 to 'N', and therefore it may be said that the power of the system is such that:

$$power > \sum_{TS=0}^{N} P_R * V_R * M_R$$

An equation for 'system-power', Equation 2.2.2, hence, is the following:

$$System_P \equiv power > \sum_{TS=0}^{N} P_R * V_R * M_R$$

Eq 2.2.2: System Power

In its ubiquitous-point-instant, Presence-Power-Knowledge-Harmony wholeness, Knowledge orchestrates emergences to continue to happen by leveraging the right instruments and circumstances.

Translating this into an equation, the notation, m_K, is used for system-knowledge. This em_K is such that it leverages the right strumentation and circumstance to bring about the progress that is possible. This concept of 'instrumentation' is denoted by the subscript 'I'. The concept of 'circumstance' is denoted by the subscript 'C'. Both instrumentation and rcumstance can be of a physical, vital, or mental type and this possibility is denoted by:

$$\begin{bmatrix} P_{I,C} \\ V_{I,C} \\ M_{I,C} \end{bmatrix}$$

Further, the notion that the 'knowledge' is such that it 'leverages' the right instrumentation and circumstance is depicted by:

$$Knowledge$$
$$\downarrow$$
$$Leverage$$

This act of leveraging results in a fundamental shift so that the physical-leading yields to the vital-leading, and the vital-leading yields to the mental-leading. Hence:

$$\begin{matrix} Knowledge \\ \downarrow \\ Leverage \end{matrix} \begin{bmatrix} P_{I,C} \\ V_{I,C} \\ M_{I,C} \end{bmatrix} \rightarrow \begin{bmatrix} P_L & \rightarrow & V_L \\ V_L & \rightarrow & M_L \end{bmatrix}$$

Since this behavior may exist across any Time-Space continuum an equation for system-knowledge, Equation 2.2.3, is suggested:

$$System_K \equiv TS_{0 \to N} \begin{bmatrix} Knowledge & \begin{bmatrix} P_{I,C} \\ V_{I,C} \\ M_{I,C} \end{bmatrix} & \to & \begin{bmatrix} P_L & \to & V_L \\ V_L & \to & M_L \end{bmatrix} \\ Leverage & & & \end{bmatrix}$$

Eq 2.2.3: System Knowledge

In its ubiquitous-point-instant, Presence-Power-Knowledge-Harmony wholeness, Harmony or nurturing allows emergences to continue to happen with more and more degrees of freedom coming to the surface.

The characteristic of this implicit-nurturing may be referred to as 'system-nurturing'. Like the other characteristics it is suggested to exist across a Time-Space continuum. This is depicted by:

$$TS_{0 \to N}$$

There is an action of nurturing such that any state is always advanced to a higher level. This is depicted by:

$$\coprod_{Nurturing} \begin{pmatrix} P_- & M_+ \\ V_- & V_+ \\ M_- & P_+ \end{pmatrix}$$

Hence, there is a 'union', depicted by 'U' that 'nurtures' the negatives towards their positives.

Further, there is an increasing action of nurturing such that the possibility of integration is always increased to form a larger and larger basis. This increasing basis is depicted as being modulated by the polar coordinates 'r' and 'θ', where r is the radius which increases from an initial value of '0', and 'θ' is an angle from '0' to '360'.

Hence, the equation of system-nurturing, Equation 2.2.4, is depicted as:

$$System_N \equiv TS_{0 \to N} \left(\coprod_{Nurturing} \begin{pmatrix} P_- & M_+ \\ V_- & V_+ \\ M_- & P_+ \end{pmatrix} mod\,(r, \theta) \right)$$

Eq 2.2.4: System Nurturing

It is suggested that these four characteristics exist across any system, and to denote this it is generalized that every point-instant in any system is embedded with this four-fold intelligence. In other words there is in effect a library of information represented by Equations 2.2.1 – 2.2.4 whose operation is governed by ∞ - entanglement. Further, there is a category of constructive genetic mutation, ∞ - entanglement mutation rooted in Equations 2.2.1 – 2.2.4 that may be activated to bring about change at the material level.

In effect the ubiquitous-point-instant pre-genetic ∞ - entanglement library provides insight into the dynamics native to the seed-singularity, and further to the kind of constructive ∞-entanglement genetic mutation that will subsequently be available at the material level.

The following chapters in this section explore additional categories of entangled pre-genetic libraries implicit in the seed-singularity that spawn additional categories of constructive genetic mutation at the material level. It is such constructive genetic mutation that alters the information by which various emergences exist, thereby allowing a biography toward the Second Singularity to in effect be recorded.

Note further, that libraries created at subsequent levels of the downward-strand, and the very process of genetic mutation involving the material level, are influenced by this subtlest level ∞-entanglement library.

Chapter 2.3: - The Seed-Singularity Architectural Forces Pre-Genetic K-Entanglement Library & Dynamics

The characteristics embedded in a ubiquitous-point-instant as in (2.1.3) suggest a possibility that is hard to fathom. One can only glimpse this extraordinary nature. And yet it can be suggested that this extraordinary nature is barely visible unless the right analytical lens of the sort being suggested in this book is first set up. Further, it is suggested that this extraordinary nature is responsible for a broader set of architectural forces that exist behind the visible face of things. Such architectural forces can be envisioned as emanating from a pre-genetic library existing in a reality where light travels at the speed c_K (above c, and closer to ∞, as defined in Chapter 2.1). Due to the process of K-entanglement this pre-genetic library can also be thought of as constructively influencing genetic mutation, and therefore all singularity-biographies, the process by which a singularity materializes possibility contained in the seed-singularity to morph into partial-singularities, and then into the Second Singularity.

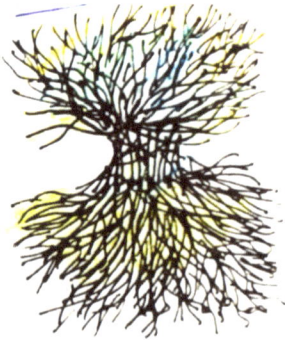

Hence, system-presence, system-power, system-knowledge, and system-nurturing that define the nature of every point in our system, become more tangible as a broader set of architectural forces that emanate from each of them.

Considering system-presence, here is a characteristic that appears to be everywhere (Malik, 2015) at the service of all the constructs that develop within it. There is a diligence and perseverance by which any opportunity for progress is seized. Further, if one considers the extraordinary detail that appears in any construct, whether an atom, a body, a planet, or a galaxy, one is struck by the high degree of perfection that surfaces in this presence.

So if one contemplates the nature of this system-presence there is a set of forces that surface. Depicting such a set as $S_{System_{Pr}}$, one can arrive at elements such as Service, Perfection, Diligence, Perseverance, amongst others, that are part of this set. Hence, the set can be described by Equation 2.3.1:

$$S_{System_{Pr}} \ni [Service, Perfection, Diligence, Perseverance, ...]$$

Eq 2.3.1: Set of System Presence

But from the point of view of the speed of light, K is a reality set up by light traveling at c_K, where $c_U < c_N < c_K < c_\infty$. Conversely assuming an inverse proportionality with 'h', as c_K is closer to ∞, h will be closer to zero, and 'matter' or form will be highly dispersed. This dispersal is presumed to be field-like so that in effect any element of the set described in (2.3.1) or in to be discussed sets (2.3.2-4) will have a field-like reality and express K-entanglement such that all emergences emanating or comprised of that element will automatically partake in an entanglement with every other unique emergence having that element as part of its foundation. K-entanglement is therefore different from ∞-entanglement and every emergence will at least have both these types of entanglements subtly coordinating or influencing its action.

Similarly, considering the characteristic of system-power, one can hypothesize that there is a family of forces that emanates from it. The kinds of forces may be thought of as Power, Courage, Adventure, Justice, amongst others. The set for system-power can hence be depicted by Equation 2.3.2:

$S_{System_P} \ni [Power, Courage, Adventure, Justice, ...]$

Eq 2.3.2: Set of System Power

Similarly, considering the system-knowledge as the root of various powers that emanate from it, one may characterize the set for system-knowledge by Equation 2.3.3:

$S_{System_K} \ni [Wisdom, Law\ Making, Spread\ of\ Knowledge\ ...]$

Eq 2.3.3: Set of System Knowledge

The set for system-nurturing is depicted by Equation 2.3.4:

$S_{System_N} \ni [Love, Compassion, Harmony, Relationship\ ...]$

Eq 2.3.4: Set of System Nurturing

Equations 2.3.1 – 4 describe potentially infinite-element sets that comprise the Architectural Force K-entanglement Pre-Genetic Library. This entangled library provides further insight into the dynamics implicit in the seed-singularity. Further, there is a category of constructive genetic mutation, K-entanglement mutation that may be activated to bring about change at the material level. As already suggested, this category of genetic mutation will also influence the narrative that defines any singularity-biography.

Chapter 2.4: The Seed-Singularity Organizational-Uniqueness Pre-Genetic N-entanglement Library & Dynamics

The seed-singularity mathematical model that leads from unity to unique diversity as is being discussed in this Section, is perhaps justified by observation of phenomena. Such observation generates hypotheses that every organization, whether an atom, cell, person, team, corporation, market, or country is unique and that this uniqueness can be specified in terms of elements of the derived light-based sets for power, knowledge, presence, and nurturing.

At the sub-atomic level, hence, Nobel Laureate Wolfgang Pauli's 'Pauli Exclusion Principle' states that no two similar fermions, which include fundamental particles with half-integer spin such as protons, neutrons, and electrons, can occupy the same quantum states simultaneously (Pauli, 1964). Spin has to do with the angle that the particle has to rotate through before being symmetrical with its original state. Half-integer spin particles need to rotate through 720 degrees before being symmetrical with their original state. The implication of the Pauli Exclusion Principle is that fundamental structure and consequently stability comes into being at the atomic level, which as is evident in the Periodic Table also allows the separation of function related to form. This stability related to the underlying structure of atoms implies the basis of uniqueness and diversity. In the absence of the Exclusion Principle matter would just be a dense soup (Hawking, 1988) with particles occupying overlapping space.

At the observable level uniqueness is evident from the immense diversity of distinct species on earth (Mora, 2011) estimated to be over 2 million, and further the uniqueness of every member of each species. This member-level uniqueness is suggested by the difference in non-coding regions of the DNA that may vary in their sequence by about 1 to 4 percent, which in turn result in unique protein binding sequences of each human (Snyder, 2010), as an example, which in turn results in unique observable qualities.

At the astronomical level Einstein's Special Theory of Relativity (Einstein, 1995) suggests that every coordinate system potentially has its own space-time rendering as opposed to there being one absolute space and time. This implies the notion of uniqueness as an implicit property of space.

The four properties explored in Chapter 2.1 and elaborated in Chapter 2.2 define the source of that uniqueness. From this source emanate 4 sets of forces that suggest the boundaries of that uniqueness as explored in Chapter 2.3.

Assuming then that the fount of uniqueness is system-presence, a general equation for organizations that belong to the family of system-presence can be derived. Such uniqueness can be depicted as Sig_x where the subscript 'x' refers to the source family, and 'Sig' or signature to 'uniqueness'. Hence the uniqueness of an organization in the family of system-presence would be notated by Sig_{Pr}.

In line with the development of properties of a point and the precipitating architectural forces as discussed in Chapter 2.2 and 2.3 respectively, an approach to constructing such uniqueness is to assume a primary factor X that drives the uniqueness that belongs to the set $S_{System_{Pr}}$. Further, assume that the uniqueness is qualified by a number of secondary factors Y that may belong to any of the 4 sets - $S_{System_{Pr}}, S_{System_P}, S_{System_K}, S_{System_N}$. The primary factor X would have a greater weightage than any of the secondary factors Y. The weightage of X hence could be depicted by the number 'a', and the weightage of Y a number $'b_{0-n}'$, such that a > b. Further, the secondary element can repeat from '0 – n' times, and is hence depicted as $\overline{Yb_{0-n}}$.

The equation, Equation 2.4.1, hence for a unique organization derived from the family of system-presence is:

$$Sig_{Pr} = Xa + \overline{Yb_{0-n}} \ where \begin{bmatrix} X \in [S_{System_{Pr}}] \\ Y \in [S_{System_{Pr}}, S_{System_P}, S_{System_K}, S_{System_N}] \\ a, b \ are \ integers; a > b \end{bmatrix}$$

Eq 2.4.1: System Presence Based Unique Organization

But from the point of view of the speed of light, N is a reality set up by light traveling at c_N, where $c_U < c_N < c_K < c_\infty$. Conversely assuming an inverse proportionality with 'h', as c_N is closer to but greater than c_U, h_U will be less than h, and 'matter' or form will be unable to accumulate as it does at U, instead tending to disperse like a wave. The wave-like nature implies an entanglement, N-entanglement, such that any emergence will always be unique. Uniqueness could not be unless there was a dynamic such as N-entanglement in place.

Similarly, an equation, Equation 2.4.2, for a unique organization derived from the family of system-power is:

$$Sig_P = Xa + \overline{Yb_{0-n}} \ where \begin{bmatrix} X \in [S_{System_P}] \\ Y \in [S_{System_{Pr}}, S_{System_P}, S_{System_K}, S_{System_N}] \\ a, b \ are \ integers; a > b \end{bmatrix}$$

Eq 2.4.2: System Power Based Unique Organization

An equation, Equation 2.4.3, for a unique organization derived from the family of system-knowledge is:

$$Sig_K = Xa +$$

$$\overline{Yb_{0-n}} \quad where \quad \begin{bmatrix} X \in [S_{System_K}] \\ Y \in [S_{System_{Pr}}, S_{System_P}, S_{System_K}, S_{System_N}] \\ a, b \ are \ integers; a > b \end{bmatrix}$$

Eq 2.4.3: System Knowledge Based Unique Organization

An equation, Equation 2.4.4, for a unique organization derived from the family of system-nurturing is:

$$Sig_N = Xa +$$

$$\overline{Yb_{0-n}} \quad where \quad \begin{bmatrix} X \in [S_{System_N}] \\ Y \in [S_{System_{Pr}}, S_{System_P}, S_{System_K}, S_{System_N}] \\ a, b \ are \ integers; a > b \end{bmatrix}$$

Eq 2.4.4: System Nurturing Based Unique Organization

The four preceding equations can be generalized by Equation 2.4.5:

$$Sig = Xa +$$

$$\overline{Yb_{0-n}} \quad where \quad \begin{bmatrix} X \in [S_{System_{Pr}}, S_{System_P}, S_{System_K}, S_{System_N}] \\ Y \in [S_{System_{Pr}}, S_{System_P}, S_{System_K}, S_{System_N}] \\ a, b \ are \ integers; a > b \end{bmatrix}$$

Eq 2.4.5: Generalized Equation for Unique Organization

Equations 2.4.1 – 5 describe potentially infinite-element sets that comprise the Organizational-Uniqueness Pre-Genetic N-entanglement Library.

The N-entanglement level, like the ∞-entanglement and K-entanglement levels, is also hypothesized as being that part of the backbone of subtle-DNA, where distinct elements of the deeper fourfold-based sets combines, to create unique seeds. These seeds are the primary organizing force at the center of any and every organization and shed light into a key dynamic of the seed-singularity. Hence the subtle-library behind any emergence of matter and life derives primarily from the action on the N-entanglement level, and further, there is a category of constructive genetic mutation, N-entanglement mutation, that may be activated to bring about change at the material level.

The next two chapters turn away from the structural or space-aspect of genetics implicit in the seed-singularity, and focus on the emergent, mutational, or time-aspect of genetics implicit in any light-based partial-singularity.

Hence Chapter 2.5 will focus on possible mutation-based narrative that describes dynamics related to partial-singularities. Essentially the deep structural basis of genetics hinted at by the ∞-entanglement, K-entanglement, and N-entanglement levels, also becomes the basis for mutational-emergence. In other words, the possibilities inherent in these levels can change emergence through genetic-type mutation to prescribe or narrate a biography of light-based partial-singularities.

Note that massively significant mutations such as took place on the SRY gene and led to the evolution of human from the chimpanzee species (Ridley, 1999) are architected by a layer of light traveling faster than c and likely involved one of the three entanglement schemes just described - ∞-entanglement, K-entanglement, or N-entanglement.

Chapter 2.6 will focus on the further mathematical systematization of the mutational-sequence discussed in Chapter 2.5, thus more generally describing the pathway of emergence for any partial-singularity.

Chapter 2.5: A Possible Mutational-Sequence in Partial-Singularities

While the uniqueness of organizations as represented by the Signature is a seed, like any seed there is a process for its emergence (Kaufmann, 1995; Portugali, 2012; Yates, 2012), and the uniqueness will often be hidden or very much behind the scene until certain conditions are fulfilled (Malik, 2009). What this implies is that possibility may emerge through mutation, and further some of that mutation may either tap into the subtle-libraries created through the structures already set up by the ∞-entanglement, K-entanglement, and N-entanglement levels, or by interaction with these layers, as will be further explored in Chapter 2.6.

Note that the libraries so set up by the ∞-entanglement, K-entanglement, and N-entanglement levels would be quite different from the Library of Babel, referred to in Beinhocker's Origin of Wealth (Beinhocker, 2006). The Library of Babel is imagined to contain all the possible 500-page books in the English language and would be vastly larger than the universe. A vast majority of these books would be gibberish with characters strung together in random fashion. By contrast the subtle-libraries at the N, K, or ∞ layers are a play on qualities or functions related to the four-foldness implicit in Light. These libraries potentially contain an infinite number of positive and useful functions that will be subtly available to inform any play involving genetic-type information.

The implicit nature of Time and Space suggest a universal developmental model that provides a cue as to the process for emergence. In this model the four sets of architectural forces and the combination of their elements form a pool in space, as already discussed, from which possibility arises. Possibility itself is unique from point to point and is governed by the Equation for Uniqueness

(Equations 2.4.1 through 2.4.5) described in the previous chapter.

In other words, Space contains seeds, as suggested in Section 1, and the implication of the discussion in the previous chapters in Section 2, is that seeds can be thought of as the result of the superposition of ever-present ∞ -entanglement, K-entanglement, and N-entanglement libraries. Thus it could be said that Space is filled with multiple levels of superposed entanglements in static form.

Time on the other hand appears to be the working out of the possibilities implicit in these superposed entanglements. The inevitable trajectory implicit in these superposed entanglements will typically follow a mutational-sequence to be described in this chapter. What is variable is the amount of time for the mutational-sequence to express itself, as this will depend on the strength of the different forces or influences from each of the different layers so set up by light, that are active.

In equation form Space could be described by Equation 2.5.1:

$Space = STATIC(superposition$

$(\infty - entanglement, K - entanglement, N - entanglement))$

Eq 2.5.1: Space

In equation form Time could be described by Equation 2.5.2:

$Time = DYNAMIC(superposition$

$(\infty - entanglement, K - entanglement, N - entanglement))$

Eq 2.5.2: Time

Hence it is observed that initially the mutational-sequence takes a 'physical' form, moving on to a 'vital' form, and then onto a 'mental' form, as will be illustrated in greater detail subsequently. Relating these forms to the equation-segment in (2.1.11), the shaded lines in Illustration 2.5.1 suggest which set of dynamics tend to be more active:

$$c_\infty : [Pr, Po, K, H]$$
$$(\downarrow R_{C_K} = f(R_{C_\infty}))$$
$$c_K : [S_{Pr}, S_{Po}, S_K, S_H]$$
$$(\downarrow R_{C_N} = f(R_{C_K}))$$
$$c_N : f(S_{Pr} \times S_{Po} \times S_K \times S_H)$$
$$(\downarrow R_{C_U} = f(R_{C_N}))$$
$$c_U : [P, V, M, C]$$
$$\Uparrow$$
$$c_{0:[D,W,I,C]}$$

Illustration 2.5.1: Active Layers in Light-Matrix Influencing Initial P-V-M Stages of Mutational Sequence

Once the characteristics implicit in each of the Physical, Vital, and Mental phases are assimilated, then the mutational-sequence takes on an 'integral' form. In this case the more active sets of dynamics are suggested by the shaded lines in Illustration 2.5.2:

$$c_\infty: [Pr, Po, K, H]$$
$$(\downarrow R_{C_K} = f(R_{C_\infty}))$$
$$c_K: [S_{Pr}, S_{Po}, S_K, S_H]$$
$$(\downarrow R_{C_N} = f(R_{C_K}))$$
$$c_N: f(S_{Pr} \times S_{Po} \times S_K \times S_H)$$
$$(\downarrow R_{C_U} = f(R_{C_N}))$$
$$c_U: [P, V, M, C]$$
$$\Uparrow$$
$$c_{0:[D,W,I,C]}$$

Illustration 2.5.2: Active Layers in Light-Matrix Influencing Integral Stage of Mutational Sequence

The integral form is a threshold phase, and allows the uniqueness suggested by the Signature to emerge in fuller force or in its 'force' form.

The final phase is the 'contextual form' that allows the signature to act with impunity within a considered context. The dynamics active in the contextual form are suggested by the shaded lines in Illustration 2.5.3:

$$c_\infty: [Pr, Po, K, H]$$
$$(\downarrow R_{C_K} = f(R_{C_\infty}))$$
$$c_K: [S_{Pr}, S_{Po}, S_K, S_H]$$
$$(\downarrow R_{C_N} = f(R_{C_K}))$$
$$c_N: f(S_{Pr} \times S_{Po} \times S_K \times S_H)$$
$$(\downarrow R_{C_U} = f(R_{C_N}))$$
$$c_U: [P, V, M, C]$$
$$\Uparrow$$
$$c_{0:[D,W,I,C]}$$

Illustration 2.5.3: Active Layers in Light-Matrix Influencing Contextual Stage of Mutational Sequence

Mathematically, if an organization exists at the physical phase, it may be suggested that its signature or uniqueness is modulated by the constant 'π'. π is the seed of a circle or sphere and can be thought of as defining behavior that is tightly bound. Within such a tightly bound volume it will likely not even be apparent what the uniqueness of an organization necessarily is. Assuming the uniqueness to be defined by the derived equation *Sig*, the physical-level (P) behavior can be described by the following equation-segment where 'mod' signifies modulated-by:

P: $Sig * mod\ (\pi)$

If an organization exists at the vital level, it may be suggested that its uniqueness is modulated by the Euler-constant 'e'. e is at the root of exponential behavior. The vital, by definition, is about assertive and aggressive growth the symbol of which is 'e'. Hence vital-level (V) modulation (represented by 'mod') can be described by the following equation-segment:

V: $Sig * mod\ (e)$

If an organization exists at the mental level, it may be suggested that its uniqueness is modulated by the Gaussian Distribution 'G'. G summarizes rational behavior with a key direction followed by most, and directions more on the edge followed by outliers. Mental-level dynamics are arguably quite similar, and it can be suggested are best modeled by such a distribution. Mental-level (M) modulation (mod) can hence be described by the following equation-segment:

$M: Sig * mod\ (G)$

The physical, the vital, and the mental levels as descriptive of the nature of light at c as discussed in Section 1, are also at the root of orientations that emerge later in which patterns of perceiving, being, behaving are set in certain ways. Each pattern has its purpose and its limitation and it can be argued that being able to learn from each orientation and yet being able to move beyond that, is the next logical step in any developmental model. The integral level hence, is about being able to leverage each of the patterns that naturally arise at the three preceding levels at will, and about further, being able to integrate these and arrive at new ways of perceiving and being.

Mathematically such behavior may be represented as being an integrative function ($\int x$) where 'x' is the ability to move between the patterns emanating from G, e, π, at will, represented by $\overline{G, e, \pi}$. Integral-level (I) modulation (mod) of uniqueness (Sig) can hence be represented by the following equation-segment:

$I:\ Sig * mod\ \left(\int \overline{G, e, \pi} \right)$

The condition of overcoming any fixed and limiting patterns is the prerequisite for the emergence of 'Force' or for entering into the force-level. At this level the uniqueness behind the particular development being considered can emerge in its purity and become a truly creative dynamic. This aspect of creativity that is in a sense not bound by circumstance may be represented by the constant 'c', the speed of light in a vacuum, which is an upper limit of the layer that systems practically operate in. This is also likely the level at which N-entanglement is overtly active. Force-level (F) modulation (mod) of uniqueness (Sig) can hence be represented by the following equation-segment:

102

$F: Sig * mod\ (c)$

Once the signature of an organization arises and continues to exercise itself in its purity, it achieves contextual-mastery (C) and is able to enforce itself as though the context it is acting in, that can vary in scale and complexity, were all of the same substance as itself. This is likely the level at which K-entanglement is active. This equality may be represented by the integrative function '$\int = 1$'. The equation-segment that notates this contextual-level (C) modulation (mod) applied to organizational uniqueness (Sig) is hence:

$$C: Sig * mod\ \left(\int = 1\right)$$

Piecing all the equation-segments together the equation for the emergence of uniqueness (Sig_E), where 'X' can be any of the discussed modulations at the respective development-model levels (P, V, M, I, F, C), is hence summarized by Equation 2.5.3:

$$Sig_E = X \begin{vmatrix} C: Sig * mod\ \left(\int = 1\right) \\ F: Sig\ mod\ (c) \\ I: Sig\ mod\ \left(\int \overline{G, e, \pi}\right) \\ M: Sig * mod\ (G) \\ V: Sig * mod\ (e) \\ P: Sig * mod\ (\pi) \end{vmatrix}$$

Eq 2.5.3: Partial-Singularity Mutational-Sequence or Emergence of Uniqueness

What is implied by (2.5.3) is that there is a bias in the phenomenon of mutation so that an implicit sequence is followed such that more and more of the uniqueness of an organization may emerge. Material change is therefore tightly tied in with implicit function. But this has to be if all is just a play of Light.

Further, the mutational-sequence also implies an upward-strand to the ubiquitous subtle-DNA. The general mathematics related to the upward-strand, and therefore to a key aspect of the dynamics of partial-singularities will be further explored in the next chapter.

Chapter 2.6: The Systematization of Partial-Singularity Mutational-Sequences

So far the inherent creativity in Light as summarized by four overarching properties in the nature of a point and its attendant '∞ -entanglement' has been considered. Further, how this deep fount of creativity is present everywhere, and how light-based sets with their attendant 'K-entanglement' dynamics that further materialize the range of creative forces has also been considered. These architectural forces elaborate the possibility inherent in any system. Leveraging these sets of forces by virtue of 'N-entanglement' an equation for the uniqueness of an organization, regardless of scale, was also arrived at.

In some sense the precipitation of creativity from the barely perceptible nature of the ubiquitous point, to how this reveals a play of forces, to how organizations take their seed and grow from these forces, giving insight too into the dynamics of entanglement and superposition in Space and Time, has been traced. It is proposed that such creativity is what also spawns subtle-libraries that elaborate the structure of subtle-DNA, also providing insight into key light-based-singularity dynamics.

In Chapter 2.5 the process of a possible mutational-sequence arising in time was explored, thereby also beginning to articulate the language of partial-singularity biography. While such a sequence is intricately tied with the precipitating layers of light, and in fact can be thought of as a reversal of the downward precipitation, thereby reinforcing the notion a subtle-DNA since DNA at the material level is composed of two tightly related strands in opposite directions, this chapter will further explore a systematization of such a mutational-sequence.

Such an upward-strand is organic in nature suggesting likely paths or mutational-sequences determined by what sets of influences are active. Yet the possibilities are wholly determined by the downward-strand and the overarching organizational principles put in place as light projects itself at varying constant speeds.

Hence, starting with the physical, which recall is suggested as being a projection of Light's property of Presence, an equation, Equation 2.6.1, is summarized as:

$$Physical = \begin{bmatrix} M_3 & \to & System_{Pr} \\ & (\uparrow F \to I) \\ M_2 & \to & S_{System_{iPr}} \\ & (\uparrow Sig \to F) \\ M_1 & \to & Sig_P \\ & (\uparrow > P_P) \\ U & \to & Physical_U \end{bmatrix} TC \to Physical_T$$

$$Where \begin{bmatrix} Physical_U & \ni & [inertia, lethargy, status\ quo, ...] \\ Physical_T & \ni & [adaptability, durability, strength, ...] \end{bmatrix}$$

Eq 2.6.1: Possible Mutational-Sequence in Physical-Type Systems

Essentially this equation is laying out the conditions of moving from the untransformed or negative physical state represented by $Physical_U$ to the transformed or positive physical state represented by $Physical_T$.

The matrix should be read from the bottom to the top:

$$\begin{bmatrix} M_3 \rightarrow System_{Pr} \\ (\uparrow F \rightarrow I) \\ M_2 \rightarrow S_{System_{Pr}} \\ (\uparrow Sig \rightarrow F) \\ M_1 \rightarrow Sig_P \\ (\uparrow > P_P) \\ U \rightarrow Physical_U \end{bmatrix}$$

Hence, at the bottom is the starting point ' $U \rightarrow Physical_U$ ' which identifies the default or untransformed (U) level of the physical. The next row up, $(\uparrow > P_P)$, states that when the patterns of the untransformed physical (P_P) have been overcome (>), movement to the next level ($\uparrow$) is facilitated. Breaking through to the next level, $M_1 \rightarrow Sig_P$, allows its dynamics to become active. Hence, the signature or uniqueness of the physical (Sig_P) becomes active at meta-level 1 (M_1). As this signature becomes more like a Force ($Sig \rightarrow F$), the conditions for breakthrough ($\uparrow$) to the next level are achieved. This next level is referred to as meta-level 2 (M_2), and indicates that the architectural forces represented by the set of system-presence ($S_{System_{Pr}}$) have become more consciously active. When this Force becomes Integral ($F \rightarrow I$) then the conditions for breakthrough ($\uparrow$) to the next level are achieved. The next level is notated as M_3 for meta-level 3, and the dynamics here indicate that the equation for system-presence becomes active. Becoming active basically means that the respective meta-level dynamic begins to act at the once 'untransformed' level (U) further modifying it. Modification or transformation began when M_1 became active. Transformation is accelerated when M_2 becomes active, and even further accelerated when M_3 becomes active.

The rate of the transformation can be better envisioned when considering action of the Transformation Circle, or TC. The TC can be thought of as 4 concentric circles, with M_3 at the center. M_3 is surrounded by M_2, which is surrounded by M_1. The outer circle is U. If TC is considered to be a clock, than at time 't = 0', the 'physical' can be thought of as being entirely in U. The clock starts ticking only when some initial patterns P_P are overcome $(>P_P)$. From this point on as time proceeds the conditions for breakthrough become riper, and a sinusoidal wave begins to integrate more of the concentric circles together. The sinusoid wave (sin) is itself modulated by an euler function, e^x, where 'x' is determined by the strength to overcome patterns (↑) which will likely vary over time but will likely tend to be positive once the clock has started ticking because of the naturally building creative expression with progressive movement. Being that the limit is the outer boundary of the concentric circles, there is further modulation by π until the 4 concentric circles have been integrated. TC, hence, may be represented by Equation 2.6.2:

$$TC \equiv (> P_P) \rightarrow \mathrm{mod}\,(\sin, e^x, \pi)$$

Eq 2.6.2: Transformation Circle

Hence, the initial nature of the physical that may be characterized by the set comprising of elements such as, lethargy, acceptance of the status quo, amongst other such elements, is represented by:

$$(Physical_U \ni [inertia, lethargy, status\ quo, ...])$$

This transforms into a physical more characterized by elements such as adaptability, durability, strength, and so on. That is:

$$(Physical_T \ni [adaptability, durability, strength, ...])$$

This transformation represents the inherent creativity-dynamic driving any mutation sequence within physical-type systems.

Such transformation as discussed in the previous chapters will implicitly involve the mechanisms of superposition and entanglement as emergence takes place.

Similarly, the equation for the 'Vital', Equation 2.6.3, which recall is suggested as being a projection of Light's property of Power, also shows the built-in transformation that represents the innovation-dynamic within the vital:

$$Vital = \begin{bmatrix} M_3 \rightarrow System_P \\ (\uparrow F \rightarrow I) \\ M_2 \rightarrow S_{System_P} \\ (\uparrow Sig \rightarrow F) \\ M_1 \rightarrow Sig_V \\ (\uparrow > P_V) \\ U \rightarrow Vital_U \end{bmatrix} TC \rightarrow Vital_T \,,$$

$$Where \begin{bmatrix} Vital_U \ni [aggression, self\ centeredness, exploitation, ...]] \\ Vital_T \ni [energy, support, adventure, enthusiasm, ...] \end{bmatrix}$$

Eq 2.6.3: Possible Mutational-Sequence of Vital-Type Systems

The equation for the 'Mental', Equation 2.6.4, which recall is suggested as being a projection of Light's property of Knowledge, is similarly summarized as:

$$Mental = \begin{bmatrix} M_3 \;\rightarrow\; System_S \\ (\uparrow F \;\rightarrow\; I) \\ M_2 \;\rightarrow\; S_{System_S} \\ (\uparrow Sig \;\rightarrow\; F) \\ M_1 \;\rightarrow\; Sig_M \\ (\uparrow > P_M) \\ U \;\rightarrow\; Mental_U \end{bmatrix} TC \;\rightarrow\; Mental_T$$

$$Where \begin{bmatrix} Mental_U \;\ni\; [fixation, fundamentalism, fragmentation, \dots] \\ Mental_T \;\ni\; [understanding, imagination, inspiration, \dots] \end{bmatrix}$$

Eq 2.6.4: Possible Mutational-Sequence of Mental-Type Systems

The equation for the 'Integral', Equation 2.6.5, suggested as being a projection of Light's property of Harmony, is similarly summarized as:

$$Integral = \begin{bmatrix} M_3 \;\rightarrow\; System_N \\ (\uparrow F \;\rightarrow\; I) \\ M_2 \;\rightarrow\; S_{System_N} \\ (\uparrow Sig \;\rightarrow\; F) \\ M_1 \;\rightarrow\; Sig_I \\ (\uparrow > P_I) \\ U \;\rightarrow\; Integral_U \end{bmatrix} TC \;\rightarrow\; Integral_T$$

$$Where \begin{bmatrix} Integral_U \;\ni\; [possession, usurpation, hidden\ agendas, \dots] \\ Integral_T \;\ni\; [appreciation, shift\ POV, MPV, synthesis, \dots] \end{bmatrix}$$

Eq 2.6.5: Possible Mutational-Sequence of Integral-Type Systems

The preceding equations can be generalized by Equation 2.6.6:

$$Innovation_{orientation-x}$$

$$= \begin{bmatrix} M_3 \to System_X \\ (\uparrow F \to I) \\ M_2 \to S_{System_X} \\ (\uparrow Sig \to F) \\ M_1 \to Sig_x \\ (\uparrow > P_x) \\ U \to x_U \end{bmatrix} TC \to x_T, where \begin{bmatrix} x_U \ni [...] \\ x_T \ni [...] \end{bmatrix}$$

Eq 2.6.6: Generalized Mutational-Sequence Equation

In this generalized equation, $Innovation_{orientation-x}$, refers to the inherent innovation within a specific orientation. Orientation refers to the physical, the vital, the mental, or the integral.

Further, the notion of a core-matrix can be summarized by the following equation, Equation 2.6.7:

$$Core\_matrix = \begin{bmatrix} M_3 \to System_X \\ (\uparrow F \to I) \\ M_2 \to S_{System_X} \\ (\uparrow Sig \to F) \\ M_1 \to Sig_x \\ (\uparrow > P_x) \\ U \to x_U \end{bmatrix}$$

Eq 2.6.7: Core Matrix

Equations 2.6.1 through 2.6.7, hence, further systematize the dynamics of mutation that are integral to partial-singularities. While these equations are tied to the time-dimension, they also define logical conditions that will allow more of the possibility resident in any seed to emerge. These conditions, as will be explored in more detail later in the book, also influence the process of quantization whereby further possibility can emerge in material reality typified by light traveling at c.

SECTION 3: FURTHER ELABORATION OF PARTIAL-SINGULARITY MATHEMATICS

The previous section outlined some dynamics of light-based-singularities. In particular these included the hi-level dynamics for the seed-singularity that occurs in the greater-than-c layers of light in the downward strand of subtle-DNA, and the partial-singularities that occur in the U-layer typified by light moving at c, also in the downward-strand. Partial-singularity dynamics, as suggested, are highly dependent on dynamics that occur in the upward-strand.

The downward-strand is caused by light slowing down in quantized-decelerations, as it were, progressively concretizing more of the information in light. The upward-strand is envisioned as prescribing a time-variable sequence wholly determined by the nature of the layers in the downward-strand. The time-variability is due to the result of the interplay of the active influences emanating from the layers of light in the downward-strand. But further, each of the layers of light spawn subtle-libraries of pre-genetic information. These subtle-libraries are effectively infinitely large encoding many different possibilities. Such genetic-type information is the language in which biographies of light-based singularities are written. Pre-genetic information is the language that describes the seed-singularity. Genetic and post-genetic information, as will be explored in further sections, is the language in which partial-singularity biographies are written.

This section will illustrate some of the interplay involving the subtle-libraries and a first projection of their possibilities in a Space-Time-Energy-Gravity construct. The Space-Time-Energy-Gravity construct is positioned as being fundamental in altering genetic information at the material level. This will be elaborated in subsequent sections. Further, numerous other constructs that have their logic determined in 'four-base logic-encoding ecosystems' (FBLEE) introduced in Section 1, envisioned as existing in the quantum layer antecedent to the material layer, will also be explored in subsequent sections. It is such ecosystems that can change due to interplay with the material layer. In the interplay or call from 'below' in which emergences find that their forms are inadequate in expressing the fullness of Light, as it were, there can be thought to be a response from 'above', from more of the fullness that Light is, that may precipitate one of the infinite functions already created at subtle-libraries at the ∞-entanglement, K-entanglement, and N-entanglement levels. Such a call and response system can be thought of as operating like a lock-and-key mechanism.

As such, this section will reinterpret some basic equations and existing quantum mechanics interpretations based on the Cosmology of Light view adopted in this treatise. Further, it will focus on the derivation of the Light-Space-Time Emergence equation leveraged in subsequent quantization analyses. The Light-Space-Time Emergence equation models the basis for quantization suggesting emergent reality for all phenomena from Light. Schrodinger's wave equation and Heisenberg's uncertainty principle are also interpreted from the point of view of the Light-based Interpretation of quantum mechanics central to this treatise, to reinforce the notion that even when considered from these points of view the existence of multiple layers of light is feasible. The chapter on quantization of space, time, matter and gravity, models how these phenomena are related to Light and subsequently also models how these fundamentals work together to potentially impact genetic-type information.

This section therefore elaborates partial-singularity dynamics in which more comprehensive emergences manifest materially to also begin to increase partial-singularity "spheres of influence".

Chapters in this section will focus on:

- Equation for Light-Space-Time Emergence
- Speed of Light and Quanta
- Interpreting Schrodinger's Equation
- Interpreting Heisenberg's Uncertainty Principle
- A Deeper Look at Quantization of Space, Time, Matter, and Gravity
- Effect of Levels of Light on Genetic Mutation
- Application of Qualified Determinism to Genetic Mutation

Equation 2.6.6, the generalized equation for mutational-sequence, can be restated as an evolving form true for all time, as in Equation 3.1.1:

$$Innovation_{orientation-x}$$
$$= \left(\begin{bmatrix} M_3 \to System_X \\ (\uparrow F \to I) \\ M_2 \to S_{System_X} \\ (\uparrow Sig \to F) \\ M_1 \to Sig_x \\ (\uparrow > P_x) \\ U \to x_U \end{bmatrix} TC \to x_T , where \begin{bmatrix} x_U \ni [...] \\ x_T \ni [...] \end{bmatrix} \right)_{\langle x_U | x_T \rangle}$$

Eq 3.1.1: Evolving Form of Generalized Equation of Mutational-Sequence

The added notation of $\langle x_U | x_T \rangle$ implies that the output of the previous iteration of the equation of innovation, x_T, where the subscript T implies relatively-transformed, now becomes the input, x_U, for the next iteration of the equation, where U implies relatively-untransformed. Hence through time there is greater and greater transformation that pushes experienced reality to greater and greater levels of functional-richness.

But further, given that quanta is proposed to be a doorway to deeper worlds of Light, that in fact allow aspects of those worlds or layers to become active at the surface layer U, the question is when have those aspects become active in manifest time. The following timeline based on generally accepted models of universal history (Particle Data Group, 2015) suggests when. Note too that the subsequent sections of exploring the generation of pre-genetic and genetic information as part of the developing biography of light-based singularities at the levels of the electromagnetic spectrum, matter, and life, will explore in far greater detail some of the statements made in the following timeline:

- At time, $t \leq 0$ seconds, only M_3 is active, and then remains active for all $t < \infty$. Recall that M_3 represents the four-fold reality present in every ubiquitous-point-instant.

- At time, $0 \geq t > \infty$, M_2 the set of architectural forces continually gets added to, thereby increasing the size of the sets of forces.

- At time, $t \geq 0$, space, time, energy, gravity, the first clear expression of the four-fold order, emerges. This emergence marks the generation of the Big-Bang partial-singularity, as will be explored in subsequent chapters. This first expression is significant because it sets in motion the interplay between the antecedent quantum-layer and the layer where matter will materialize. Note that the antecedent quantum-layer likely houses the ever-enhanced four-base logic-encoding ecosystems (FBLEE) critical to genetics and evolution.

- At time, $0 > t \geq 10^{-36}$ seconds, the equation of Innovation, $Innovation_{orientation-x}$, is such that M_1 also becomes active. The activation of M_1 begins to result in unique expressions or signatures of the set of architectural forces, and in this case in the reality of the essentially ubiquitous electromagnetic-spectrum (EM Spectrum) as a vehicle of the four-fold order that expressed itself in all that existed and in all that unfolded from that point in time on. This marks the generation of the electromagnetic spectrum partial-singularity.

- At time, $t \sim 10^{-10}$ seconds, fundamental particles emerge as an essential material basis of the four architectural forces that frame all further development. As in the case of the EM Spectrum this implies the activity of M_1, and then also of U. This also marks the generation of the quantum particle partial-singularity.

- At time, $t \sim 3 \times 10^5$ years light atoms emerge, and at time $t \sim 10^9$ years heavier atoms in the stars emerge. These also imply the continued activity of M_1 and U and the generation of atom-based partial-singularities with increasing spheres of influence.

- At time, $t \sim 13.8 \times 10^9$ years, a further clear expression of the same fourfold order as the bases of an even more complex organization, that of cellular life and all that is founded on it comes into being. This time-point will be represented by the notation $t \sim E_{Cell}$, where 'E' stands for emergence. This too implies the activity of M_1. Note that with the generation of the living cell partial-singularity the sets of architectural forces specified by M_2 continue to increase the number of elements they comprise of as the complex interaction between the layers continues.

- At time $t > 13.8 \times 10^9$ years, human-beings, and more complex social organizations emerge. Here TC acts with an implicit direction of operation from U to M_3. This time-point will be represented by $t \sim E_{Human}$ and marks the generation of human-based partial-singularities.

Note that the emergence of space-time-energy-gravity, and subsequently the generation of the electromagnetic spectrum partial-singularity, quantum particles partial-singularity, and atoms-based partial-singularities, implies that the pre-genetic information in their attendant FBLEE must also be present in some form in genes as appear later with the advent of cellular life. This must be the case since the logic of cosmic fundamentals, such as space-time-energy-gravity, and the very basis of matter, as in quantum particles, atoms, and so on, has to be deeply ingrained in all things, inanimate and animate. This feature of subsequent emergences containing biographies for all previous emergences is fundamental to light-based emergences and to the dynamics of any light-based-singularity. A vast and actionable information-base therefore animates any light-based emergence. Note that this is in contrast to any AI-based emergence that is in this view a fragmented development that is not based on the persistent quantum computation and generation of genetic-type information that animates every light-based emergence.

Based on the aforementioned timeline and description Equation 3.1.2 for Emergence true of any space-time scale may be generalized as the following:

$Emergence_{space-time}$

$$= \begin{vmatrix} \begin{bmatrix} M_3 \to System_X \\ (\uparrow F \to I) \\ M_2 \to S_{System_X} \\ (\uparrow Sig \to F) \\ M_1 \to Sig_x \\ (\uparrow > P_x) \\ U \to x_U \end{bmatrix}_{Space} \\ \begin{bmatrix} M_3 : -\infty \leq t \leq \infty \\ \downarrow \\ M_2 : 0 \geq t > \infty \\ \downarrow \\ M_1 : 0 > t > \infty \\ \downarrow \\ U \to \begin{array}{l} t \leq E_{Cell}; TC: M_3 \to U \\ t \sim E_{Human}; TC: U \to M_3 \end{array} \\ TC \to x_T, where \begin{bmatrix} x_U \ni [...] \\ x_T \ni [...] \end{bmatrix} \end{bmatrix}_{Time} \end{vmatrix}_{\langle x_U | x_T \rangle}$$

Eq 3.1.2: Space-Time Emergence

An implication of this equation, brought out more explicitly through the elaboration of the 'Time' component, is that the layers U, M_1, M_2, and M_3 exist simultaneously. Further the advancement of time also causes the generation of partial-singularities with increasing spheres of influence. Adding the Light-Matrix derived in Chapter 2.1 enhances Equation 3.1.2 to the Light-Space-Time Emergence form as represented by:

$Emergence_{light-space-time} =$

$$\left\| \begin{bmatrix} \begin{array}{c} c_\infty : [Pr, Po, K, H] \\ (\downarrow R_{C_K} = f(R_{C_\infty})) \\ c_K : [S_{Pr}, S_{Po}, S_K, S_H] \\ (\downarrow R_{C_N} = f(R_{C_K})) \\ c_N : f(S_{Pr} \times S_{Po} \times S_K \times S_H) \\ (\downarrow R_{C_U} = f(R_{C_N})) \\ c_U : [P, V, M, C] \\ \Uparrow \\ c_{0:[D,W,I,C]} \end{array} \end{bmatrix}_{Light} \begin{bmatrix} \begin{array}{c} M_3 \rightarrow System_X \\ (\uparrow F \rightarrow I) \\ M_2 \rightarrow S_{System_X} \\ (\uparrow Sig \rightarrow F) \\ M_1 \rightarrow Sig_X \\ (\uparrow > P_x) \\ U \rightarrow x_U \end{array} \end{bmatrix}_{Space} \right.$$

$$\left. \begin{bmatrix} \begin{array}{c} M_3 : -\infty \leq t \leq \infty \\ \downarrow \\ M_2 : 0 \geq t > \infty \\ \downarrow \\ M_1 : 0 > t > \infty \\ \downarrow \\ U \rightarrow \begin{array}{l} t \leq E_{Cell}; \text{TC: } M_3 \rightarrow U \\ t \sim E_{Human}; \text{TC: } U \rightarrow M_3 \end{array} \end{array} \end{bmatrix}_{Time} \begin{bmatrix} \\ TC \rightarrow x_T \\ \\ \end{bmatrix} \right\| \langle x_U | x_T \rangle$$

Eq 3.1.3: *Light-Space-Time Emergence or Partial-Singularity Generator*

In (3.1.3) there is a 1:1 mapping between the Light and

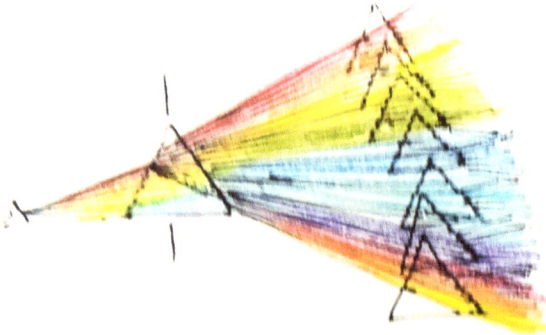

Space matrices in that M_3 reflects the ever-present C_∞, M_2 reflects C_K, M_1 reflects C_N, and U reflects C_U. Hence there are fundamental mathematical symmetries at play that suggests that the same underlying reality is only being perceived from a different point of view. The Time matrix simply gives estimates of the time at which each of the layers became active. But also (3.1.3) can, in this analyses, be thought of as the equation that generates partial-singularities. Hence, (3.1.3) is also a partial-singularity generator equation.

Chapter 3.2: Speed of Light and Quanta

As discussed conceptually in Section 1 and also mathematically in Chapter 2.1, since c is finite and therefore there is past, present, and future implied by it, this also implies that at U a point has to become quanta. This is implicit in the notion of finiteness. Since light takes a finite amount of time to get from A to B, a "unit" of light can be thought of as related to the finite time to traverse that. Quanta at the subatomic level can be thought of as related to this finite time and distance for a unit of light to be expressed.

Planck's discovery that energy at the subatomic level requires a minimum threshold 'quanta' to express itself therefore makes sense. It is to be noted though that Planck's treatment of quanta was more as a mathematical convenience that allowed the derivation of an equation that explained the curve of radiation wave-lengths at varying temperatures of a heated black-body (Isaacson, 2008). Einstein though postulated quanta as a fundamental property of light itself, rather than as something that arose in the interaction of light with matter as Planck thought. Einstein's theory produced a law of the photoelectric effect where the energy of emitted electrons would depend on the frequency of light. Einstein received the Nobel Prize for this discovery (Isaacson, 2008).

Summarizing, if c is the upper limit of the layer U, then it makes sense that the lower limit h (Planck's constant) should be inversely proportional to c. Hence:

$$h \propto \frac{1}{c}$$

This relationship is substantiated by combining two well-known equations: the first is the electromagnetic equation connecting speed of light with wavelength and frequency, and the second is Einstein's photoelectric equation connecting energy with frequency of light:

(1) $C = \nu\lambda$
(2) $E = h\nu$

Yields:

$$h = \frac{E\lambda}{C}$$

About h, H.A. Lorentz the Dutch scientist has commented in The Science of Nature (Lorentz, 1925): "We have now advanced so far that this constant not only furnishes the basis for explaining the intensity of radiation and the wavelength for which it represents a maximum, but also for interpreting the quantitative relations existing in several other cases among the many physical quantities it determines. I shall mention a few only, namely the specific heat of solids, the photo-chemical effects of light, the orbits of electrons in the atom, the wavelengths of the lines of the spectrum, the frequency of the Roentgen rays which are produced by the impact of electrons of given velocity, the velocity with which gas molecules can rotate, and also the distances between the particles which make up a crystal. It is no exaggeration to say that in our picture of nature nowadays it is the quantum conditions that hold matter together and prevent it from completely losing its energy by radiation."

So just as c sets up the past-present-future experience and reality of U, h suggests that this experience will take place in shells of matter. In the absence of the limit h, as pointed out by Lorentz, only radiation, and no matter would exist. This 'past-present-future-matter' dynamic reinforces the notion of the four-foldness implicit in the nature of Light as already discussed in Sections 1 and 2.

The suggested variance in the speed of light by meta-layer may also throw some further light on the quantum realm. First, summarizing:

1. At U the speed of light in a vacuum, c_U, is finite at 186,000 miles/sec. This finiteness creates the reality and experience of past-present-future, and further a sense of fragmentation and separation. Further, assuming that c_U is a fundamental upper-limit at U, the inverse of it, $\frac{1}{c_U}$, must define some fundamental lower limit at U. This is indeed the case as Planck's constant, h, can be perceived as being proportional to this. 'h' allows for matter to be sustained, as it fundamentally limits the dispersion of energy as suggested by Lorentz.

2. At M_3, the speed of light, c_{M_3}, is suggested as being ∞ miles/sec. This allows a reality of 'oneness' and the possibility of a suggested fourfold-intelligence existing in every ubiquitous-point-instant as already discussed.

3. As also already suggested the quantum world, here designated by Q, because it is at boundary of U, accesses and interrelates with the meta-levels. As such, the speed of light, c_Q, will appear as a hybrid as in the following figure. Note though that it is really the speed of light at the native or resident layer that becomes active, and that this is simply being represented as c_Q for convenience:

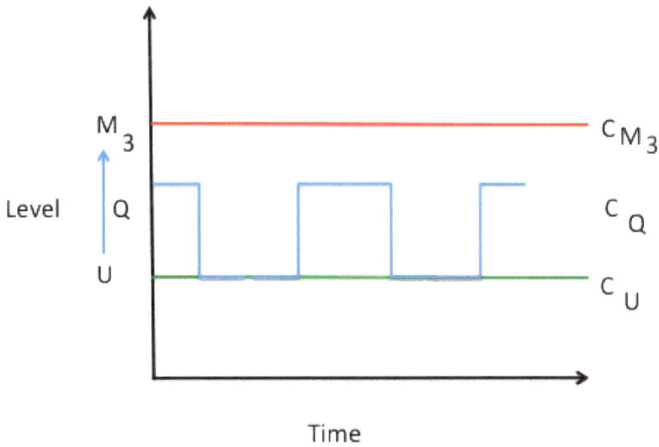

Figure 3.2.1 Speed of Light at Quantum Level

Note that research on the speed of light also indicates that it may go faster than c_U. While the speeds suggested currently through experimental research, and summarized below, may be only incrementally higher than c_U the notion that c_U can be exceeded appears to be put in place:

1. The Heisenberg Uncertainty Principle (to be discussed subsequently in Chapter 3.4) already suggests that photons can travel at any speed, even exceeding c_U, for short periods.
2. Notion of different space-time realities, also known as meta-levels in this treatise, suggests that light can travel differently in a layer different from the four-dimensional space-time that apparently defines our observable world (Hawking, 1988).
3. In his book QED Feynman (Feynman, 1985) says "...there is also an amplitude for light to go faster (or slower) than the conventional speed of light. You found out in the last lecture that light doesn't go only in straight lines; now, you find out that it

doesn't go only at the speed of light! It may surprise you that there is an amplitude for a photon to go at speeds faster or slower than the conventional speed, c." In research conducted at Humboldt University (Chown, 1990), Scharnhorst has made calculations using the theory of quantum electrodynamics to reveal the possible existence of "faster-than-light" photons. This is known as the Scharnhorst effect.

4. As reviewed in Chapter 2.1, Perkowitz makes the point that the theory of relativity does not disallow particles already moving at c or greater.

The point is that the reality at Q is going to be different than the reality at U. This should be apparent from considering the relation of c_x, where 'x' is '∞', 'k', 'n', 'u', and 0 respectively, to the consequent reality as elaborated in Chapter 2.1. In Q the fundamental lower limit, h, which allows matter to sustain itself, is itself going to fluctuate. Hence, as X tends to M_3, c_x will tend to infinity, and h will become a fraction of itself. As it becomes a fraction of itself the quantization effect will be lowered, and matter will get dispersed more and more easily to in effect take on a wave-like, or field-like, or even block-like appearance as also suggested in Chapter 2.1.

Chapter 3.3: Schrodinger's Equation & Multiple Layers of Light

Schrodinger's equation, which seeks to model how a quantum state of a quantum system changes with time, or in other words seeks to model matter as a wave rather than as a particle (Stewart, 2012), is depicted in Equation 3.3.1:

$$i\frac{h}{2\pi}\frac{\partial}{\partial x}\psi = \hat{H}\psi$$

Eq 3.3.1: Schrodinger's Equation

ψ depicts a wave form and can be thought of as a probable cloud of possible states. $\hat{H}$ is the Hamiltonian operator which is a focusing function, and in its essence what the equation may be suggesting is that the way a wave form changes over time is equivalent to some expressible state of the possibilities inherent in the cloud of possible states.

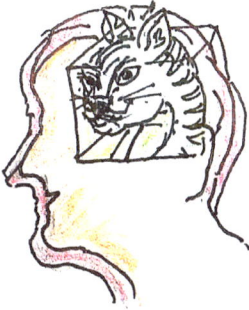

But the cloud of possible states is another way of saying that behind the layer U, form is represented in another way than at U. If the existence of the meta-levels, and in this case, of Sig_x at M_1, is considered possible, then it is far more reasonable to admit that form is configured by function and the very dynamics of what may appear to have been random, as some interpretations of quantum phenomena suggest (Wimmel, 1992), may now appear to be far more logical. There is now more *context* to interpreting observation at the quantum level.

127

Interestingly Schrodinger himself had misgivings about the applicability of this equation that seemed to apply at the quantum level, to the macro-world (Stewart, 2012). To bring his misgivings to light he invented a thought experiment concerning a cat. This cat would be in a superposed state in a quantum black-box. A radioactive particle, a decaying-particle detector, and a flask of poison, were the other inhabitants of the black-box. At some point the particle will decay, be detected, and as in the thought-experiment at that point, triggered by the decaying particle, the poison in the flask would be released. The cat would then die. But in the meanwhile the cat would be in superposed states of being both dead and alive. Only when the box was opened would the wave function collapse and a single definite state emerge. This thought experiment is also considered the origin of the notion of 'superposition'. The mathematical model presented in this book though has already interpreted 'superposition' differently as discussed previously.

Schrodinger was hoping to highlight the absurdity of the application of having a cat in both a dead and alive state at the macro-level. Instead physicists found this thought experiment to be sensible, even at the macro-level, and began to generalize the quantum theory based on this. Hence, for example the idea of superposition at the physical level began to be thought of as real. It is interesting to note that in his lectures on Schrodinger's equation Feynman (Gottlieb, 2013) has stated: "Where did he get that [equation] from? Nowhere. It is not possible to derive it from anything you know. It came out of the mind of Schrödinger".

Considering Schrodinger's equation in the light of the discussion on Q in Chapter 3.2, 'i' is a complex number and suggests the interplay of two dimensions, one being real, and one being 'imaginary'. But the 'imaginary' dimension could be thought of as none other than the meta-levels implicit in the mathematical model in this treatise, and suggested to be real at Q. Further, $\frac{h}{2\pi}$ is in line with the suggestion also made in Chapter 3.2 that h will have to become a fraction of itself as c increases. Hence, the change in the wave function, $\frac{\partial}{\partial x}\psi$, is intimately related to i and $\frac{h}{2\pi}$, and perhaps more fully makes sense when considered in the context of i x $\frac{h}{2\pi}$ x $\frac{\partial}{\partial x}\psi$, which has to be the case when dealing with the integration of dynamics of multiple levels of light.

Further, the change in the wave function, $\frac{\partial}{\partial x}\psi$, is related to $\hat{H}\psi$, and suggests that there is some system "energy", represented by the Hamiltonian, $\hat{H}$, that when applied to the existing wave, ψ, will indicate how the wave will be expressed going forward.

But as discussed, at Q the dynamics of Sig_X become real, and in fact is a fundamental organizing principle for all organization at U, and starting at the space-dimension of h.

Chapter 3.4: Heisenberg's Uncertainty Principle & Multiple Layers of Light

In his book, The Little Book of String Theory, Princeton University's Gubser (Gubser, 2010) describes the effect on approaching absolute zero temperature on molecules. He takes the example of water molecules and relates that one cannot make the water molecules colder than absolute zero, -273.15 Celsius, because there is no more thermal energy to suck out at that temperature. However, quantum uncertainty, the phenomenon which relates the momentum and location of electrons in atoms necessitates that the water molecules will still vibrate. Gubser suggests this by considering Heisenberg's uncertainty relation, reproduced in Equation 3.4.1:

$$\Delta p \; X \; \Delta x \; \geq \; \frac{h}{4\pi}$$

Eq 3.4.1: Heisenberg's Uncertainty Relation

In Equation 3.4.1, Δp is the uncertainty in a particle's momentum, Δx is the uncertainty in the particle's location, and h is the Planck's constant. In frozen water crystals it is precisely known where the water molecules are, and therefore Δx is fairly small. This means that Δp has to be considerably larger, and therefore that the water molecules are still vibrating even though they are at absolute zero. This innate vibration, known as 'quantum zero-point' energy, expresses the phenomenon of quantum fluctuations.

The Planck's constant order of magnitude (10^{-34}) though, suggests the boundary between U and M_1 and the quantum fluctuations, the uncertainty relation, and the quantum zero-point energy could be an expression of the essential Signature function, Sig_x, that is posited as a key formative force behind organization at U. In this interpretation the thermal energy describes the essential energy at U, while the uncertainty relation may suggest the phenomenon of function-precipitation from other layers of Light, "physically" linking M_1 and U. In this case it may be suggested that integration of meta-levels with the surface level, I_U^M, is indicated by the uncertainty relation, as in Equation 3.4.2:

$$I_U^M \;\rightarrow\; \Delta p \; X \; \Delta x \;\geq\; \frac{h}{4\pi}$$

Eq 3.4.2: Integration of Levels (Leveraging Heisenberg's Uncertainty Relation)

But further, it may also be suggested that the uncertainty principle itself is only valid at U, and that too, because of the finiteness of c. This finiteness as already suggested

implies h, which implies that if the position of a particle is going to be observed by shining light on it, the light has to have at least a quantum of energy. But to determine the position of a particle accurately, light of a shorter wavelength would have to be used (Hawking, 1988) which would have to have a minimum amount of energy, which in turn would interfere with the velocity and hence momentum of the particle. The uncertainty in measuring the momentum could therefore be thought of as a consequence of the finiteness of the speed of light, c.

If c_U were to approach c_Q though, which as suggested in Chapter 3.2 could be anywhere between c and ∞ miles per second, the quantum would be smaller and the uncertainty in measuring position or momentum would be reduced. At c_{M_3} there would be no uncertainty since light would accurately tell both position and momentum definitively.

Hence, the uncertainty principle may be further qualified, as in Equations 3.4.3, 3.4.4, and 3.4.5:

$$@C_U:\ \Delta p\ X\ \Delta x\ \geq\ \frac{h}{4\pi}$$

Eq 3.4.3: Uncertainty Principle at U

$$@C_Q:\ \Delta p\ X\ \Delta x\ \rightarrow\ 0$$

Eq 3.4.4: Uncertainty Principle at Q

$$@C_{M_3}:\ \Delta p\ X\ \Delta x\ =\ 0$$

Eq 3.4.5: Uncertainty Principle at M3

The notion of position and momentum becoming finite at U also may imply that space, time, and quanta are emergent rather than absolute properties, as also suggested in Section 1. This is also the conclusion of Arkani-Hamed of the Institute of Advanced Studies in the following thought experiment (Wolchover, 2013):

'Locality says that particles interact at points in space-time. But suppose you want to inspect space-time very closely. Probing smaller and smaller distance scales requires ever higher energies, but at a certain scale, called the Planck length, the picture gets blurry: So much energy must be concentrated into such a small region that the energy collapses the region into a black hole, making it impossible to inspect. "There's no way of measuring space and time separations once they are smaller than the Planck length," said Arkani-Hamed. "So we imagine space-time is a continuous thing, but because it's impossible to talk sharply about that thing, then that suggests it must not be fundamental — it must be emergent."

Unitarity says the quantum mechanical probabilities of all possible outcomes of a particle interaction must sum to one. To prove it, one would have to observe the same interaction over and over and count the frequencies of the different outcomes. Doing this to perfect accuracy would require an infinite number of observations using an infinitely large measuring apparatus, but the latter would again cause gravitational collapse into a black hole. In finite regions of the universe unitarity can therefore only be approximately known.'

Chapter 3.5: A Deeper Look at Quantization of Space, Time, Energy, and Gravity

The Light-Space-Time Emergence equation (3.1.3) derived in Chapter 3.1, and reproduced here for convenience, offers some insight into the process of quantization. Reproducing the equation:

$Emergence_{light-space-time} =$

$$
\begin{Vmatrix}
\begin{bmatrix}
\begin{array}{c}
c_\infty: [Pr, Po, K, H] \\
(\downarrow R_{C_K} = f(R_{C_\infty})) \\
c_K: [S_{Pr}, S_{Po}, S_K, S_H] \\
(\downarrow R_{C_N} = f(R_{C_K})) \\
c_N: f(S_{Pr} \times S_{Po} \times S_K \times S_H) \\
(\downarrow R_{C_U} = f(R_{C_N})) \\
c_U: [P, V, M, C] \\
\Uparrow \\
c_{0:[D,W,I,C]}
\end{array}
\end{bmatrix}_{Light}
\begin{bmatrix}
M_3 \rightarrow System_X \\
(\uparrow F \rightarrow I) \\
M_2 \rightarrow S_{System_X} \\
(\uparrow Sig \rightarrow F) \\
M_1 \rightarrow Sig_x \\
(\uparrow > P_x) \\
U \rightarrow x_U
\end{bmatrix}_{Space} \\
\begin{bmatrix}
U \rightarrow
\begin{array}{c}
M_3 : -\infty \leq t \leq \infty \\
\downarrow \\
M_2 : 0 \geq t > \infty \\
\downarrow \\
M_1 : 0 > t > \infty \\
\downarrow \\
t \leq E_{Cell}; TC: M_3 \rightarrow U \\
t \sim E_{Human}; TC: U \rightarrow M_3
\end{array}
\end{bmatrix}_{Time}
\quad TC \rightarrow x_T
\end{Vmatrix}
\langle x_U | x_T \rangle
$$

The Light-Matrix, the top left-hand matrix, suggests that there are particular kinds of quantization that could occur. Along the vertical realm these can be thought of as inter-relating one layer of the matrix with the previous layers, or c_0 with c_U. Hence the kind of quantization that is relevant to the physical layer, U, would relate c_U with c_x where $x \in (N, K, \infty, 0)$, one of the possible speeds of light as per this mathematical treatise. This quantization can be represented by h_U, where h stands for Planck's

constant. Note that in this model there are several fundamental quantization possible that may inter-relate a specific light-layer with other layers of light. For the sake of simplicity it will be assumed that h_U is the quanta experienced in the inter-relation between U and the collective-set of light-layers behind or above U.

In looking at the Light-Matrix it is also clear that each subsequent layer below the top layer, describing the reality set up by an infinite speed, is emergent. Hence Space, Time, Energy, and Gravity, that arise when Light slows down from c_N to c_U, can be thought of as emergent phenomena. The emergence itself can be thought of as a function of the Light-Matrix. At layer U, and as discussed in Chapter 1.2, Space is related to Light's property of Knowledge, Time is related to Light's property of Power, Matter or Energy is related to Light's property of Presence, and Gravity is related to Light's Property of Harmony or Nurturing. Each of these will emerge in a particular way and it is possible that there are multiple h_U's.

Hence, the Planck's constant for matter or energy, which we are already familiar with from past scientific discoveries, can be depicted as h_{UPr}, since it is related to Presence (Pr). But similarly, quantization for space, time, and gravity could potentially be governed by other similar constants, referred to as h_{UK}, h_{UP}, and h_{UH}, and related to Knowledge (K), Power (P), and Harmony (H), respectively. Since the relationship between these constants, in absolute terms, is uncertain, this can be represented by the equality-inequality as depicted by Equation 3.5.1:

$$h_{UPr} \lesseqgtr h_{UK} \lesseqgtr h_{UP} \lesseqgtr h_{UH}$$

Eq 3.5.1: Equality-Inequality Relationship Between Different 'Planck' Constants

Further, the "quantization-window", that is, the window that quanta provides to layers of light behind U, as it were, potentially allows the precipitation of, or inter-relation with, or creation of a cohesive and compelling meta-function or signature as modeled by the c_N (or M_1) layer. This quantization-window is positioned as being key in allowing a phenomenon of "quantum-certainty" to occur. Quantum certainty can be thought of as allowing space-time-energy-gravity quantization to occur and is described in detail in the book Quantum Certainty (Malik, 2017e), part of the 6-book Cosmology of Light series (Malik, 2017-2018).

Subsequent Sections in this book will describe several examples of the activation of the multi-dimensional quantization that ensures change at the material-level initiated at FBLEE at the quantum-levels. Such change at the FBLEE and material levels is also the basis of genetic mutation. If the mutation is influenced by the layers set up by c_K, c_N, or c_∞, then the mutation is constructive because the

136

material outcomes of Light are being more integrated with Light itself. If the mutation is influenced by c_0, then the mutation is destructive because the material outcomes of Light are being further separated or fragmented from Light itself.

Such quantization requires patterns to be overcome, and so long as these are, this will result in the phenomenon of quantum-certainty. The 'Time' matrix, in (3.1.3) reproduced above, indicates the general default direction under which such quantum-certainty can occur at Layer U. Generally at the pre-human level it may proceed more automatically by the dynamics of the system itself, as represented by the segment: $t \sim E_{Cell}$; TC: $M_3 \rightarrow U$. Beyond this level it will generally occur through the action of a cohesive will, as represented by the segment: E_{Human}; TC: $U \rightarrow M_3$. The opening of this quantization-window is modeled by the 'Space' matrix and will happen when habitual patterns are overcome so that 'will' or 'want' becomes cohesive.

Such inter-dependence between human want or will and quantum dynamics is a possible basis of Human-Quantum computational models, as discussed in greater detail in 'The Emperor's Quantum Computer' (Malik, 2018c) and is envisioned to play a critical role in genetic mutation and evolution.

The specific quantization that are occurring along the space, time, energy/matter, and gravity dimensions are modeled by the following equations, that are all based on the Sig_x equations derived in Chapter 2.4. As a reminder (2.4.5) is reproduced here for convenience:

$$Sig = Xa + \overline{Yb_{0-n}}$$

$$where: \begin{bmatrix} X \in [S_{System_{Pr}}, S_{System_P}, S_{System_K}, S_{System_N}] \\ Y \in [S_{System_{Pr}}, S_{System_P}, S_{System_K}, S_{System_N}] \\ a, b \ are \ integers; a > b \end{bmatrix}$$

The Quantization and Structure of Space

The quantization of Space, which holds the seeds of knowledge of all that will emerge, hence, is modeled in the following way as specified by Equation 3.5.2:

$$Space_{quantization} = h_{UK}(Xa + \overline{Yb_{0-n}})$$

$$where: \begin{bmatrix} X \in [S_{System_K}] \\ Y \in [S_{System_{Pr}}, S_{System_P}, S_{System_K}, S_{System_N}] \\ a, b \text{ are integers}; a > b \end{bmatrix}$$

Eq 3.5.2: Space Quantization

In this model 'space' as an emergent property of Light, is structured by infinite seeds of knowledge. But because the emergence is taking place in a layer of reality generally itself structured by a finite speed of light, c, it has to be quantized. The quantization assures that the knowledge is not dissipated, but can accumulate to create seeds, and therefore, the structure of space itself. But such quantization as suggested earlier, also ensures that something of the possibility available at light traveling at a higher speed is materialized at the layer created when light slows down. In this treatise the only reason that Light slows down is to allow such materialization to occur. Such a structure of space appears consistent with recent models on the structure of space as explored in Rovelli's 'Reality Is Not What It Seems' (Rovelli, 2017). It is assumed that there is some kind of 'Planck's constant' in effect, that is modeled as being specific to the way knowledge or space may be quantized – hence, h_{UK}.

The structure of Space itself would be the summation of infinite seeds, as in Equation 3.5.3, Structure of Space:

$$Space_{Structure} = \sum_{i,j=1}^{\to \infty} h_{UK}(X_i a + \overline{Y_j b_{0-n}})$$

$$where: \begin{bmatrix} X_i \in [S_{System_K}] \\ Y_j \in [S_{System_{Pr}}, S_{System_P}, S_{System_K}, S_{System_N}] \\ a, b, i, j \ are \ integers; a > b \end{bmatrix}$$

Eq 3.5.3: Structure of Space

The Structure of Time

The quantization of Time, which holds the inevitability of the seeds or knowledge emerging in a phased maturity, hence, is modeled in the following way as specified by Equation:

$$Time_{quantization} = h_{UP}\left(Xa + \overline{Yb_{0-n}}\right)$$

$$where: \begin{bmatrix} X \in [S_{System_P}] \\ Y \in [S_{System_{Pr}}, S_{System_P}, S_{System_K}, S_{System_N}] \\ a, b \ are \ integers; a > b \end{bmatrix}$$

Eq 3.5.4: Time Quantization

In this model 'time' as an emergent property of Light, is structured by an inevitable process of maturity, which due to its inevitability, is related to power. But because the emergence is taking place in a layer of reality generally itself structured by a finite speed of light, c, it has to be quantized. The quantization assures that the power is not dissipated, but can accumulate to express phased maturity, and therefore, the structure of time itself. It is assumed that there is some kind of 'Planck's constant' in effect, that is modeled as being specific to the way power or time may be quantized – hence, h_{UP}.

The infinity of Time itself would be the summation of the maturation of infinite seeds, as in Equation 3.5.5, Infinity of Time:

$$Time_{Infinity} = \sum_{i,j=1}^{\to \infty} h_{UP}\left(X_i a + \overline{Y_j b_{0-n}}\right)$$

$$where: \left[\begin{array}{c} X_i \in [S_{System_P}] \\ Y_j \in [S_{System_{Pr}}, S_{System_P}, S_{System_K}, S_{System_N}] \\ a, b, i, j \ are\ integers; a > b \end{array}\right]$$

Eq 3.5.5: Infinity of Time

Energy & Matter

Energy, which through a process of containment can result in Matter, hence, is modeled in the following way as specified by Equation 3.5.6:

$$Energy_{quantization} = h_{UPr}\left(Xa + \overline{Yb_{0-n}}\right)$$

$$where: \left[\begin{array}{c} X \in [S_{System_{Pr}}] \\ Y \in [S_{System_{Pr}}, S_{System_P}, S_{System_K}, S_{System_N}] \\ a, b \ are\ integers; a > b \end{array}\right]$$

Eq 3.5.6: Energy Quantization

In this model 'energy' as an emergent property of Light, results in the reality of matter. But because the emergence is taking place in a layer of reality generally itself structured by a finite speed of light, c, it has to be quantized. The quantization assures that the energy is not dissipated, but can accumulate to create matter. Planck's constant is referred to as - h_{UPr}.

Gravity

Gravity, which holds seemingly distinct objects in the layer of reality created by c together in a harmony, hence, is modeled in the following way as specified by Equation 3.5.7:

140

$$Gravity_{quantization} = h_{UH}\left(Xa + \overline{Yb_{0-n}}\right)$$

$$where: \begin{bmatrix} X \in [S_{System_N}] \\ Y \in [S_{System_{Pr}}, S_{System_P}, S_{System_K}, S_{System_N}] \\ a, b \text{ are integers}; a > b \end{bmatrix}$$

Eq 3.5.7: Gravity Quantization

In this model 'gravity' as an emergent property of Light, results in a harmonious collectivity of seemingly independent objects. But because the emergence is taking place in a layer of reality generally itself structured by a finite speed of light, c, it has to be quantized. The quantization assures that the harmony is not dissipated, but can accumulate to express more and more complex collectivities on large-scale. It is assumed that there is some kind of 'Planck's constant' in effect, that is modeled as being specific to the way harmony or gravity may be quantized – hence, h_{UH}.

The preceding analysis also suggests that Space is a repository of pre-genetic information or of pre-genetic libraries. Time, Gravity, and Energy work with Space to bring about material possibilities. Change to any material possibility requires therefore the action of space, time, energy, and gravity. At the margin there is a fourfold quantization that brings about such change.

Such composite space-time-energy-gravity quantization is hence a key light-based singularity dynamic. In particular it is a key dynamic in the unfolding biography of any partial-singularity as will be further discussed in the next chapter.

Chapter 3.6: The Effect of Levels of Light on Genetic Mutation

The Light-Space-Time Emergence or Partial-Singularity Generation equation (3.1.3) being iterative can be used to model emergence from simpler to more complex four-fold manifestations. In other words (3.1.3) suggests a computational approach to the development of the universe. Pre-genetic and genetic information may be thought of as the output of such computation. This computational approach involves multiple layers of reality as suggested by (3.1.3) and is driven more by a process of qualified determinism, to be explored in more detail in the next chapter, than probability and statistics. This notion of qualified determinism is in contrast to the prevalent probabilistic view that has been erected as the cornerstone of quantum theory, and to MIT's Seth Lloyd's Copenhagen-like quantum superposition, probability-based computational approach to the development of the universe as described in his book 'Programming the Universe' (Lloyd, 2007).

An example of the Light-Space-Time based computation will be elaborated in Section 4, to provide an overview of

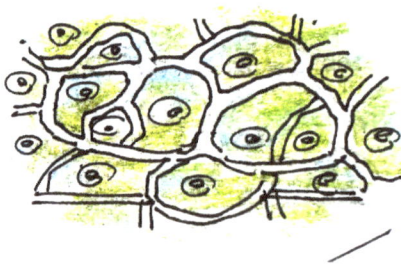

the process by which pre-genetic and genetic information is created. Sections 5, 6, 7, and 8 will further leverage the Light-Space-Time based computation to discuss in more detail how such information changes, or how mutation occurs. However, in keeping with the current quantum-analysis paradigm of using wavefunction to explore outcomes at a particular space and time, this chapter will develop two general approaches summarized by equations, to understanding the effect that different layers or levels of light may have on genetic mutation. While the natural line of development given the mathematical treatise explored in this book is to proceed with such an understanding using (3.1.3), the wavefunction depiction will be elaborated to suggest how (3.1.3) can be leveraged to provide further insight into a probability-based quantum-analytical approach.

(3.1.3) is reproduced below for convenience and a more concise form of it, the Simplified Version of Light-Space-Time Emergence, Equation 3.6.1, shall be utilized through the rest of this chapter.

$Emergence_{light-space-time} =$

$$
\left| \begin{bmatrix} \begin{array}{c} c_\infty: [Pr, Po, K, H] \\ (\downarrow R_{C_K} = f(R_{C_\infty})) \\ c_K: [S_{Pr}, S_{Po}, S_K, S_H] \\ (\downarrow R_{C_N} = f(R_{C_K})) \\ c_N: f(S_{Pr} \times S_{Po} \times S_K \times S_H) \\ (\downarrow R_{C_U} = f(R_{C_N})) \\ c_U: [P, V, M, C] \\ \Uparrow \\ c_{0:[D,W,I,C]} \end{array} \end{bmatrix}_{Light} \begin{bmatrix} M_3 \rightarrow System_X \\ (\uparrow F \rightarrow I) \\ M_2 \rightarrow S_{System_X} \\ (\uparrow Sig \rightarrow F) \\ M_1 \rightarrow Sig_x \\ (\uparrow > P_{x)} \\ U \rightarrow x_U \end{bmatrix}_{Space} \\ \begin{bmatrix} U \rightarrow \begin{array}{c} M_3: -\infty \le t \le \infty \\ \downarrow \\ M_2: 0 \ge t > \infty \\ \downarrow \\ M_1: 0 > t > \infty \\ \downarrow \\ t \le E_{Cell}; TC: M_3 \rightarrow U \\ t \sim E_{Human}; TC: U \rightarrow M_3 \end{array} \end{bmatrix}_{Time} \quad TC \rightarrow x_T \end{array} \right| \langle x_U | x_T \rangle
$$

Simplifying each of the main matrices in (3.1.3), hence, yields Equation 3.6.1, the Simplified Version of the Light-Space-Time Emergence equation:

$Emergence_{light-space-time} = |[L][S][T]TC \rightarrow x_T| \langle x_U | x_T \rangle$

Eq. 3.6.1: Simplified Version of Light-Space-Time Emergence

Effect of Levels of Light on Mutation Using Simplified Version of Light-Space-Time Emergence

As detailed in Chapter 2.6 on the systematization of the mutational-sequence, while relatively transformed organizations have opened to the active influence of meta-levels, relatively untransformed organizations remain under the influence of untransformed physical, vital, mental, and integral dynamics at U. This modeling can also be applied to micro-level dynamics such as mutation. Keep in mind that genetic-mutation as a level of granularity is important because this is the script or language in which partial-singularity biographies are written. Hence, as discussed in Chapter 2.6, the notion of relatively transformed organizations will translate to that of constructive-mutation. The notion of relatively untransformed organizations will translate to that of destructive-mutation.

Equations 3.6.2 through 3.6.4 summarize three types of mutation using (3.6.1). These are the destructive-mutation with only untransformed bases active, random-mutation with mixed bases, and constructive-mutation with bases relatively transformed due to the active influence of meta-levels, respectively. Hence:

$$Destructive - Mutation = \left\| |[L][S][T]TC \rightarrow x_T|_{\langle x_U | x_T \rangle} \right\|_U$$

Eq. 3.6.2: Destructive-Mutation

$$Random - Mutation = \left\| |[L][S][T]TC \rightarrow x_T|_{\langle x_U | x_T \rangle} \right\|_{U \& M_x}$$

Eq. 3.6.3: Random-Mutation

$$Consgtructive - Mutation = \left\| |[L][S][T]TC \rightarrow x_T|_{\langle x_U | x_T \rangle} \right\|_{M_x}$$

Eq. 3.6.4: Constructive-Mutation

Now the question is to what extent will the antecedent, and therefore the more permanent four-base logic-encoding ecosystem (FBLEE) be altered? For in the model based on Light it is this category of library that is important in heredity or the longer-term scheme of things. Destructive or random mutation may alter local copies of libraries, such as exist in DNA at the cellular-level, but it is only constructive mutation that can alter the deeper level subtle libraries that are preserved in a different kind of way in longevity.

This has to be the case since there is an in-built buffer due to the quantization that occurs as light slows down. The conditions for entering into that region where light is traveling faster must require a special, non-trivial set of conditions to be active. The conditions for traversing this in-built buffer will be explored in greater detail in subsequent chapters. Traversing the buffer implies that meta-levels are active, and that they have to become active in a special way for FBLEE to be affected.

Equation 3.6.5, Potential Effect of Levels of Light on Genetic-Type Information, summarizes how this may happen:

$$Potential\ Effect\ of\ Levels\ of\ Light\ on\ Genetic$$
$$-\ Type\ Information\ =$$

$$\begin{bmatrix} STATIC \; \langle |[L][S][T]TC \rightarrow x_T|_{\langle x_U|x_T \rangle} \\ \times \\ \Big((Y > U : Z_Q) \lor (Y \leq U : Z_F) \lor (Y = U : Z_R) \Big) \\ \ni \\ \Big(\begin{matrix} Z \in \mathbb{U} \, (\, Space, Time, Energy, Gravity) \\ Q : Quantization; F : Fragmentation; R : Random \end{matrix} \Big) \end{bmatrix} \rightarrow h \ni$$

$$h \in \left(\begin{matrix} [Q] : Constructive \; zone, \\ [Q] : Constructive \; zone \land Constructive \; mutation, \\ [F] : Destructive \; mutation, \\ [R] : Random \; mutation \end{matrix} \right)$$

Eq. 3.6.5: Potential Effect of Levels of Light on Genetic-Type Information

Line 1 from the top in the matrix is simply a static form of (3.6.1) the Simplified Light-Space-Time Emergence equation. The static form is designated by 'STATIC', and implies that fundamental operations true of (3.6.1) are being highlighted in (3.6.5). In other words (3.6.1) already has all the operations highlighted in (3.6.5) in it, but by 'freezing' it by making it static, the essential dynamics leading to possible mutations at the genetic level can more clearly be highlighted.

Line 1 is then subjected ($\times$) to a determination of the dominant levels of light that may be active, designated by Line 2, ' $\Big((Y > U : Z_Q) \lor (Y \leq U : Z_F) \lor (Y = U : Z_R) \Big)$ '. Unpacking this, 'Y > U' implies meta-levels are active and as a result it is possible that Z_Q is going to take place (the subscript 'Q' implies quantum-level action). In other words the "quantization-window" referred to in the previous chapter has been opened. This also implies activation and potential change of FBLEE. The call from below, as it were, may invoke some function that already exists in the subtle-libraries 'above', so that some already existing function may influence FBLEE through ∞ - entanglement, K-entanglement, or N-entanglement. This may be thought of as a key-and-lock mechanism, where a deep enough visceral urge from below acting as the key, opens an entangled lock to alter FBLEE as per the visceral urge. The alteration of FBLEE is what is referred to as "quantum-certainty" in the previous chapter because such certain alteration is happening at the quantum-level. '$Y \leq U$' implies that only the untransformed levels are active, and therefore also the sub-level where the speed of light is 0 is active and as a result Z_F is going to take place (the subscript 'F' implies 'fragmentation'). '$Y = U$' implies that all levels are active and as a result Z_R is going to take place (the subscript 'R' implies 'random').

Line 3 elaborates the significance of Z_Q, Z_F, and Z_R. Hence Z is the union of potential quantum-operations of space, time, energy, and gravity, designated by '$Z \in \mathbb{U}\,(Space, Time, Energy, Gravity)$'. But the nature of the operations is designated by '$Q: Quantization; F: Fragmentation; R: Random$'. Z_Q, then, implies that the full quantization originating from updated four-base logic-encoding ecosystems (FBLEE)

can take place, and will result in lasting material change at the Genetic-Type level. Z_F implies that the essential set will be fragmented and that only libraries at the level of local cellular-level DNA or precipitated material-fabric logic can potentially be altered. Z_R also precludes full quantization, and that some partial local-library constructive or destructive mutation may take place.

The '$\rightarrow h \ni h \in$' segment resolves the outcome of the operations implied by Lines 1 – 3, suggesting that the outcome will be 'h' such that ($\ni$) 'h' is an element ($\in$) of the set specified by the members '[Q]: *Constructive zone*', '[Q]: *Constructive zone AND Constructive mutation*', '[F]: *Destructive mutation*', and '{R}: *Random mutation*'. '[Q]:

147

Constructive zone' implies that the in-built buffer has been crossed and that access to the deeper four-base logic-encoding ecosystem (FBLEE) has been granted. '[Q}' in this segment implies that there is the possibility that full-quantization as specified by Line 3 of the previous matrix can take place. Access to this zone is a prerequisite for constructive mutation to occur, as designated by the element   '   [Q}: *Constructive zone* ∩ *Constructive mutation* ', which implies that full-quantization is going to take place and will result in material change. The '[F]' specifies the relationship between 'Fragmentation' in Line 3 of the previous matrix and destructive mutation. The '[R]' specifies the relationship between 'Random' in Line 3 of the previous matrix and random mutation.

Effect of Levels of Light on Mutation Using Wavefunction Form of Light-Space-Time Emergence

In Chapter 3.3 we were introduced to Schrodinger's wavefunction equation (3.3.1), reproduced below for convenience, which seeks to model matter as a wave rather than as a particle:

$$i\frac{h}{2\pi}\frac{\partial}{\partial x}\psi = \hat{H}\psi$$

This wavefunction equation deals with dynamics behind the surface, and particularly of as many as infinite potential superposed states that then collapse into a material possibility in time and space at U. Such superposition is another way of saying that there is wholeness behind the surface. In (3.6.1) such wholeness is represented by L, S, T, the Light, Space, and Time matrices respectively.

As explained by Neil Turok in his book The Universe Within (Turok, 2012), Euler's formula reproduced below, can be used to model many naturally occurring phenomena because of its sinusoidal oscillation between narrow bounds as x increases. Reproducing Euler's formula:

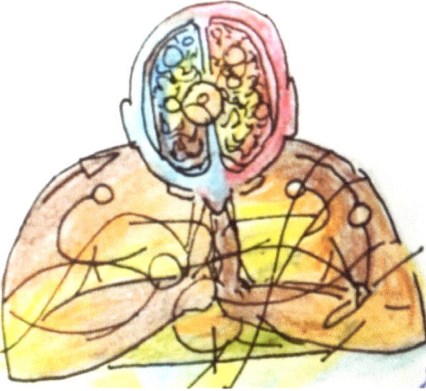

$$e^{ix} = \cos x + i \sin x$$

A key characteristic of this equation is that the sum of the squares of the ordinary and complex parts, on the right side of the equation, is one. In many contemporary interpretations of quantum theory this ensures that the probabilities for all possible outcomes add up to one. Dealing in probabilities become important when considering as many as infinite superposed states.

Further, in modeling material reality a modified notation of the Schrodinger wavefunction as interpreted by Feynman is leveraged. This version features the integral sign, $\int$, meaning that all terms to the right of it have to be summed up for all space and time till the moment when the wavefunction is required to be known.

Hence, combining (3.6.1) with Feynman's interpretation of the Schrodinger wavefunction, with the Euler formula yields the following equations, 3.6.6-11 for genetic mutation:

$$\psi_{destructive-mutation} = \left|\left| \int e^{i \int |[L][S][T]|} \right|\right|_U$$

Eq. 3.6.6: Wavefunction for Destructive-Mutation

Note that in (3.6.6) the formulation specifying iteration, $\langle x_U | x_T \rangle$, becomes redundant since iteration is implied by the integral across space and time, and hence is removed. Further, (3.6.6) suggests that so long as the basis of an organization is untransformed, specified by the U following the vertical-brackets, the outcome is going to be one of destructive mutation, specified by $\psi_{destructive-mutation}$.

Further, as specified by Equation 3.6.7, the Probability-View of Destructive-Mutation, the driver of such destructive-mutation is going to be either the untransformed physical (P_U), the untransformed vital (V_U), the untransformed mental (M_U), or the untransformed integral (I_U):

$$|\psi_{destructive-mutation}|^2 = P_U^2 + V_U^2 + M_U^2 + I_U^2 = 1$$

Eq. 3.6.7: Probability-View for Destructive-Mutation

As specified by (3.6.7) the probability that any of these bases will be leveraged in destructive mutation adds up to one. Note that the physical, the vital, the mental, and integral were discussed in Chapter 2.6 focusing on the systematization of the mutational-sequence.

Equation 3.6.8, Wavefunction for Random-Mutation, suggests the mixed bases for organizations, as specified by dynamics of both the untransformed (U) and the meta-levels (M_x), which therefore results in random mutation.

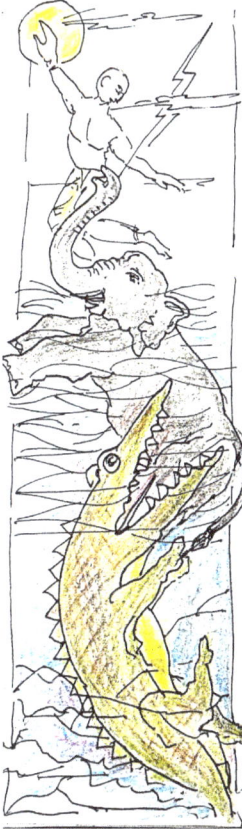

$$\psi_{ransom-mutation} = \left| \int e^{i \int |[L][S][T]|} \right|_{U \& M_x}$$

Eq. 3.6.8:   Wavefunction for Random-Mutation

As specified by Equation 3.6.9, Probability-View for Random-Mutation, the probability that any of the untransformed (U) and transformed (T) bases will be leveraged in random mutation adds up to one:

$$|\psi_{random-mutation}|^2$$
$$= P_U^2 + P_T^2 + V_U^2 + V_T^2$$
$$+ M_U^2 + M_T^2 + I_U^2 + I_T^2$$
$$= 1$$

Eq. 3.6.9: Probability View for Random-Mutation

Equation 3.6.10, Wavefunction for Constructive-Mutation, suggests some transformed bases for an organization, as specified by dynamics of the meta-levels (M_x), which therefore can result in a constructive mutation.

$$\psi_{constructive-mutation} = \left| \int e^{i \int |[L][S][T]|} \right|_{M_x}$$

Eq. 3.6.10: Wavefunction for Constructive-Mutation

As specified by Equation 3.6.11, Probability-View for Constructive-Mutation, the probability that the transformed (T) bases will be leveraged in constructive mutation adds up to one:

$$|\Psi_{constructive-mutation}|^2 = P_T^2 + V_T^2 + M_T^2 + I_T^2 = 1$$

Eq. 3.6.11: Probability View for Constructive-Mutation

Now the question is how would genetic information be impacted based on the levels of light active, and what might an equation that captured that look like? Equation 3.6.12, Potential Effect of Levels of Light on Genetic-Type Information (Wavefunction Form), is such an equation:

$$\Psi_{Potential\ Effect\ of\ Levels\ of\ Light\ on\ Pre\ Genetic\ or\ Genetic\ Information}$$

$$=$$

$$\begin{bmatrix} \left(\left| \int\int e^{i\int [L][S][T]} \right|_{Y=(U,U\&M_x,M_x)} \right) \\ \times \\ \left((Y > U: Z_Q) \vee (Y \le U: Z_F) \vee (Y = U: Z_R) \right) \\ \ni \\ \left(\begin{array}{c} Z \in \mathbb{U}\ (Space, Time, Energy, Gravity) \\ Q: Quantization; F: Fragmentation; R: Random \end{array} \right) \end{bmatrix} \rightarrow$$

$$h \ni h \in \left(\begin{array}{c} Constructive\ zone, \\ Constructive\ zone \wedge Constructive\ mutation, \\ Destructive\ mutation, \\ Random\ mutation \end{array} \right)$$

Eq. 3.6.12: Potential Effect of Levels of Light on Genetic-Type Information (Wavefunction Form)

Taking each line in the equation separately: Line 1 from the top combines (3.6.6), (3.6.8), and (3.6.10) to essentially state that the wave-function for effect on genetic information while governed by the Light-Space-Time matrices in (3.1.3) will vary based on the bases that are active, hence $Y = (U, U\&M_x, M_x)$, where Y equates to the family of bases that may be active: untransformed (U), a mix (U & M_x), or M_x. Since the primary part of Line 1 falls under the $\int$ it means that the operation is iterative since dynamics are summed up for space and time to that point where the wavefunction is sought.

Line 1 is then subjected (×) to a determination of the dominant levels of light that may be active, designated by Line 2, ' $\left((Y > U : Z_Q) \vee (Y \le U : Z_F) \vee (Y = U : Z_R) \right)$ '. Unpacking this, '$Y > U$' implies meta-levels are active and as a result it is possible that Z_Q is going to take place. This also implies activation and potential change of FBLEE. The call from below, as it were, may invoke some function that already exists in the subtle-libraries 'above', so that some already existing function may influence FBLEE through ∞ -entanglement, K-entanglement, or N-entanglement. As discussed earlier this may be thought of as a key-and-lock mechanism, where a deep enough visceral urge from below acting as the key, opens an entangled lock to alter FBLEE as per the visceral urge. '$Y \le U$' implies that only the untransformed levels are active, and therefore also the sub-level where the speed of light is 0 is active and as a result Z_F is going to take place. '$Y = U$' implies that all levels are active and as a result Z_R is going to take place.

Line 3 elaborates the significance of Z_Q, Z_F, and Z_R. Hence Z is the union of potential quantum-operations of space, time, energy, and gravity, designated by '$Z \in \mathbb{U} (Space, Time, Energy, Gravity)$'. But the nature of the operations is designated by '$Q: Quantization; F: Fragmentation; R: Random$'. Z_Q, then, implies that the full quantization originating from updated four-base logic-encoding ecosystems (FBLEE) can take place, and will result in lasting material change at the Genetic-Type level. Z_F implies that the essential set will be fragmented and that only libraries a the level of local cellular-level DNA or precipitated material-fabric logic can potentially be altered. Z_R also precludes full quantization, and that some partial local-library constructive or destructive mutation may take place.

The '→ $h \ni h \in$' segment resolves the outcome of the operations implied by Lines 1 – 3, suggesting that the outcome will be 'h' such that (∋) 'h' is an element (∈) of the set specified by the members '[Q]: *Constructive zone*', '[Q}: *Constructive zone AND Constructive mutation*', '[F}; *Destructive mutation*', and '{R}: *Random mutation*'. '[Q]: *Constructive zone*' implies that the in-built buffer has been crossed and that access to the deeper four-base logic-encoding ecosystem (FBLEE) has been granted. '[Q}' in this segment implies that there is the possibility that full-quantization as specified by Line 3 of the previous matrix can take place. Access to this zone is a prerequisite for constructive mutation to occur, as designated by the element '[Q}: *Constructive zone* ∩ *Constructive mutation* ', which implies that full-quantization is going to take place and will result in material change. The '[F]' specifies the relationship between 'Fragmentation' in Line 3 of the previous matrix and destructive mutation. The '[R]' specifies the relationship between 'Random' in Line 3 of the previous matrix and random mutation.

Chapter 3.7: Application of Qualified Determinism to Genetic Mutation

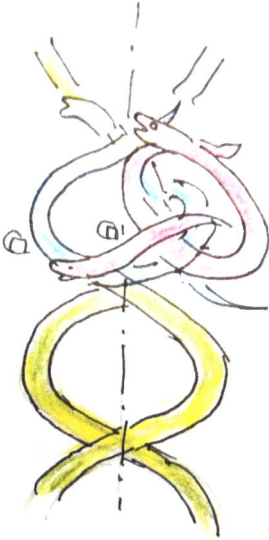

The previous chapter explored the possible impact of different layers or levels of light on genetic mutation. Recall that genetic mutation results in altering or creating new genetic-type information, which is positioned as being the biographical language of any partial-singularity. The exploration was at a high level and broadly considered a mapping between active layers of light and the type of mutation that may result. Further, the exploration was framed in terms of the light-space-time matrices, central to the mathematics in this treatise, and additionally in terms of a wavefunction form, central to the probabilistic foundation in quantum theory. Consistent with the mathematics in this treatise though, this chapter further elaborates a non-probabilistic approach by application of a process of Qualified Determinism to genetic mutation. In this process structural forces influencing mutation are disaggregated into vertical and horizontal components. The vertical component is resolved by determination of the most active layer of light. The horizontal component is resolved by determination of the most influential base of change amongst the physical, vital, mental, and integral possibilities.

This chapter, hence, explores a mathematical notion of qualified determinism - Dynamic Interaction (DI) that has a 'vertical' and a 'horizontal' component. The vertical component is designated as DI_V and the horizontal component as DI_H. Several equations to capture the

155

inherent dynamism manifest as the likely mutational-sequence along the upward-strand have already been derived in Chapter 2.6. These included equations for the mutational-sequence propelled by the physical, the vital, the mental, and the integral bases. The derived equations propose a model to give insight into how on-going mutation occurs by changing the fundamental states that a microorganism, as an instance of organization in general, is subject to. Several scientists, such as (Prigogine, 1977) and others, are proposing that a system can bifurcate in unpredictable ways to create an emergent property that cannot be predicted. DI is going to propose that in fact there is a 'qualified determinism' as opposed to randomness that occurs (Malik et al, 2017).

DI is going to propose an alternative to the paradigm of statistics and probability thought to govern quantum and other nano- and micro-level objects.

This qualified determinism is the result of the relative strengths of the levels within the core-matrix as summarized by Equation 2.6.7 and reproduced here for convenience:

$$\begin{bmatrix} M_3 \rightarrow System_X \\ (\uparrow F \rightarrow I) \\ M_2 \rightarrow S_{System_X} \\ (\uparrow Sig \rightarrow F) \\ M_1 \rightarrow Sig_x \\ (\uparrow > P_x) \\ U \rightarrow x_U \end{bmatrix}$$

The application of the vertical component of the new function being proposed, DI_V, to this core matrix will yield the nature or 'strength' of the state (x) or orientation under consideration. If the untransformed or U layer is strongest, implying that the habitual patterns that keep an organization locked into its untransformed way of operation are still very active, then the nature of the output of DI_V, notated by x-state, will be x_U. If the

156

habitual patterns have been overcome then the strength of the x-state increases since it is the dynamics of M_1 or Sig_x that are now active. In this case the x-state will be Sig_x. If the unique 'signature' has become a 'force', then the conditions for activation of M_2 have been put in place and the x-state will be even higher, S_{System_X}. The architectural forces active in M_2 are by definition more powerful than Sig_x that is a derivation of a set of such architectural forces. If the 'force' so acting becomes impersonal so that an organizational ego-state is overcome, then the x-state will have the most strength and is characterized by $System_X$ active at M_3. Hence, DI_V applied to a core-matrix will yield the 'strength' in terms of the x-dynamic that is active. Recall also, that each of these upward-strand levels is in fact determined by the parallel light-level that exists in the downward-strand. This is illustrated by the following equation, Equation 3.7.1, which can be considered to be a deductive proof in the context of this model:

$$DI_V \begin{bmatrix} M_3 \rightarrow System_X \\ (\uparrow F \rightarrow I) \\ M_2 \rightarrow S_{System_X} \\ (\uparrow Sig \rightarrow F) \\ M_1 \rightarrow Sig_x \\ (\uparrow > P_x) \\ U \rightarrow x_U \end{bmatrix} =>$$

$$\text{x-state} \in (x_U, Sig_x, S_{System_X}, System_X)$$

$$Where: Strength(System_X) > Strength(S_{System_X})$$
$$> Strength(Sig_x) > Strength(x_U)$$

Eq 3.7.1: Illustrating Action of Dynamic Interaction – vertical component

What is to be noted here is that while the action of DI_V yields a relative strength and therefore a 'single' value for the core- or x-matrix under consideration yet each x-matrix in itself could have an infinite number of possibilities. This should be clear in looking at how x_U, Sig_x, S_{System_X}, and $System_X$, were initially defined.

Hence, taking the example where x = physical:

$$Physical_U \ni [inertia, lethargy, status\ quo, ...]$$

As can be seen $Physical_U$, defined in Chapter 2.6, is practically speaking already an infinite set with qualities similar to the ones specified.

Similarly, Sig_{Pr}, defined in Chapter 2.4, also has an infinite variation:

$$Sig_{Pr}$$
$$= Xa$$
$$+ \overline{Yb_{0-n}} \quad where \begin{bmatrix} X \in [S_{System_{Pr}}] \\ Y \in [S_{System_{Pr}}, S_{System_P}, S_{System_K}, S_{System_N}] \\ a, b\ are\ integers; a > b \end{bmatrix}$$

$S_{System_{Pr}}$, defined in Chapter 2.3, is also an infinite set with forces of the nature specified in the following equation:

$$S_{System_{Pr}} \ni [Service, Perfection, Diligence, Perseverance, ...]$$

And recall that in Chapter 2.2, $System_{Pr}$ has been defined as:

$$System_{Pr}$$
$$\equiv TS_{0 \to N} \begin{bmatrix} Presence \\ \downarrow \\ Opportunity \end{bmatrix} \begin{bmatrix} P_L & \to & V_L \\ V_L & \to & M_L \end{bmatrix} \& Container \begin{bmatrix} System_P \\ System_K \\ System_N \end{bmatrix} \end{bmatrix}$$

So in essence DI_V is really giving us a summary assessment of the 'level' of the x-matrix under consideration with all its infinite potentiality. An example will follow shortly.

The other component of DI, as suggested earlier in this chapter, is the horizontal component, DI_H. Just as DI_V yields a summary assessment of the level that an x-matrix is operating at, similarly DI_H yields a summary assessment of the direction that a system or organization under consideration is going to continue its development in, considering the physical, the vital, the mental, and the integral bases to be the choices.

Assuming that any organization or system is inherently unique, as this mathematical model proposes, and as the phenomena of entanglement proves, and assuming that the infinite sets of x_U and S_{System_X} applied across the physical, vital, mental, and integral bases or orientations respectively will account for any state that a nano- or micro-organization can experience, then at a certain point in time any such organization under consideration is going to have a direction-bias in one of the possible physical, vital, mental, or integral directions. Hence, DI_H will yield the summary direction that is going to lead an organization into its future given the current states active in it.

This summary direction is going to be yielded by considering the relative strengths of the separate core x-matrices – the physical, the vital, the mental, and the integral. The assumption is that there will be one core-matrix that will be stronger than the others.

Hence, as an example, first applying DI_V across all four x-matrices may, for example, yield the following results, with the strongest level within each x-matrix highlighted and bolded:

$$\begin{bmatrix} System_{Pr} \\ \mathbf{S_{System_{Pr}}} \\ Sig_P \\ Physical_U \end{bmatrix} \begin{bmatrix} System_P \\ S_{System_P} \\ Sig_V \\ \mathbf{Vital_U} \end{bmatrix} \begin{bmatrix} System_K \\ S_{System_K} \\ Sig_M \\ \mathbf{Mental_U} \end{bmatrix} \begin{bmatrix} System_N \\ S_{System_N} \\ \mathbf{Sig_I} \\ Integral_U \end{bmatrix}$$

Since by definition the strength of $System_x$ is greater than S_{System_x}, which is greater than Sig_x, which is greater than x_U, applying DI_H, as in Equation 3.7.2, across these x-matrices, as in the example following it will then yield the strongest direction, which in this example is the Physical:

$$DI_H \left(\begin{bmatrix} System_{Pr} \\ S_{System_{Pr}} \\ Sig_P \\ Physical_U \end{bmatrix} \begin{bmatrix} System_P \\ S_{System_P} \\ Sig_V \\ Vital_U \end{bmatrix} \begin{bmatrix} System_K \\ S_{System_K} \\ Sig_M \\ Mental_U \end{bmatrix} \begin{bmatrix} System_N \\ S_{System_N} \\ Sig_I \\ Integral_U \end{bmatrix} \right)$$
$$= Orientation_{Strongest}$$

Eq 3.7.2: Illustrating Action of Dynamic Interaction - horizontal component

Example:

$$DI_H \left(\begin{bmatrix} System_{Pr} \\ \mathbf{S_{System_{Pr}}} \\ Sig_P \\ Physical_U \end{bmatrix} \begin{bmatrix} System_P \\ S_{System_P} \\ Sig_V \\ \mathbf{Vital_U} \end{bmatrix} \begin{bmatrix} System_K \\ S_{System_K} \\ Sig_M \\ \mathbf{Mental_U} \end{bmatrix} \begin{bmatrix} System_N \\ S_{System_N} \\ \mathbf{Sig_I} \\ Integral_U \end{bmatrix} \right)$$
$$= Physical$$

Hence, DI function will yield the following direction of mutation, as in Equation 3.7.3, where 'x_matrix' is used interchangeably with 'orientation' or 'bases':

$$Mutation\_Dir = DI \left(\begin{bmatrix} M_3 \to System_{Pr} \\ (\uparrow F \to I) \\ M_2 \to S_{System_{Pr}} \\ (\uparrow Sig \to F) \\ M_1 \to Sig_P \\ (\uparrow > P_P) \\ U \to Physical_U \end{bmatrix} \begin{bmatrix} M_3 \to System_P \\ (\uparrow F \to I) \\ M_2 \to S_{System_P} \\ (\uparrow Sig \to F) \\ M_1 \to Sig_V \\ (\uparrow > P_V) \\ U \to Vital_U \end{bmatrix} \atop \begin{bmatrix} M_3 \to System_S \\ (\uparrow F \to I) \\ M_2 \to S_{System_S} \\ (\uparrow Sig \to F) \\ M_1 \to Sig_M \\ (\uparrow > P_M) \\ U \to Mental_U \end{bmatrix} \begin{bmatrix} M_3 \to System_N \\ (\uparrow F \to I) \\ M_2 \to S_{System_N} \\ (\uparrow Sig \to F) \\ M_1 \to Sig_I \\ (\uparrow > P_I) \\ U \to Integral_U \end{bmatrix} \right) \to$$

$$x\_matrix_{strongest} @ level_{strongest}$$

Eq 3.7.3: Direction of Mutation

Generalizing, as in Equation 3.7.4, where Mutation_Dir is direction of mutation:

$$Mutation\_Dir = DI \left(\begin{bmatrix} M_3 \to System_X \\ (\uparrow F \to I) \\ M_2 \to S_{System_X} \\ (\uparrow Sig \to F) \\ M_1 \to Sig_X \\ (\uparrow > P_P) \\ U \to x_U \end{bmatrix}_{x=p,v,m,i} \right) \to$$

$$x\_matrix_{strongest} @ level_{strongest}$$

Eq 3.7.4: Generalized Equation for Direction of Mutation

Hence, this mathematical model is suggesting that any situation, including micro and nano-level situations, rather than having a random outcome, has a 'qualified deterministic' outcome. In the introduction to his book "Where is Science Going?" (Planck, 1933), James Murphy points out that the reason Planck spent so much of his time giving lectures on causation was because of the trend of physicists at the time, which has continued to the modern day, to overthrowing the principle of causation following the development of quantum theory, which he felt was misplaced. "Planck would claim", he wrote, "and so would Einstein, that it is not the principle of causation itself which has broken down in modern physics, but rather the traditional formulation of it." Murphy also quotes James Jeans (Jeans, 1932) to suggest the issue associated with causation and determinism: "Einstein showed in 1917 that the theory founded by Planck appeared, at first sight at least, to entail consequences far more revolutionary than mere discontinuity", and here he is referring to the finding that radiant energy is not emitted in a continuous flow, but in integral quantities, or quanta, which can be expressed in integral numbers. Continuing: "It appeared to dethrone the law of causation from the position it had therefore held as guiding the course of the natural world. The old science had confidently proclaimed that nature could follow only one road, the road which was mapped out from the beginning of time to its end by the continuous chain of cause and effect; state A was inevitably succeeded by state B. So far the new science has only been able to say that state A may be followed by state B or C or D or by innumerable other states. It can, it is true, say that B is more likely than C, C than D, and so on; it can even specify the relative probabilities of B, C, and D. But, just because it has to speak in terms of probabilities, it cannot speak with certainty which state will follow which; this is a matter which lies on the knees of the gods – whatever gods there may be."

While under the apparent dynamics at the quantum, micro, or nano level there may appear to be randomness and a dethroning of the principle of causation, the notion of a multiplicity of levels, each having its impact on the strength of an orientation or base and further on the consequent direction from a multiplicity of possible orientations, is being suggested here as determining the direction of any system, while still allowing infinite variation in the details that may define it. Hence, the positions of Planck and Einstein are vindicated when considering Equation 3.7.4.

Further, assuming any system where multiple elements are active, connected, interdependent, and emergent, it may be possible to understand, through application of calculus, as to which level is the source for change.

Hence, where N may be source of change, the rate of change of N will resolve into one of P_U, V_U, M_U, I_U, P_T, V_T, M_T, or I_T. This may be summarized by Equation 3.7.5, where y is either U or T:

$$\frac{dN}{dt} \rightarrow \begin{bmatrix} P_U & P_T \\ V_U & V_T \\ M_U & M_T \\ I_U & I_T \end{bmatrix} \rightarrow x_y, where\ y \in (U,T)$$

Eq 3.7.5: Establishing the Nature of the Change

If T, implying that the action of one of the meta-levels has caused the transformation or mutation, then application of one of the following integrals will determine which level is the likely source for change.

Hence, for M_1, if the integral of $\frac{\partial(x_U \rightarrow x_T)}{\partial t}$ across a limited area 'a' in the vicinity of the change, is greater than some threshold value $Threshold_{Signature}$, then the signature dynamics are likely the source of change, where 'a' can be thought of as some kind of gene-footprint. This is summarized by Equation 3.7.6:

$$\int_0^a \frac{\partial(x_U \rightarrow x_T)}{\partial t} > Threshold_{Signature}$$

Eq 3.7.6: Signature Dynamics as the Source of Change

For M_2, if the integral of $\frac{\partial(x_U \rightarrow x_T)}{\partial t}$ across a larger area 'b' extending beyond the vicinity of the change, is greater than some threshold value $Threshold_{Architectural\ Forces}$, then the architectural forces are likely the source of change, , where 'b' can be thought of as some kind of gene-footprint larger than 'a'. This is summarized by Equation 3.7.7:

164

$$\int_0^b \frac{\partial (x_U \rightarrow x_T)}{\partial t} > Threshold_{ArchitecturalForces}$$

Eq 3.7.7: Architectural Forces as the Source of Change

For $M_{3''}$ if the double integral of $\frac{\partial (x_U \rightarrow x_T)}{\partial t}$ across the system specified by 'A', and across some time 't', is greater than some threshold value $Threshold_{System\ Property}$, then the system properties are likely the source of change, , where 'A' can be thought of as some kind of gene-footprint larger than both 'a' and 'b'. This is summarized by Equation 3.7.8:

$$\int_0^t \int_0^A \frac{\partial (x_U \rightarrow x_T)}{\partial t} > Threshold_{SystemProperty}$$

Eq 3.7.8: System Properties as the Source of Change

SECTION 4: OVERVIEW OF DEVELOPMENT OF PRE-GENETIC CODE, MATERIAL-FABRIC, GENETIC CODE, AND POST-GENETIC CODE IN LIGHT-BASED SINGULARITIES

If the mathematics presented in the previous sections is true, then the pre-genetic and genetic code that imbues everything becomes the result of continuous, real-time Light-Space-Time Matrix based computations implicitly involving quantum-levels.

This section therefore briefly outlines how it may be that Light-Space-Time Matrix based computations generates a range of code from an era pre-dating the Big Bang to modern day Global Civilization, to whatever else may emerge.

The implications of this are that everything that exists is imbued with code. It is just the way the code is housed that changes in forms recognized as living. This also implies that there is likely code within all living cells that has to do with space, time, energy, gravity, the functioning of the electromagnetic spectrum, and all the stages of material elaboration that precede the emergence of a living cell.

It is the sharing of such large swathes of code that precede an emergent form, with the emergent form, that typifies the notion of a light-based singularity. In fact it is a "singularity" because the most current emergent form has all previous code embedded or available to it thereby inherently abiding with all "laws" that have thus far emerged in previous singularities. This also implies the notion that the seed-singularity itself will have had a progressive biography resulting in partial-seed-singularities along the way, until the seed-singularity has itself emerged.

Note that the sections following this will go into more detail to elaborate the process by which quantization incrementally changes material reality through alteration of pre-material-fabric, four-base logic encoding ecosystem, pre-genetic material-fabric, genetic, and post-genetic code.

Chapter 4.1: Generation of Pre-Big Bang to Big Bang Pre-Genetic Code and Activation of Material-Fabric in Light-Based Singularities

The Light-Matrix component of the Light-Space-Time Emergence equation (3.1.3), reproduced below for convenience, essentially provides an algorithm by which pre-Big Bang existence proceeds to a reality of the Big Bang due to precipitating light. The process generates pre-Big Bang pre-genetic code, and with the Big Bang culminates in the activation of the material-fabric through the initialization of space-time-energy-gravity pre-genetic code.

$$Light - Space - Time\ Emergence_T =$$

$$\left| \begin{matrix} \begin{bmatrix} c_\infty : [Pr, Po, K, H] \\ (\downarrow R_{C_K} = f(R_{C_\infty})) \\ c_K : [S_{Pr}, S_{Po}, S_K, S_H] \\ (\downarrow R_{C_N} = f(R_{C_K})) \\ c_N : f(S_{Pr} \times S_{Po} \times S_K \times S_H) \\ (\downarrow R_{C_U} = f(R_{C_N})) \\ c_U : [P, V, M, C] \\ \Uparrow \\ c_{0:[D,W,I,C]} \end{bmatrix}_{Light} \begin{bmatrix} M_3 \rightarrow System_X \\ (\uparrow F \rightarrow I) \\ M_2 \rightarrow S_{System_X} \\ (\uparrow Sig \rightarrow F) \\ M_1 \rightarrow Sig_x \\ (\uparrow > P_x) \\ U \rightarrow x_U \end{bmatrix}_{Space} \\ \begin{bmatrix} M_3 : -\infty \le t \le \infty \\ \downarrow \\ M_2 : 0 \ge t > \infty \\ \downarrow \\ M_1 : 0 > t > \infty \\ \downarrow \\ U \rightarrow \begin{matrix} t \le E_{Cell} ; TC: M_3 \rightarrow U \\ t \sim E_{Human} ; TC: U \rightarrow M_3 \end{matrix} \end{bmatrix}_{Time} TC \rightarrow x_T \end{matrix} \right| \langle x_U | x_T \rangle$$

Partial-Seed-Singularity Pre-Big Bang Pre-Precipitation ∞-Entanglement Pre-Material-Fabric Pre-Genetic Code

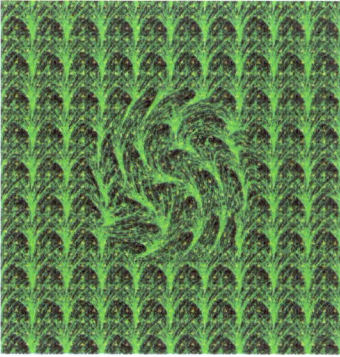

Before the Big Bang, as per discussions in Section 1 and 2, it can be assumed that Light precipitated in stages to the point where it slowed down to c. Equation 4.1.1 highlights the pre-precipitation pre-material-fabric pre-genetic code. Hence (3.1.3) collapses to the active highlighted portion only.

$Light - Space - Time\ Emergence_T =$

$$
\left|\left|\begin{array}{l}
\begin{bmatrix}
\boldsymbol{c_\infty} : [\boldsymbol{Pr, Po, K, H}] \\
\left(\downarrow R_{C_K} = f\left(R_{C_\infty}\right)\right) \\
c_K : [S_{Pr}, S_{Po}, S_K, S_H] \\
\left(\downarrow R_{C_N} = f\left(R_{C_K}\right)\right) \\
c_N : f(S_{Pr}\ x\ S_{Po}\ x\ S_K\ x\ S_H) \\
\left(\downarrow R_{C_U} = f\left(R_{C_N}\right)\right) \\
c_U : [P, V, M, C] \\
\Uparrow \\
c_{0 : [D, W, I, C]}
\end{bmatrix}_{Light}
\begin{bmatrix}
M_3 \rightarrow System_X \\
(\uparrow F \rightarrow I) \\
M_2 \rightarrow S_{System_X} \\
(\uparrow Sig \rightarrow F) \\
M_1 \rightarrow Sig_x \\
(\uparrow > P_x) \\
U \rightarrow x_U
\end{bmatrix}_{Space} \\
\begin{bmatrix}
M_3 : -\infty \le t \le \infty \\
\downarrow \\
M_2 : 0 \ge t > \infty \\
\downarrow \\
M_1 : 0 > t > \infty \\
\downarrow \\
U \rightarrow \begin{array}{l} t \le E_{Cell}; TC : M_3 \rightarrow U \\ t \sim E_{Human}; TC : U \rightarrow M_3 \end{array}
\end{bmatrix}_{Time}
\quad TC \rightarrow x_T \\
\Rightarrow
\end{array}\right.\right| \langle x_U | x_T \rangle
$$

169

$$\left(System_{P_T} \equiv TS_{0 \to N} \begin{bmatrix} Presence \\ \downarrow \\ Opportunity \end{bmatrix} \begin{bmatrix} P_L & \to & V_L \\ V_L & \to & M_L \end{bmatrix} \& \ Container \begin{bmatrix} System_P \\ System_K \\ System_N \end{bmatrix} \right)$$

$$System_P \equiv power > \sum_{TS=0}^{N} P_R * V_R * M_R$$

$$System_K \equiv TS_{0 \to N} \begin{bmatrix} Knowledge \\ \downarrow \\ Leverage \end{bmatrix} \begin{bmatrix} P_{I,C} \\ V_{I,C} \\ M_{I,C} \end{bmatrix} \to \begin{bmatrix} P_L & \to & V_L \\ V_L & \to & M_L \end{bmatrix}$$

$$System_N \equiv TS_{0 \to N} \left(\coprod_{Nurturing} \begin{pmatrix} P_- & M_+ \\ V_- & V_+ \\ M_- & P_+ \end{pmatrix} mod \ (r, \theta) \right)$$

Eq 4.1.1: Partial-Seed-Singularity Pre-Big Bang Pre-Precipitation ∞-Entanglement Pre-Material-Fabric Pre-Genetic Code

Note the equation-segments following the ' ⇒ ' were derived in Chapter 2.2, The Seed-Singularity Ubiquitous-Point-Instant Pre-Genetic ∞-Entanglement Library. The generated pre-genetic code belongs to the pre-material-fabric genre since it will remain in a subtle domain exercising its influence through ∞-entanglement where possible. Further, the generated pre-genetic code also narrates the biography for a composite singularity in that it is whole and complete, even though it is describing only a part of the seed-singularity. Hence, strictly speaking (4.1.1) is describing a partial-seed-singularity.

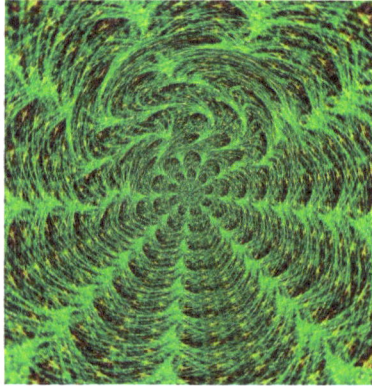

Partial-Seed-Singularity Pre-Big Bang Pre-Precipitation Zero-Speed Pre-Material-Fabric Pre-Genetic Code

As discussed in the Chapter 2.1, The Light-Based-Singularity Superstructure, Light could project itself at speed zero. This would create a reality opposite to that in its native state. This reality could have been projected prior to the gradual precipitation from light traveling infinitely fast, that is the corner stone of this treatise on Light. The reality created when light exists at zero speed, as already discussed in Sections 2 and 3, can be seen to play a significant part in destructive mutation, and is seen to influence the reality modeled by the layer U, where light moves at c. Because obstinate patterns can be better explained with a 'speed zero' reality in the background, it will be assumed that this pre-exists the precipitation of Light down to c.

Equation 4.1.2 highlights the pre-precipitation zero-speed pre-material-fabric pre-genetic code. Hence (3.1.3) collapses to the active highlighted portion only.

$$Light - Space - Time\ Emergence_T =$$

$$\left[\left[\begin{array}{l} c_\infty: [Pr, Po, K, H] \\ \left(\downarrow R_{C_K} = f\left(R_{C_\infty}\right)\right) \\ c_K: [S_{Pr}, S_{Po}, S_K, S_H] \\ \left(\downarrow R_{C_N} = f\left(R_{C_K}\right)\right) \\ c_N: f(S_{Pr} \times S_{Po} \times S_K \times S_H) \\ \left(\downarrow R_{C_U} = f\left(R_{C_N}\right)\right) \\ c_U: [P, V, M, C] \\ \Uparrow \\ C_{0:[D,W,I,C]} \end{array}\right]_{Light} \begin{bmatrix} M_3 \rightarrow System_X \\ (\uparrow F \rightarrow I) \\ M_2 \rightarrow S_{System_X} \\ (\uparrow Sig \rightarrow F) \\ M_1 \rightarrow Sig_X \\ (\uparrow > P_x) \\ U \rightarrow x_U \end{bmatrix}_{Space}\right.$$

$$\left. U \rightarrow \begin{bmatrix} M_3: -\infty \leq t \leq \infty \\ \downarrow \\ M_2: 0 \geq t > \infty \\ \downarrow \\ M_1: 0 > t > \infty \\ \downarrow \\ t \leq E_{Cell}; TC: M_3 \rightarrow U \\ t \sim E_{Human}; TC: U \rightarrow M_3 \end{bmatrix}_{Time} \quad TC \rightarrow x_T \right]_{\langle x_U | x_T \rangle}$$

$$\Rightarrow$$

$$\left\langle \begin{array}{l} Pre - Big - Bang \ Pre - Precipitation \ \infty - Entanglement \\ Pre - Material - Fabric \ Pre - Genetic \ Code \end{array} \right\rangle +$$

$$(c_0: [Darkness, Weakness, Ignorance, Chaos])$$

Eq 4.1.2:   Partial-Seed-Singularity Pre-Big Bang Pre-Precipitation Zero-Speed Pre-Material-Fabric Pre-Genetic Code

Note that the last equation-segment following the '$\Rightarrow$' was derived in Chapter 2.1, The Light-Based-Singularity Superstructure. This equation is stating that when light is projected at zero-speed the only active code-segments are due to ∞-entanglement as discussed in the previous sub-section and zero-speed of light. The fact that code segments due to ∞-entanglement are still active implies that this too can be thought of as a partial-seed-singularity.

Partial-Seed-Singularity Pre-Big Bang K-Entanglement Pre-Material-Fabric Pre-Genetic Code

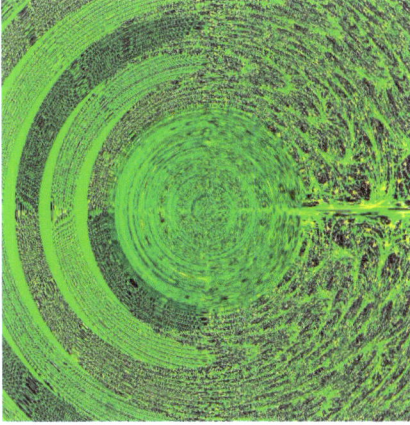

The first stage of precipitation prior to the Big Bang is modeled by K-entanglement as discussed in Chapter 2.3, The Seed-Singularity Architectural Forces K-Entanglement Pre-Genetic Library & Dynamics. Equation 4.1.3 highlights the pre-material-fabric pre-genetic code active at this point. Hence (3.1.3) collapses to the active highlighted portion only.

$$Light - Space - Time\ Emergence_T =$$

173

$$\left\| \begin{array}{l} \begin{bmatrix} \begin{array}{c} c_\infty : [Pr, Po, K, H] \\ \left(\downarrow R_{C_K} = f(R_{C_\infty}) \right) \\ c_K : [S_{Pr}, S_{Po}, S_K, S_H] \\ \left(\downarrow R_{C_N} = f(R_{C_K}) \right) \\ c_N : f(S_{Pr} \times S_{Po} \times S_K \times S_H) \\ \left(\downarrow R_{C_U} = f(R_{C_N}) \right) \\ c_U : [P, V, M, C] \\ \Uparrow \\ c_{0:[D,W,I,C]} \end{array} \end{bmatrix}_{Light} \begin{bmatrix} M_3 \to \text{System}_X \\ (\uparrow F \to I) \\ M_2 \to S_{\text{System}_X} \\ (\uparrow Sig \to F) \\ M_1 \to Sig_x \\ (\uparrow > P_x) \\ U \to x_U \end{bmatrix}_{Space} \\ \begin{bmatrix} U \to \begin{array}{c} M_3 : -\infty \le t \le \infty \\ \downarrow \\ M_2 : 0 \ge t > \infty \\ \downarrow \\ M_1 : 0 > t > \infty \\ \downarrow \\ t \le E_{Cell}; TC: M_3 \to U \\ t \sim E_{Human}; TC: U \to M_3 \end{array} \end{bmatrix}_{Time} \quad TC \to x_T \\ \Rightarrow \end{array} \right\|_{\langle x_U | x_T \rangle}$$

$$\left\langle \begin{array}{c} Pre - Big - Bang\ Pre - Precipitation\ \infty - Entanglement \\ Pre - Material - Fabric\ Pre - Genetic\ Code \end{array} \right\rangle +$$

$$\left\langle \begin{array}{c} Pre - Big - Bang\ Pre - Precipitation\ Zero - Speed \\ Pre - Material - Fabric\ Pre - Genetic\ Code \end{array} \right\rangle +$$

$$\left(\begin{array}{l} \sum S_{System_{Pr}} \ni [Service, Perfection, Diligence, Perseverance, \dots] \\ \quad \sum S_{System_P} \ni [Power, Courage, Adventure, Justice, \dots] \\ \quad \quad \sum S_{System_K} \ni [Wisdom, Law\ Making, Spread\ of\ Knowledge \dots] \\ \quad \quad \quad \sum S_{System_N} \ni [Love, Compassion, Harmony, Relationship \dots] \end{array} \right)$$

Eq 4.1.3: Partial-Seed-Singularity Pre-Big Bang K-entanglement Pre-Material-Fabric Pre-Genetic Code

Note that the last equation-segment following the '$\Rightarrow$' was derived in Chapter 2.3, The Seed-Singularity Architectural Forces K-Entanglement Pre-Genetic Library

& Dynamics. The '$\sum x$' signifies all the possibilities for the 'x' equation-segment. All segments following the '$\Rightarrow$' are active as pre-material-fabric pre-genetic code at this point. These segments also imply wholeness and the formation of a partial-seed-singularity.

Seed-Singularity Pre-Big Bang N-Entanglement Pre-Material-Fabric Pre-Genetic Code

The second stage of precipitation prior to the Big Bang is modeled by N-entanglement as discussed in Chapter 2.4, The Seed-Singularity Organizational Uniqueness N-Entanglement Pre-Genetic Library & Dynamics. Equation 4.1.4 highlights the pre-material-fabric pre-genetic code active at this point. Hence (3.1.3) collapses to the active highlighted portion only.

$$Light - Space - Time\ Emergence_T =$$

$$\left[\begin{array}{c} \left[\begin{array}{c} c_\infty: [Pr, Po, K, H] \\ \left(\downarrow R_{C_K} = f(R_{C_\infty})\right) \\ c_K: [S_{Pr}, S_{Po}, S_K, S_H] \\ \left(\downarrow R_{C_N} = f(R_{C_K})\right) \\ c_N: f(S_{Pr} \times S_{Po} \times S_K \times S_H) \\ \left(\downarrow R_{C_U} = f(R_{C_N})\right) \\ c_U: [P, V, M, C] \\ \Uparrow \\ c_{0:[D,W,I,C]} \end{array}\right]_{Light} \begin{array}{c} \left[\begin{array}{c} M_3 \rightarrow System_X \\ (\uparrow F \rightarrow I) \\ M_2 \rightarrow S_{System_X} \\ (\uparrow Sig \rightarrow F) \\ M_1 \rightarrow Sig_x \\ (\uparrow > P_x) \\ U \rightarrow x_U \end{array}\right]_{Space} \end{array} \\ \left[\begin{array}{c} M_3 : -\infty \le t \le \infty \\ \downarrow \\ M_2 : 0 \ge t > \infty \\ \downarrow \\ M_1 : 0 > t > \infty \\ \downarrow \\ U \rightarrow \begin{array}{l} t \le E_{Cell}; TC: M_3 \rightarrow U \\ t \sim E_{Human}; TC: U \rightarrow M_3 \end{array} \end{array}\right]_{Time} \quad TC \rightarrow x_T \end{array}\right] \langle x_U | x_T \rangle$$

$$\Rightarrow$$

$$\left\langle \begin{array}{c} Pre - Big - Bang \; Pre - Precipitation \; \infty - Entanglement \\ Pre - Material - Fabric \; Pre - Genetic \; Code \end{array} \right\rangle +$$

$$\left\langle \begin{array}{c} Pre - Big - Bang \; Pre - Precipitation \; Zero - Speed \\ Pre - Material - Fabric \; Pre - Genetic \; Code \end{array} \right\rangle +$$

$$\left\langle \begin{array}{c} Pre - Big - Bang \; K - Entanglement \; Pre - Material - Fabric \\ Pre - Genetic \; Code \end{array} \right\rangle$$
$$+$$

$$\left(\begin{array}{c} \sum Sig = Xa + \overline{Yb_{0-n}} \\ where \left[\begin{array}{c} X \in [S_{System_{Pr}}, S_{System_P}, S_{System_K}, S_{System_N}] \\ Y \in [S_{System_{Pr}}, S_{System_P}, S_{System_K}, S_{System_N}] \\ a, b \; are \; integers; a > b \end{array}\right] \end{array} \right)$$

Eq 4.1.4: Seed-Singularity Pre-Big Bang N-Entanglement Pre-Material-Fabric Pre-Genetic Code

Note that the last equation-segment following the '⇒' was derived in Chapter 2.4, The Seed-Singularity

Organizational Uniqueness N-Entanglement Pre-Genetic Library & Dynamics. The '$\sum x$' signifies all the possibilities for the 'x' equation-segment. The fact that all antecedent layers of Light are now active implies the emergence of the seed-singularity.

Big Bang Partial-Singularity Pre-Material-Fabric and Material-Fabric Pre-Genetic Code

Assuming Light then projects itself or precipitates as suggested from its state of ∞ to that of c, this creates the Big Bang. The process is modeled as yielding two equations.

Equation 4.1.5, Big Bang Pre-Material-Fabric Pre-Genetic Code, highlights the operative portions. The important point is that this equation generates the additional space-time-energy-gravity code-segment as a four-base logic-encoding ecosystem depicted by the sub-script 'FBLEE' at the right-base of the newly generated code-segment, which then immediately activates the material-fabric in Equation 4.1.6.

Hence (4.1.5):

$$Light - Space - Time\ Emergence_T =$$

177

$$\left[\begin{array}{l} \begin{array}{c} c_\infty : [Pr, Po, K, H] \\ \left(\downarrow R_{C_K} = f(R_{C_\infty})\right) \\ c_K : [S_{Pr}, S_{Po}, S_K, S_H] \\ \left(\downarrow R_{C_N} = f(R_{C_K})\right) \\ c_N : f(S_{Pr} \times S_{Po} \times S_K \times S_H) \\ \left(\downarrow R_{C_U} = f(R_{C_N})\right) \\ c_U : [P, V, M, C] \\ \Uparrow \\ c_{0:[D,W,I,C]} \end{array} \end{array}\right]_{Light} \quad \left[\begin{array}{c} M_3 \;\to\; System_X \\ (\uparrow F \to I) \\ M_2 \;\to\; S_{System_X} \\ (\uparrow Sig \to F) \\ M_1 \;\to\; Sig_x \\ (\uparrow > P_x) \\ U \;\to\; x_U \end{array}\right]_{Space}$$

$$\left[U \;\to\; \begin{array}{c} M_3 : -\infty \le t \le \infty \\ \downarrow \\ M_2 : 0 \ge t > \infty \\ \downarrow \\ M_1 : 0 > t > \infty \\ \downarrow \\ t \le E_{Cell}; TC: M_3 \to U \\ t \sim E_{Human}; TC: U \to M_3 \end{array}\right]_{Time} \quad TC \;\to\; x_T$$

$$\langle x_U | x_T \rangle$$

$\Rightarrow$

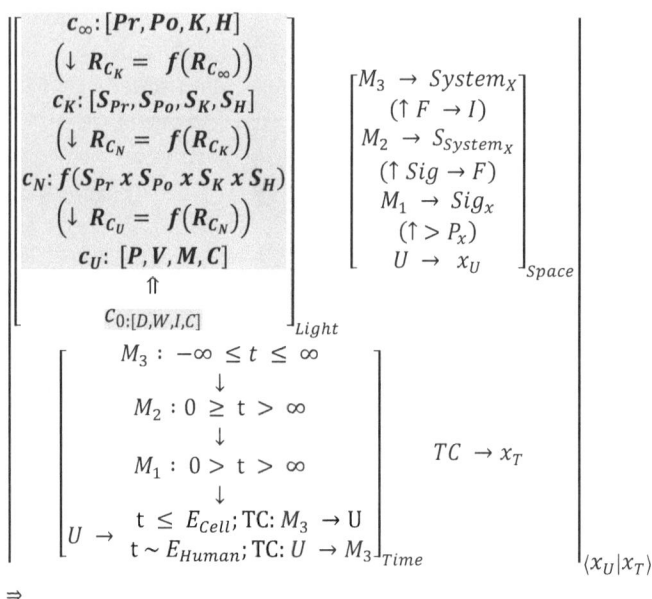

$$\left\langle \begin{array}{c} Pre-Big-Bang\ Pre-Precipitation\ \infty-Entanglement \\ Pre-Material-Fabric\ Pre-Genetic\ Code \end{array}\right\rangle +$$

$$\left\langle \begin{array}{c} Pre-Big-Bang\ Pre-Precipitation\ Zero-Speed \\ Pre-Material-Fabric\ Pre-Genetic\ Code \end{array}\right\rangle +$$

$$\left\langle \begin{array}{c} Pre-Big-Bang\ K-Entanglement\ Pre-Material-Fabric \\ Pre-Genetic\ Code \end{array}\right\rangle +$$

$$\left\langle \begin{array}{c} Pre-Big-Bang\ N-Entanglement \\ Pre-Material-Fabric\ Pre-Genetic\ Code \end{array}\right\rangle +$$

178

$$\left(\begin{array}{c} Space_{quantization} \;=\; h_{UK}\left(Xa + \overline{Yb_{0-n}}\,\right) \\ where: \begin{bmatrix} X \in [S_{System_K}] \\ Y \in [S_{System_{Pr}}, S_{System_P}, S_{System_K}, S_{System_N}] \\ a, b \; are \; integers; a > b \end{bmatrix} \\ Time_{quantization} \;=\; h_{UP}\left(Xa + \overline{Yb_{0-n}}\,\right) \\ where: \begin{bmatrix} X \in [S_{System_P}] \\ Y \in [S_{System_{Pr}}, S_{System_P}, S_{System_K}, S_{System_N}] \\ a, b \; are \; integers; a > b \end{bmatrix} \\ Energy_{quantization} \;=\; h_{UPr}\left(Xa + \overline{Yb_{0-n}}\,\right) \\ where: \begin{bmatrix} X \in [S_{System_{Pr}}] \\ Y \in [S_{System_{Pr}}, S_{System_P}, S_{System_K}, S_{System_N}] \\ a, b \; are \; integers; a > b \end{bmatrix} \\ Gravity_{quantization} \;=\; h_{UH}\left(Xa + \overline{Yb_{0-n}}\,\right) \\ where: \begin{bmatrix} X \in [S_{System_N}] \\ Y \in [S_{System_{Pr}}, S_{System_P}, S_{System_K}, S_{System_N}] \\ a, b \; are \; integers; a > b \end{bmatrix} \end{array}\right)_{FBLEE}$$

Eq 4.1.5: Big Bang Partial-Singularity Pre-Material-Fabric Pre-Genetic Code

Equation 4.1.5 may be thought of as being active as light

approaches c. Hence the generated FBLEE-segment exists at the quantum-levels and can be thought of as exhibiting entanglement, referred to as FBLEE-entanglement (FBLEEE). As light slows down to c the multiple quantization discussed in Section 3 becomes active and spawns the material-fabric, which may be thought of as the gestalt or the emergent property due to the combined space-time-energy-gravity precipitation.

179

Space, Time Energy, and Gravity have therefore a specific meaning and significance in the context of the material universe. At the macro-level as discussed in Chapter 3.5 through acting as the mechanisms by which potentially infinite number of seeds are interrelated and mature, they define essential cosmic parameters by which the universe operates.

At the micro-level their action is essential in materializing pre-genetic code, genetic code, and other codes that may also cause different kinds of materialization in the future. In living cells such code is known to exist in DNA. In forms prior to the living cell, while Science may not yet have acknowledged that such code exists and similarly must be housed in a material structure, yet this must be the case.

In this treatise such a material structure housing pre-genetic code is referred to as the 'material-fabric'. Material-fabric is envisioned as existing at the interface of the antecedent quantum-layer and matter. Note though that the Big Bang space-time-energy-gravity pre-genetic code is still created first as a four-base logic encoding ecosystem assumed to exist in the antecedent quantum-layer and depicted by the FBLEE subscript in (4.1.5) as just discussed.

Through further application of (3.1.3) this FBLEE pre-genetic code will be housed in the material-fabric as illustrated by Equation 4.1.6:

$$Light - Space - Time\ Emergence_{Big\ Bang} =$$

$$\Rightarrow \begin{bmatrix} \begin{bmatrix} c_\infty: [Pr, Po, K, H] \\ \left(\downarrow R_{C_K} = f(R_{C_\infty}) \right) \\ c_K: [S_{Pr}, S_{Po}, S_K, S_H] \\ \left(\downarrow R_{C_N} = f(R_{C_K}) \right) \\ c_N: f(S_{Pr} \times S_{Po} \times S_K \times S_H) \\ \left(\downarrow R_{C_U} = f(R_{C_N}) \right) \\ c_U: [P, V, M, C] \\ \Uparrow \\ c_{0:[D,W,I,C]} \end{bmatrix}_{Light} \quad \begin{bmatrix} M_3 \to System_X \\ (\uparrow F \to I) \\ M_2 \to S_{System_X} \\ (\uparrow Sig \to F) \\ M_1 \to Sig_x \\ (\uparrow > P_x) \\ U \to x_{Big\ Bang[FBLEE]} \end{bmatrix}_{Space} \\ \\ U \to \begin{bmatrix} M_3: -\infty \le t \le \infty \\ \downarrow \\ M_2: 0 \ge t > \infty \\ \downarrow \\ M_1: 0 > t > \infty \\ \downarrow \\ t \le E_{Cell}; TC: M_3 \to U \\ t \sim E_{Human}; TC: U \to M_3 \end{bmatrix}_{Time} \quad TC \to x_{Big\ Bang} \end{bmatrix} \langle x_U | x_T \rangle$$

$$\Rightarrow$$

$$\begin{pmatrix} Space_{quantization} = h_{UK}\left(Xa + \overline{Yb_{0-n}}\right) \\ where: \begin{bmatrix} X \in [S_{System_K}] \\ Y \in [S_{System_{Pr}}, S_{System_P}, S_{System_K}, S_{System_N}] \\ a, b\ are\ integers; a > b \end{bmatrix} \\ Time_{quantization} = h_{UP}\left(Xa + \overline{Yb_{0-n}}\right) \\ where: \begin{bmatrix} X \in [S_{System_P}] \\ Y \in [S_{System_{Pr}}, S_{System_P}, S_{System_K}, S_{System_N}] \\ a, b\ are\ integers; a > b \end{bmatrix} \\ Energy_{quantization} = h_{UPr}\left(Xa + \overline{Yb_{0-n}}\right) \\ where: \begin{bmatrix} X \in [S_{System_{Pr}}] \\ Y \in [S_{System_{Pr}}, S_{System_P}, S_{System_K}, S_{System_N}] \\ a, b\ are\ integers; a > b \end{bmatrix} \\ Gravity_{quantization} = h_{UH}\left(Xa + \overline{Yb_{0-n}}\right) \\ where: \begin{bmatrix} X \in [S_{System_N}] \\ Y \in [S_{System_{Pr}}, S_{System_P}, S_{System_K}, S_{System_N}] \\ a, b\ are\ integers; a > b \end{bmatrix} \end{pmatrix}_{MF}$$

Eq 4.1.6: Big Bang Partial-Singularity Material-Fabric Pre-Genetic Code

Not the change in several notations in (4.1.6) as compared with previous iterations of (3.1.3). These include the

181

highlighted portions in 4.1.6, the change in subscripts T and U, and the additional subscript 'MF' after the generated code-segment.

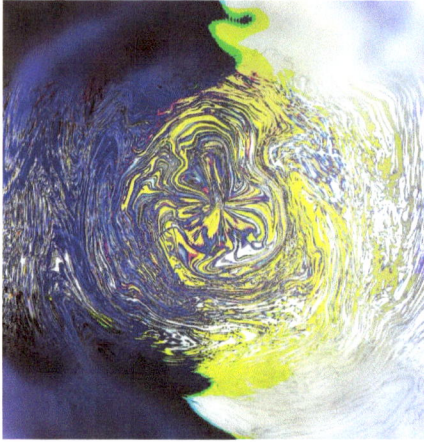

Hence, all the portions of the Light-Matrix are highlighted as in (4.1.5). But in addition in the Time-Matrix, the highlighted 't $\leq E_{Cell}$; TC: $M_3 \rightarrow$ U' indicates that since the space-time-energy-gravity construct precedes the emergence of the cell, all meta-layers will be active in the Space-Matrix. Hence, the entire space-matrix is also highlighted indicating that quantization is active and therefore precipitation into a material-fabric is inevitable. This materialization into the material-fabric is depicted by the subscript 'MF' following the generated code-segment. Specifically also, the U has been changed to 'Big Bang [FBLEE]' and the T to 'Big Bang' indicating the passage through the action of (4.1.6) from FBLEE to MF.

Note that in future sections the actions specific to (4.1.5) and (4.1.6) will generally be combined into one equation only. This will signify that FBLEE code generation is implicit and will always precede MF code generation. The combined equation will always have the U and T subscripts named.

An observation of (4.1.5) and (4.1.6) suggest the vast amount of pre-genetic information that has already been generated, and also suggests the means by which this information can be materialized. The Big Bang is therefore a watershed event that puts into play a potentially endless evolution materializing more and more possibility that exists in Light. The Big Bang hence also signals the start of a new narrative putting into effect all future partial-singularity biographies. Hence (4.1.6) is named 'Big Bang Partial-Singularity…' implying the generation of material-fabric code.

While (4.1.6) does not explicitly show all the pre-material-fabric code-segments, these exist in antecedent layers of Light and will continue to exercise influence through both entanglement and through being part of the source material that generates FBLEE code-segments. The means for materialization of any generated FBLEE code-segment is the fourfold space-time-gravity-energy quantization that is itself generated as pre-genetic code that must exist in every iota of the universe as part of its material-fabric.

Note also that the Big Bang is a significant event that in the model of emergence depicted in this book alters the possibilities of evolution. If the only operative layer besides c_∞ was c_0 then the nature of emergence and evolution would be radically different than with all the layers as depicted by the Light-Matrix in (3.1.3). As discussed each of these light layers implies the presence of libraries of pre-genetic pre-material-fabric information, and a much quicker and potentially richer evolution than would be possible if only c_0

183

existed. Ironically, though, it appears that in the contemporary rendering of physics and the notion of emergence as described in mainstream complex adaptive systems literature evolution is viewed as originating from c_0 only. Note that fragmented-singularities, such as any AI-based singularity, can be thought of as existing in the same manner as any creation based only on c_0.

Chapter 4.2: Post-Big Bang Generation of Material-Fabric, Genetic Code, and Post-Genetic Code in Light-Based Singularities

As implied in Chapter 4.1 it is only with (4.1.6) that the emergence through matter, life, and mega-organization can proceed. The occurrence of space-time-energy-gravity quantization means that for pre-cellular structures the mechanism for affecting the fabric of material existence, or material-fabric, and for cellular structures and beyond, the means for affecting genetic code and any possible post-genetic code is now in place. In other words the mechanism for narrating the biography of any partial-singularity is now in place because there is a meaningful way to record change in genetic-type information in more material structure. Such quantization puts into place the mechanism for the evolution of the universe as will be elaborated in subsequent sections.

This chapter will offer a high-level overview of various fourfold code projections that accelerate material evolution. Hence, if we trace the potential in Light working from the basis of an initial electromagnetic spectrum to the current basis of a possible sustainable global civilization, and beyond, some of the iterations the equation may have gone through can be envisioned as follows in Equation 4.2.1 – 8. Each iteration generates a new code-segment and in totality these are added to existing code-segments in the material-fabric in the case of pre-genetic information, to DNA in the case of genetic information, and to some hybrid or unknown structure in the case of post-genetic information. Each generated code-segment being seamlessly added to existing code-segments creates

185

a new partial-singularity. In effect therefore, there can be an infinite number of partial-singularities. Note that these remain singularities precisely because there is wholeness in that all the code is active all the time. The generated code is universal and true for any emergences. This differs from fragmented-singularities in that while the code associated with fragmented-singularities may appear to be universal in effect they are not, as they apply only to the fragmented-objects they are associated with. As suggested in the previous chapter fragmented-singularities can perhaps be best modeled as existing on an essential c_0 substrate completely separated from any other layer of Light.

Note that the following equations and their outputs are only snapshots and always there will be a vaster number of iterations that will be required to capture the enormity of any kind of genetic-type information that exists in reality.

Further, these iterations do not necessarily imply that one code-segment output is dependent on a previous one. Space-Time-Energy-Gravity logic embedded in the material-fabric allows quantization to take place. Similarly the emergence of other code-segments allows other necessary operations to take place and accelerates what is possible in the material universe. But emergence itself may equally be dictated by the different kinds of entanglement and the vast variety of information in pre-material-fabric subtle-libraries.

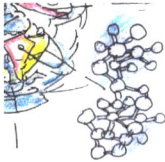

In Equation 4.2.1, the starting point, x_U, is 'EM Spectrum', and the ending point, x_T, is 'Quantum Particle'. x_U is highlighted, and x_T is bolded. As suggested by (4.1.6) the naming of subscripts implies that the code generated is actually housed in the material-fabric through a similar 2-step process as captured by (4.1.5) and (4.1.6). Hence the FBLEE code-segments for quantum particles will already have been generated at the quantum-levels. The output is the pre-genetic code for quantum particles that will be added to the existing space-time-energy-gravity and electromagnetic spectrum pre-genetic code in the material-fabric. In this model such code embedded into the material-fabric is assumed to be operative and binding on constructs that arise within the universe. Such

187

code also changes the possibilities that can arise within the material universe. Hence what has materialized is a complete and whole partial-singularity.

Hence:

$Light - Space - Time\ Emergence_{Quantum\ Particles} =$

$$
\left| \begin{array}{l} \left[\begin{array}{c} c_\infty : [Pr, Po, K, H] \\ \left(\downarrow R_{C_K} = f\left(R_{C_\infty}\right)\right) \\ c_K : [S_{Pr}, S_{Po}, S_K, S_H] \\ \left(\downarrow R_{C_N} = f\left(R_{C_K}\right)\right) \\ c_N : f(S_{Pr} \times S_{Po} \times S_K \times S_H) \\ \left(\downarrow R_{C_U} = f\left(R_{C_N}\right)\right) \\ c_U : [P, V, M, C] \\ \Uparrow \\ C_{0:[D,W,I,C]} \end{array} \right]_{Light} \begin{array}{c} \left[\begin{array}{c} M_3 \rightarrow System_X \\ (\uparrow F \rightarrow I) \\ M_2 \rightarrow S_{System_X} \\ (\uparrow Sig \rightarrow F) \\ M_1 \rightarrow Sig_X \\ (\uparrow > P_x) \\ U \rightarrow x_{EM\ Spectrum} \end{array} \right]_{Space} \end{array} \\ \left[U \rightarrow \begin{array}{c} M_3 : -\infty \le t \le \infty \\ \downarrow \\ M_2 : 0 \ge t > \infty \\ \downarrow \\ M_1 : 0 > t > \infty \\ \downarrow \\ t \le E_{Cell}; TC: M_3 \rightarrow U \\ t \sim E_{Human}; TC: U \rightarrow M_3 \end{array} \right]_{Time} \quad TC \rightarrow x_{Quantum\ Particles} \end{array} \right| \langle x_U | x_T \rangle
$$

$\Rightarrow$

$\langle Space - Time - Energy - Gravity\ Material - Fabric\ Pre$
$\qquad - Genetic\ Code \rangle +$

$\langle Electromagnetic\ Spectrum\ Material - Fabric\ Pre - Genetic\ Code \rangle +$

$Quantum - Particle\ Material - Fabric\ Pre - Genetic\ Code$

188

Eq 4.2.1: Quantum Particle Partial-Singularity Material-Fabric Pre-Genetic Code

Note that Section 5 will discuss the computation leading to the emergence of the electromagnetic spectrum and its associated pre-genetic code segments, while Section 6 will discuss the computations leading to the emergence of matter, including quantum particles, and the associated pre-genetic code segments.

Atoms Partial-Singularity Material-Fabric Pre-Genetic Code Iteration

In Equation 4.2.2, Atoms Partial-Singularity Material-Fabric Pre-Genetic Code, the starting point, x_U , is 'Quantum Particles', and the ending point, x_T, is 'Atoms'. This iteration of the Light-Space Time Emergence equation suggests therefore that the generated code-segments for atoms will be added to the pre-existing material-fabric code-segments or "laws" for space-time-energy-gravity, the electromagnetic spectrum, and quantum particles, hence facilitating the emergence of a new whole or partial-singularity.

Hence:

$$Light - Space - Time\ Emergence_{Atoms} =$$

$$\left\| \begin{bmatrix} \begin{matrix} c_\infty : [Pr, Po, K, H] \\ \left(\downarrow R_{C_K} = f\left(R_{C_\infty}\right) \right) \\ c_K : [S_{Pr}, S_{Po}, S_K, S_H] \\ \left(\downarrow R_{C_N} = f\left(R_{C_K}\right) \right) \\ c_N : f(S_{Pr} \ x \ S_{Po} \ x \ S_K \ x \ S_H) \\ \left(\downarrow R_{C_U} = f\left(R_{C_N}\right) \right) \\ c_U : [P, V, M, C] \\ \Uparrow \\ c_{0 : [D, W, I, C]} \end{matrix} \end{bmatrix}_{Light} \begin{bmatrix} M_3 \to System_X \\ (\uparrow F \to I) \\ M_2 \to S_{System_X} \\ (\uparrow Sig \to F) \\ M_1 \to Sig_X \\ (\uparrow > P_x) \\ U \to x_{Quantum\ Particles} \end{bmatrix}_{Space} \right.$$

$$\begin{bmatrix} M_3 : -\infty \leq t \leq \infty \\ \downarrow \\ M_2 : 0 \geq t > \infty \\ \downarrow \\ M_1 : 0 > t > \infty \\ \downarrow \\ U \to \begin{matrix} t \leq E_{Cell}; TC: M_3 \to U \\ t \sim E_{Human}; TC: U \to M_3 \end{matrix} \end{bmatrix}_{Time} \quad TC \to x_{Atoms} \Bigg|_{\langle x_U | x_T \rangle}$$

$\Rightarrow$

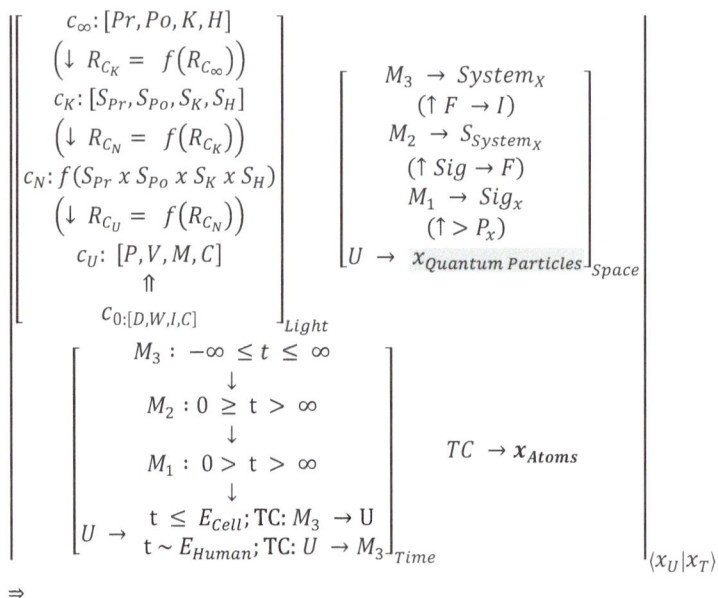

$\langle Space - Time - Energy - Gravity \ Material - Fabric \ Pre$
$\qquad\qquad - Genetic \ Code \rangle +$

$\langle Electromagnetic \ Spectrum \ Material - Fabric \ Pre - Genetic \ Code \rangle +$

$\langle Quantum - Particle \ Material - Fabric \ Pre - Genetic \ Code \rangle +$

$Atoms \ Material - Fabric \ Pre - Genetic \ Code$

Eq 4.2.2: Atoms Partial-Singularity Material-Fabric Pre-Genetic Code

The computation leading to the emergence of atoms and its associated pre-genetic code segments is discussed in greater detail in Chapter 6.3.

Molecules Partial-Singularity Material-Fabric Pre-Genetic Code Iteration

In Equation 4.2.3, Molecules Partial-Singularity Material-Fabric Pre-Genetic Code, the starting point, x_U, is 'Atoms', and the ending point, x_T, is 'Molecules', signaling the emergence of a new molecules-based partial-singularity:

$Light - Space - Time\ Emergence_{Molecules} =$

$$\left| \begin{bmatrix} \begin{bmatrix} c_\infty: [Pr, Po, K, H] \\ \left(\downarrow R_{C_K} = f(R_{C_\infty})\right) \\ c_K: [S_{Pr}, S_{Po}, S_K, S_H] \\ \left(\downarrow R_{C_N} = f(R_{C_K})\right) \\ c_N: f(S_{Pr} \times S_{Po} \times S_K \times S_H) \\ \left(\downarrow R_{C_U} = f(R_{C_N})\right) \\ c_U: [P, V, M, C] \\ \Uparrow \\ c_{0:[D,W,I,C]} \end{bmatrix}_{Light} \begin{bmatrix} M_3 \rightarrow System_X \\ (\uparrow F \rightarrow I) \\ M_2 \rightarrow S_{System_X} \\ (\uparrow Sig \rightarrow F) \\ M_1 \rightarrow Sig_x \\ (\uparrow > P_x) \\ U \rightarrow x_{Atoms} \end{bmatrix}_{Space} \\ \begin{bmatrix} M_3 : -\infty \leq t \leq \infty \\ \downarrow \\ M_2 : 0 \geq t > \infty \\ \downarrow \\ M_1 : 0 > t > \infty \\ \downarrow \\ U \rightarrow \begin{matrix} t \leq E_{Cell}; TC: M_3 \rightarrow U \\ t \sim E_{Human}; TC: U \rightarrow M_3 \end{matrix} \end{bmatrix}_{Time} \quad TC \rightarrow x_{Molecules} \end{bmatrix} \right| \langle x_U | x_T \rangle$$

$\Rightarrow$

$\langle Space - Time - Energy - Gravity\ Material - Fabric\ Pre\\ - Genetic\ Code \rangle +$

$\langle Electromagnetic\ Spectrum\ Material - Fabric\ Pre - Genetic\ Code \rangle +$

$\langle Quantum - Particle\ Material - Fabric\ Pre - Genetic\ Code \rangle +$

$\langle Atoms\ Material - Fabric\ Pre - Genetic\ Code \rangle +$

$Molecules\ Material - Fabric\ Pre - Genetic\ Code$

191

Eq 4.2.3: Molecules Partial-Singularity Material-Fabric Pre-Genetic Code

Cells Partial-Singularity Genetic Code Iteration

In Equation 4.2.4, Cells Partial-Singularity Genetic Code, the starting point, x_U, is 'Molecules', and the ending point, x_T, is 'Cells':

$$Light - Space - Time\ Emergence_{Cells} =$$

$$
\left|
\begin{array}{c}
\left[
\begin{array}{c}
c_\infty : [Pr, Po, K, H] \\
\left(\downarrow R_{C_K} = f(R_{C_\infty}) \right) \\
c_K : [S_{Pr}, S_{Po}, S_K, S_H] \\
\left(\downarrow R_{C_N} = f(R_{C_K}) \right) \\
c_N : f(S_{Pr}\ x\ S_{Po}\ x\ S_K\ x\ S_H) \\
\left(\downarrow R_{C_U} = f(R_{C_N}) \right) \\
c_U : [P, V, M, C] \\
\Uparrow \\
c_{0:[D,W,I,C]}
\end{array}
\right]_{Light}
\left[
\begin{array}{c}
M_3 \rightarrow System_X \\
(\uparrow F \rightarrow I) \\
M_2 \rightarrow S_{System_X} \\
(\uparrow Sig \rightarrow F) \\
M_1 \rightarrow Sig_x \\
(\uparrow > P_x) \\
U \rightarrow\ x_{Molecules}
\end{array}
\right]_{Space} \\
\left[
\begin{array}{c}
M_3 : -\infty \le t \le \infty \\
\downarrow \\
M_2 : 0 \ge t > \infty \\
\downarrow \\
M_1 : 0 > t > \infty \\
\downarrow \\
U \rightarrow \begin{array}{c} t \le E_{Cell}; TC: M_3 \rightarrow U \\ t \sim E_{Human}; TC: U \rightarrow M_3 \end{array}
\end{array}
\right]_{Time}
TC \rightarrow x_{Cells}
\end{array}
\right|_{\langle x_U | x_T \rangle}
$$

$$\Rightarrow$$

$$\langle Space - Time - Energy - Gravity\ Material - Fabric\ Pre - Genetic\ Code \rangle +$$

$$\langle Electromagnetic\ Spectrum\ Material - Fabric\ Pre - Genetic\ Code \rangle +$$

$$\langle Quantum - Particle\ Material - Fabric\ Pre - Genetic\ Code \rangle +$$

$$\langle Atoms\ Material - Fabric\ Pre - Genetic\ Code \rangle +$$

$$\langle Molecules\ Material - Fabric\ Pre - Genetic\ Code \rangle + LSTE \langle ... \rangle$$
$$+ Cells\ Genetic\ Code$$

Eq 4.2.4: Cells Partial-Singularity Genetic Code

The computation leading to the emergence of cells and associated code segments is discussed in greater detail in Chapter 7.2. Note that there are likely many iterations of this equation between molecules and cells and this in general is depicted by '*LSTE* $\langle ... \rangle$' in (4.2.4), where LSTE signifies iterations(s) of the Light-Space-Time Emergence equation. Also note that an assumption is made that all the previous material-fabric code-segments are intimately active at the level of cellular genetic code. This may be through the device of entanglement, or perhaps even by the material-fabric code segments being embedded in cellular genetic code.

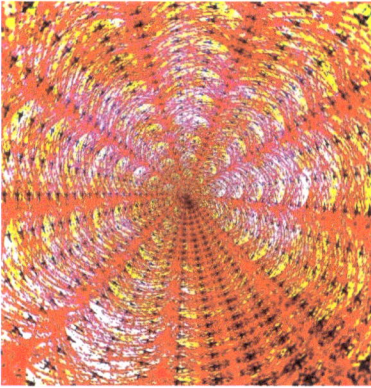

Humans Partial-Singularity Genetic Code Iteration

In Equation 4.2.5, Humans Partial-Singularity Genetic Code, the starting point, x_U, is 'Cells', and the ending point, x_T, is 'Humans'. Clearly there will be a vast number of iterations before humans emerge which also covers vast tracts of the plant and animal kingdoms that will cause an appropriate divergence in genetic code. Further the development of microbiomes – that contain a multitude of bacteria and other microbes – will have been developed and shared between a vast number of species (Young, 2016).

193

Further, somewhere between the cell and the human, the emergence of vision is said to have sparked an explosion of evolution (Parker, 2003). This is not surprising if in fact we exist due to a Cosmology of Light as is being suggested in this book. It should be natural that

instrumentation to perceive light, such as vision, would then generate an explosion in evolution.

Since any genetic code will be the materialization of four-base logic-encoding ecosystems (FBLEE) as a result of the Light-Matrix computation, there will also be a degree of entanglement to such ecosystems. Such entanglement will be different from ∞-entanglement, K-entanglement, and N-entanglement, and as suggested in the last chapter, will be referred to as FBLEE-entanglement (FBLEEE). Such entanglement is due to four-base logic-encoding ecosystems existing at the quantum level. FBLEE-entanglement will allow code-segment developments to be shared or accessed on an inter-species level, and are depicted by 'FBLEEE $\langle \dots \rangle$', as in Equation 4.2.5.

Hence:

$$Light - Space - Time \; Emergence_{Humans} =$$

$$\begin{Vmatrix} \begin{bmatrix} c_\infty : [Pr, Po, K, H] \\ \left(\downarrow R_{C_K} = f\left(R_{C_\infty}\right) \right) \\ c_K : [S_{Pr}, S_{Po}, S_K, S_H] \\ \left(\downarrow R_{C_N} = f\left(R_{C_K}\right) \right) \\ c_N : f(S_{Pr} \; x \; S_{Po} \; x \; S_K \; x \; S_H) \\ \left(\downarrow R_{C_U} = f\left(R_{C_N}\right) \right) \\ c_U : [P, V, M, C] \\ \Uparrow \\ c_{0:[D,W,I,C]} \end{bmatrix}_{Light} & \begin{bmatrix} M_3 \to System_X \\ (\uparrow F \to I) \\ M_2 \to S_{System_X} \\ (\uparrow Sig \to F) \\ M_1 \to Sig_x \\ (\uparrow > P_x) \\ U \to x_{Cells} \end{bmatrix}_{Space} \\ \begin{bmatrix} M_3 : -\infty \le t \le \infty \\ \downarrow \\ M_2 : 0 \ge t > \infty \\ \downarrow \\ M_1 : 0 > t > \infty \\ \downarrow \\ U \to \begin{array}{l} t \le E_{Cell}; \text{TC}: M_3 \to U \\ t \sim E_{Human}; \text{TC}: U \to M_3 \end{array} \end{bmatrix}_{Time} & TC \to x_{Humans} \end{Vmatrix} \langle x_U | x_T \rangle$$

$$\Rightarrow$$

$\langle Space - Time - Energy - Gravity \; Material - Fabric \; Pre - Genetic \; Code \rangle +$

$\langle Electromagnetic \; Spectrum \; Material - Fabric \; Pre - Genetic \; Code \rangle +$

$\langle Quantum - Particle \; Material - Fabric \; Pre - Genetic \; Code \rangle +$

$\langle Atoms \; Pre - Genetic \; Code \rangle +$

$\langle Molecules \; Material - Fabric \; Pre - Genetic \; Code \rangle + LSTE \; \langle ... \rangle +$

$\langle Cells \; Genetic \; Code \rangle + LSTE \; \langle ... \rangle + FBLEEE \; \langle ... \rangle$
$+ \; Humans \; Genetic \; Code$

Eq 4.2.5: Humans Partial-Singularity Genetic Code

If we follow the lines of emergence as will be described in the following subsection then (4.2.5) will be the starting point for potentially different trajectories of development. In one trajectory 'Humans' will evolve 'Basic Capacities of Self' and then 'Truer Individuality'. In another line of development 'Humans' will be the starting point for 'Stable Mega-Organization' culminating in 'Sustainable Global Civilization'. The gist of these can be represented by Equations 4.2.6 and 4.2.7 respectively. As in (4.2.5) FBLEEE-segments depict the sharing of code segments between these different lines of development through the process of four-base logic-encoding ecosystem entanglement at the quantum level. In the latter line of development though the advance of post-human structures signals the start of a process that will correspondingly advance a post-genetic structure.

Truer Individuality Partial-Singularity Genetic Code Iteration

Hence in Equation 4.2.6, Truer Individuality Partial-Singularity Genetic Code, the starting point, x_U , is 'Humans', and the ending point, x_T , is 'Truer Individuality'. Clearly there will be a vast number of iterations before 'Truer Individuality' emerges, that will contain 'Basic Capacities of Self' as a milestone along the way:

$Light - Space - Time\ Emergence_{Truer\ Individuality} =$

$$\left| \begin{array}{c} \begin{bmatrix} c_\infty : [Pr, Po, K, H] \\ \left(\downarrow R_{C_K} = f(R_{C_\infty}) \right) \\ c_K : [S_{Pr}, S_{Po}, S_K, S_H] \\ \left(\downarrow R_{C_N} = f(R_{C_K}) \right) \\ c_N : f(S_{Pr} \times S_{Po} \times S_K \times S_H) \\ \left(\downarrow R_{C_U} = f(R_{C_N}) \right) \\ c_U : [P, V, M, C] \\ \Uparrow \\ c_{0:[D,W,I,C]} \end{bmatrix}_{Light} \begin{bmatrix} M_3 \rightarrow System_X \\ (\uparrow F \rightarrow I) \\ M_2 \rightarrow S_{System_X} \\ (\uparrow Sig \rightarrow F) \\ M_1 \rightarrow Sig_X \\ (\uparrow > P_X) \\ U \rightarrow x_{Humans} \end{bmatrix}_{Space} \\ \begin{bmatrix} M_3 : -\infty \le t \le \infty \\ \downarrow \\ M_2 : 0 \ge t > \infty \\ \downarrow \\ M_1 : 0 > t > \infty \\ \downarrow \\ U \rightarrow \begin{array}{l} t \le E_{Cell}; TC: M_3 \rightarrow U \\ t \sim E_{Human}; TC: U \rightarrow M_3 \end{array} \end{bmatrix}_{Time} \qquad TC \rightarrow x_{Truer\ Individuality} \end{array} \right| \langle x_U | x_T \rangle$$

$\Rightarrow$

$\langle Space - Time - Energy - Gravity\ Material - Fabric\ Pre - Genetic\ Code \rangle +$

$\langle Electromagnetic\ Spectrum\ Material - Fabric\ Pre - Genetic\ Code \rangle +$

$\langle Quantum - Particle\ Material - Fabric\ Pre - Genetic\ Code \rangle +$

$\langle Atoms\ Pre - Genetic\ Code \rangle +$

$\langle Molecules\ Material - Fabric\ Pre - Genetic\ Code \rangle + LSTE \langle ... \rangle +$

$\langle Cells\ Genetic\ Code \rangle + LSTE \langle ... \rangle + FBLEEE \langle ... \rangle + Humans\ Genetic\ Code +$

$LSTE \langle ... \rangle + FBLEEE \langle ... \rangle + Truer\ Individuality\ Genetic\ Code$

Eq 4.2.6: Truer Individuality Partial-Singularity & Genetic Code

Sustainable Global Civilization FBLEE-Based Partial-Singularity Genetic Code Iteration

In equation 4.2.7, Sustainable Global Civilization Partial-Singularity Genetic Code, the starting point, x_U, is 'Humans', and the ending point, x_T, is 'Sustainable Global Civilization'. Clearly there will be a vast number of iterations before 'Sustainable Global Civilization' emerges, that will contain 'Stable Mega-Organization' as a milestone along the way. This line of development though, advancing post-human structures will initiate a process that will begin to create post-genetic code. This will be referred to as Post Genetic Code Initiation or PGCI. Further, given that such a post-genetic structure does not yet exist, the sustainable global civilization will be a FBLEE-based partial-singularity:

$$Light - Space - Time\ Emergence_{Sustainable\ Global\ Civilization} =$$

198

$$\left| \begin{bmatrix} \begin{bmatrix} c_\infty : [Pr, Po, K, H] \\ \left(\downarrow R_{C_K} = f(R_{C_\infty}) \right) \\ c_K : [S_{Pr}, S_{Po}, S_K, S_H] \\ \left(\downarrow R_{C_N} = f(R_{C_K}) \right) \\ c_N : f(S_{Pr} \times S_{Po} \times S_K \times S_H) \\ \left(\downarrow R_{C_U} = f(R_{C_N}) \right) \\ c_U : [P, V, M, C] \\ \Uparrow \\ c_{0:[D,W,I,C]} \end{bmatrix}_{Light} \quad \begin{bmatrix} M_3 \rightarrow System_X \\ (\uparrow F \rightarrow I) \\ M_2 \rightarrow S_{System_X} \\ (\uparrow Sig \rightarrow F) \\ M_1 \rightarrow Sig_x \\ (\uparrow > P_x) \\ U \rightarrow x_{Humans} \end{bmatrix}_{Space} \\[2em] \begin{bmatrix} M_3 : -\infty \leq t \leq \infty \\ \downarrow \\ M_2 : 0 \geq t > \infty \\ \downarrow \\ M_1 : 0 > t > \infty \\ \downarrow \\ U \rightarrow \begin{matrix} t \leq E_{Cell}; TC: M_3 \rightarrow U \\ t \sim E_{Human}; TC: U \rightarrow M_3 \end{matrix} \end{bmatrix}_{Time} \quad TC \rightarrow x_{Sust.Global\ Civilization} \end{bmatrix} \right| \langle x_U | x_T \rangle$$

$\Rightarrow$

$\langle Space - Time - Energy - Gravity\ Material - Fabric\ Pre - Genetic\ Code \rangle +$

$\langle Electromagnetic\ Spectrum\ Material - Fabric\ Pre - Genetic\ Code \rangle +$

$\langle Quantum - Particle\ Material - Fabric\ Pre - Genetic\ Code \rangle +$

$\langle Atoms\ Pre - Genetic\ Code \rangle + \langle Molecules\ Material - Fabric\ Pre - Genetic\ Code \rangle +$

$LSTE \langle \dots \rangle + PGGT < \dots > + \langle Cells\ Genetic\ Code \rangle + LSTE \langle \dots \rangle + FBLEEE \langle \dots \rangle +$

$\langle Humans\ Genetic\ Code \rangle + LSTE \langle \dots \rangle + FBLEEE \langle \dots \rangle +$

$\langle Stable\ Mega - Organization\ FBLEE\ Genetic\ Code \rangle + PGCI \langle \dots \rangle +$

$Sustainable\ Global\ Civilization\ Code$ FBLEE Genetic Code

The computation leading to the emergence of a sustainable global civilization is discussed in greater detail in Chapter 8.3.

Super-Matter Partial-Singularity Post-Genetic Code Iteration

In Equation 4.2.8, Super-Matter Partial-Singularity Post-Genetic Code, the starting point, x_U, is 'Molecules, and the ending point, x_T, is 'Super-Matter'. As per the mathematical model presented here all the preceding Light-Space-Time Emergence equation iterations progressively change the reality of matter through subtle quantization that allows space, time, energy, and gravity to operate differently as will be further explored in subsequent sections. The following equation is therefore an approximation in stating how super-matter may emerge:

$$Light - Space - Time\ Emergence_{Suoer-Matter} =$$

$$\left| \begin{bmatrix} \begin{bmatrix} c_\infty : [Pr, Po, K, H] \\ \left(\downarrow R_{C_K} = f(R_{C_\infty}) \right) \\ c_K : [S_{Pr}, S_{Po}, S_K, S_H] \\ \left(\downarrow R_{C_N} = f(R_{C_K}) \right) \\ c_N : f(S_{Pr} \times S_{Po} \times S_K \times S_H) \\ \left(\downarrow R_{C_U} = f(R_{C_N}) \right) \\ c_U : [P, V, M, C] \\ \Uparrow \\ c_{0:[D,W,I,C]} \end{bmatrix}_{Light} & \begin{bmatrix} M_3 \rightarrow System_X \\ (\uparrow F \rightarrow I) \\ M_2 \rightarrow S_{System_X} \\ (\uparrow Sig \rightarrow F) \\ M_1 \rightarrow Sig_x \\ (\uparrow > P_x) \\ U \rightarrow x_{Molecules} \end{bmatrix}_{Space} \\ \begin{bmatrix} M_3 : -\infty \le t \le \infty \\ \downarrow \\ M_2 : 0 \ge t > \infty \\ \downarrow \\ M_1 : 0 > t > \infty \\ \downarrow \\ U \rightarrow \begin{array}{l} t \le E_{Cell}; TC: M_3 \rightarrow U \\ t \sim E_{Human}; TC: U \rightarrow M_3 \end{array} \end{bmatrix}_{Time} & TC \rightarrow x_{Super-Matter} \end{bmatrix} \right\rangle \langle x_U | x_T \rangle$$

$\Rightarrow$

$\langle Space - Time - Energy - Gravity\ Material - Fabric\ Pre - Genetic\ Code \rangle +$

$\langle Electromagnetic\ Spectrum\ \ Material - Fabric\ Pre - Genetic\ Code \rangle +$

$\langle Quantum - Particle\ Material - Fabric\ Pre - Genetic\ Code \rangle +$

$\langle Atoms\ Material - Fabric\ Pre - Genetic\ Code \rangle +$

$\langle Molecules\ Material - Fabric\ Pre - Genetic\ Code \rangle + LSTE \langle \dots \rangle + FBLEEE \langle \dots \rangle +$

$LSTE \langle \dots \rangle + PGCI \langle \dots \rangle + Super - Matter\ Post - Genetic\ Code$

Eq 4.2.8: Super-Matter Partial-Singularity & Post-Genetic Code

The emergence of super matter is also discussed in the book Super-Matter (Malik, 2018a), with (4.2.8) included here to suggest computational realities yet to emerge. The essential action of space-time-energy-gravity quantization can cause the materialization of potentially infinite four-base logic-encoding ecosystems. It is conceivable that such a variation of space-time-energy-gravity quantization coupled with the ability to easily move back and forth between the material and antecedent realms makes the need to house genetic information in a form such as DNA incomplete. (4.2.7) suggested a process – Post Genetic Code Initiation – to capture such a dynamic, and is included in (4.2.8) as a code-segment. It is conceivable that four-base logic-encoding ecosystems (FBLEE) and even FBLEEE, or even the ∞-entanglement, K-entanglement, and N-entanglement may be able to more directly act at the material level in some composite post-genetic form. This possibility will be referred to as Post-Genetic Code.

SECTION 5: GENERATION OF THE ELECTROMAGNETIC SPECTRUM PARTIAL-SINGULARITY

Section 5 will explore the computation involved in the creation of the electromagnetic partial-singularity and its associated pre-genetic, material-fabric code.

The electromagnetic spectrum is a technical way to refer to light, essentially because amongst its range of properties it displays a tangible and simultaneous electric and magnetic or "electromagnetic" reality as well. So as light becomes more concrete to us or as the possibilities within it begin to emerge, one of the first forms it takes is as the electromagnetic spectrum. It must be the case that the previously surfaced properties of light – Presence, Power, Knowledge, and Harmony – emerge so as to define the very architecture of the electromagnetic spectrum.

Chapter 5.1, Generation of Electromagnetic Spectrum Partial-Singularity Pre-Genetic, Material-Fabric Code, summarizes the emergence of the electromagnetic spectrum in terms of the underlying Light-Space-Time Emergence equation and the process of quantization that must occur to create the logic of the electromagnetic spectrum ecosystem that precipitates into the material-fabric. So we find that the four underlying properties of Light that we call Harmony, Knowledge, Power, and Presence are of the essence of the speed with which the electromagnetic spectrum moves, the wave-range within the electromagnetic spectrum, the energy-gradient within the electromagnetic spectrum, and the mass-possibilities due to the electromagnetic spectrum, respectively. Further the description – electromagnetic – seems to have captured the Power-Harmony aspects implicit in light. In reality the electro-magnetic spectrum can likely be more completely described as electro-magnetic-wavearchetype-masspotential spectrum.

Chapter 5.1: Generation of Electromagnetic Spectrum Partial-Singularity Material-Fabric, Pre-Genetic Code

As elaborated in Section 4, the Light-Space-Time Emergence equation (3.1.3) can be used to model emergence as it proceeds from simpler four-fold to more complex four-fold manifestation. In essence (3.1.3) can also be thought of as the equation by which the biographies for any light-based-singularity, including all partial-singularities such as the electromagnetic spectrum partial-singularity are generated. Being iterative, it can be used to understand the generation of code-segments that will continue to add to Genetic-Type code thus enriching the living narrative that accompanies the materialization of infinite possibility in Light. (3.6.5), the Potential Effect of Levels of Light on Genetic-Type Information equation, basically sheds light on specific types of mutation that can occur and therefore on the nature of the code-segment that is being generated by (3.1.3). Note that (3.6.5) shows a particular working of (3.1.3), but is wholly contained within it.

This chapter elaborates the action of (3.1.3) and (3.6.5) in the creation of the code-segments that define the electromagnetic spectrum ecosystem logic.

Application of Light-Space-Time Emergence Equation

As suggested by (3.1.3) the architecture and details of the electromagnetic spectrum, and the resulting material-fabric, pre-genetic code can be seen to be the result of the application of the Light, Space, and Time matrices. This is illustrated in Equation 5.1.1, Generation of Electromagnetic Spectrum Partial-Singularity:

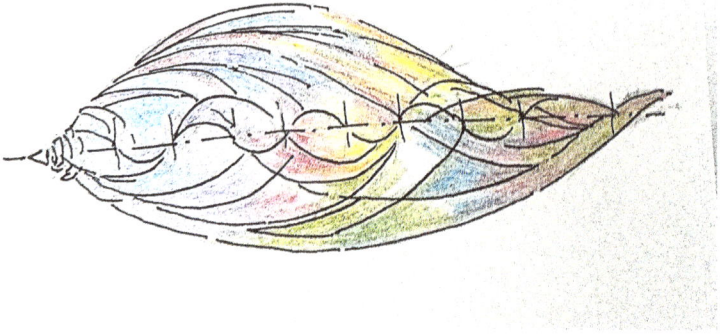

$$Light - Space - Time\ Emergence_{EM\ Spectrum} =$$

$$\Rightarrow$$

$$\begin{Vmatrix} \begin{bmatrix} \begin{matrix} c_\infty:[Pr,Po,K,H] \\ \left(\downarrow R_{C_K} = f(R_{C_\infty})\right) \\ c_K:[S_{P_T},S_{P_O},S_K,S_H] \\ \left(\downarrow R_{C_N} = f(R_{C_K})\right) \\ c_N:f(S_{P_T} \times S_{P_O} \times S_K \times S_H) \\ \left(\downarrow R_{C_U} = f(R_{C_N})\right) \\ c_U:[P,V,M,C] \\ \Uparrow \\ c_0:[D,W,J,C] \end{matrix} \end{bmatrix}_{Light} \\ \begin{bmatrix} M_3 : -\infty \le t \le \infty \\ \downarrow \\ M_2 : 0 \ge t > \infty \\ \downarrow \\ M_1 : 0 > t > \infty \\ \downarrow \\ U \to \begin{matrix} t \le E_{Cell}; TC: M_3 \to U \\ t \sim E_{Human}; TC: U \to M_3 \end{matrix} \end{bmatrix}_{Time} & \begin{bmatrix} M_3 \to System_X \\ (\uparrow F \to I) \\ M_2 \to S_{System_X} \\ (\uparrow Sig \to F) \\ M_1 \to Sig_X \\ (\uparrow > P_X) \\ U \to x_{Space-Time-Energy-Gravity} \end{bmatrix}_{Space} \\ \\ TC \to x_{EM\ Spectrum} \end{Vmatrix}_{\langle x_U | x_T \rangle}$$

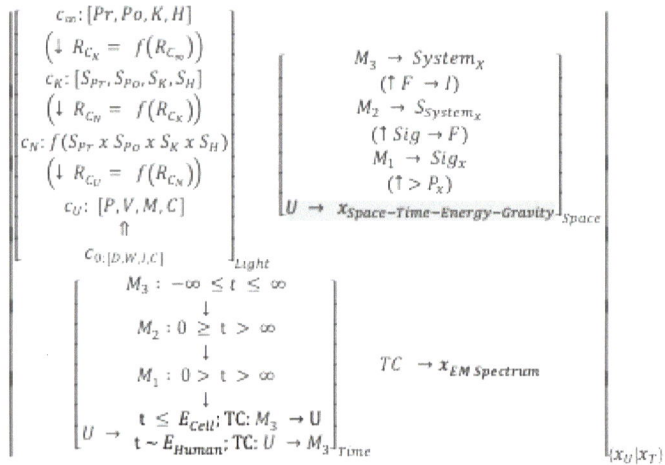

$$\langle Space - Time - Energy - Gravity\ Material - Fabric\ Pre \\ - Genetic\ Code \rangle +$$

$$Electromagnetic\ Spectrum\ Material - Fabric\ Pre - Genetic\ Code$$

Eq 5.1.1:   Generation of Electromagnetic Spectrum Partial-Singularity

Starting with the Light-Matrix, the top left-hand matrix in (5.1.5), the first line from the top, $C_\infty: [Pr, Po, K, H]$, specifies the fundamental architecture of the electromagnetic spectrum. As will be elaborated in the subsections on Harmony, Knowledge, Power, and Presence, each of these aspects are an emergence of the fundamental properties of Light at ∞. Hence, it is only Light that architects the materialization of the infinite possibility within it – this is the hallmark of any light-based-singularity. There is an underlying oneness that founds the dynamics of any light-based-singularity.

Line 3 in the Light-Matrix, $C_K: [S_{Pr}, S_{Po}, S_K, S_H]$, elaborates the sets for Presence, Power, Knowledge, and Harmony, each containing multiple elements. For example, as will be explored in the subsection on Harmony, the following elements: 'connection', 'growing into one's own', 'form bonds', are suggested to exist in the Set of Harmony or Nurturing, and are accessed to contribute to the operative reality so set up by virtue of light traveling at c.

Specifically, Line 5, $C_N: f(S_{Pr} \, x \, S_{Po} \, x \, S_K \, x \, S_H)$, suggests that unique seeds are created from a combination of the elements from all four sets, with a particular element leading or having more weight, that in effect creates the

207

distinctness of the vast variety of waves possible in the electromagnetic spectrum.

Line 6, ($\downarrow R_{C_U} = f(R_{C_N})$) , specifies quantization between the layer where the seeds are formed, and the physical layer, and as explored in Chapter 3.5 and 3.6, will result in Line 7, C_U: $[P, V, M, C]$, hence changing the material-fabric of existence. Note that as in the process describing the generation of the code-segments for space-time-energy-gravity quantization at the time of the Big Bang, the generation of the FBLEE (four-base logic-encoding ecosystem) code-segment as specified by (4.1.5) and the MF (material-fabric) code segment as specified by (4.1.6) are combined together into one equation so that the FBLEE process is apparently transparent. In reality one may say that any partial-singularity is first a FBLEE-partial-singularity and then a MF-partial-singularity.

The possibilities represented by Lines 1 through 5 hence concretize through the quantization represented by Line 6 to become the electromagnetic spectrum with its physical (related to Presence), vital (related to Power), mental (related to Knowledge), and connection (related to Harmony) aspects now existing in material reality typified by Light moving at c. Note that just as Line 6 represents a process of quantization relating the layer of reality created by Light traveling at c with the antecedent layers, so too Lines 2 and 4 as previously discussed, also represent quantization of a more subtle kind that ultimately plays a critical part in allowing the material-fabric to express infinite diversity.

Typically it is the process as captured by the Space-Matrix that will determine if Line 6 is activated. Specifically patterns at the untransformed layer, U, will need to be overcome, as specified by the second-line from the bottom of the Space-Matrix: ($\uparrow > P_x$). But as specified by the bottom-line of the Time-Matrix, reproduced below, it is assumed that only with the advent of the human-system that the automaticity of the action of meta-levels is reversed:

$$U \rightarrow \begin{array}{l} t \le E_{Cell}; \text{TC}: M_3 \rightarrow U \\ t \sim E_{Human}; \text{TC}: U \rightarrow M_3 \end{array}$$

Hence in the case of the electromagnetic spectrum system, which in this emergence is a pre-human system, the fact that patterns do not need to be overcome means that quantization happens automatically.

From the point of view of pre-genetic information this means that constructive mutation has occurred and that the available code-segments that impact the material operation of the universe has been changed. But 'constructive' implies that the operation, materially, allows the manifested universe to 'materially' come closer to the complete integrated nature of Light. Of course the emergence of the electromagnetic spectrum in this rendering occurs after the emergence of space-time-energy-gravity, and marks the beginning stages, relatively, of the long curve that evolution will go through as more and more possibility emerges from Light. These

fundamental emergences are milestones along the way marking the expansion of one edifice, a light-based edifice, which may culminate in the reality of the Second Singularity, were all conditions for such fulfillment to be achieved.

Application of Potential Effect of Levels of Light on Genetic-Type Information Equation

Even though the electromagnetic spectrum is a pre-human system and therefore the action of meta-levels are modeled as being automatic, it is nonetheless useful to review (3.6.5), Potential Effect of Levels of Light on Genetic-Type Information, to understand how the relationship with different forms of mutation has been specified:

$$Potential\ Effect\ of\ Levels\ of\ Light\ on\ Genetic$$
$$-\ Type\ Information =$$

$$\begin{bmatrix} STATIC\ \langle |[L][S][T]TC \rightarrow x_T|_{\langle x_U | x_T \rangle} \\ \times \\ \left((Y > U \colon Z_Q) \vee (Y \leq U \colon Z_F) \vee (Y = U \colon Z_R) \right) \\ \ni \\ \begin{pmatrix} Z \in \mathbb{U}\ (Space, Time, Energy, Gravity) \\ Q \colon Quantization;\ F \colon Fragmentation;\ R \colon\ Random \end{pmatrix} \end{bmatrix} \rightarrow h \ni$$

$$h \in \begin{pmatrix} [Q] \colon Constructive\ zone, \\ [Q] \colon Constructive\ zone \wedge Constructive\ mutation, \\ [F] \colon Destructive\ mutation, \\ [R] \colon Random\ mutation \end{pmatrix}$$

Line 1 from the top in the matrix is simply a static form of (3.6.1) the Simplified Light-Space-Time Emergence equation. The static form is designated by 'STATIC', and implies that fundamental operations true of (3.6.1) are being highlighted in (3.6.5). In other words (3.6.1) already has all the operations highlighted in (3.6.5) in it, but by 'freezing' it by making it static, the essential dynamics leading to possible mutations at the genetic level can more clearly be highlighted.

Line 1 is then subjected ($\times$) to a determination of the dominant levels of light that may be active, designated by Line 2, ' $\left((Y > U \colon Z_Q) \vee (Y \leq U \colon Z_F) \vee (Y = U \colon Z_R) \right)$ '. Unpacking this, '$Y > U$' implies meta-levels are active and as a result it is possible that Z_Q is going to take place (the subscript 'Q' implies quantum-level action). This also implies activation and potential change of FBLEE. The call from below, as it were, may invoke some function that already exists in the subtle-libraries 'above', so that some already existing function may influence FBLEE through ∞ -entanglement, K-entanglement, or N-entanglement. This may be thought of as a key-and-lock mechanism, where a deep enough visceral urge from below acting as the key, opens an entangled lock to alter FBLEE as per the visceral urge. 'Y $\leq$ U' implies that only the untransformed levels are active, and therefore also the sub-level where the speed of light is 0 is active and as a result Z_F is going to take place (the subscript 'F' implies 'fragmentation'). 'Y = U' implies that all levels are active and as a result Z_R is going to take place (the subscript 'R' implies 'random').

Line 3 elaborates the significance of Z_Q, Z_F, and Z_R. Hence Z is the union of potential quantum-operations of space, time, energy, and gravity, designated by ' $Z \in \mathbb{U} \, (Space, Time, Energy, Gravity)$'. But the nature of the operations, as suggested in the previous paragraph, is designated by ' $Q: Quantization; F: Fragmentation; R: Random$ '. Z_Q, then, implies that the full quantization originating from updated four-base logic-encoding ecosystems (FBLEE) can take place, and will result in lasting material change at the Genetic-Type level. Z_F implies that the essential set will be fragmented and that only libraries at the level of local cellular-level DNA or precipitated material-fabric logic can potentially be altered. Z_R also precludes full quantization, and that some partial local-library constructive or destructive mutation may take place.

The ' $\rightarrow h \ni h \in$ ' segment resolves the outcome of the operations implied by Lines 1 – 3, suggesting that the outcome will be 'h' such that ($\ni$) 'h' is an element ($\in$) of the set specified by the members '[Q]: *Constructive zone*', '[Q]: *Constructive zone AND Constructive mutation*', '[F]: *Destructive mutation*', and '{R}: *Random mutation*'. '[Q]: *Constructive zone*' implies that the in-built buffer has been crossed and that access to the deeper four-base logic-encoding ecosystem (FBLEE) has been granted. '[Q}' in this segment implies that there is the possibility that full-quantization as specified by Line 3 of the previous matrix can take place. Access to this zone is a prerequisite for constructive mutation to occur, as designated by the element '$[Q\}$: *Constructive zone* ∩ *Constructive mutation* ', which implies that full-quantization is going to take place and will result in material change. It is in this case that the FBLEE iteration will be followed by an MF iteration. The '[F]' specifies the relationship between 'Fragmentation' in Line 3 of the previous matrix and destructive mutation. The '[R]' specifies the relationship between 'Random' in Line 3 of the previous matrix and random mutation.

But as just summarized in the Time-Matrix in (5.1.1) Y is by definition greater than U and hence quantization is automatic. In terms of the electromagnetic spectrum such quantization implies that wholeness becomes fully active through specific space, time, energy, and gravity quantization to create an holistic "ecosystem" with its own "electromagnetic spectrum logic" as it were. The wholeness has now precipitated into the material-fabric and is available to be consciously and unconsciously tapped into. This "logic" or more specifically pre-genetic code is elaborated in the following sub-sections.

Generation of Electromagnetic Spectrum Partial-Singularity 'Magnetic' Material-Fabric Pre-Genetic Code

To begin with, we had already looked at how the speed of light sets up the nature of reality because of its speed. So for light traveling at c, past, present, future, and the notion of separation due to creation of islands of matter is the reality. It seems then that c architects the possibility of matter-based interaction in our system and can therefore be thought of as a projection from the property of Harmony to create the basis for a matter-based harmony.

So being, it can be suggested that the nature of the resultant interactions allows matter-based organizations, regardless of scale, to come into their own, to grow into their boundaries, and to form bonds based on the sense of being separated from other perceived organizations. This notion of forming bonds seems also to be related to the "magnetic" in electromagnetic.

In equation form, as in Equation 5.1.2 it would therefore be possible to specify the nature of reality so set up by the electromagnetic spectrum (EM_Spectrum) moving at c:

$$EM\_Spectrum_{Speed} = \frac{Xa+}{Yb_{0-n}} \quad where \begin{bmatrix} X \in [S_{System_N}] \\ Y \in [S_{System_{Pr}}, S_{System_P}, S_{System_K}, S_{System_N}] \\ a, b \ are \ integers; a > b \end{bmatrix}$$

Eq 5.1.2: Speed of EM Spectrum

The notion of 'connection', 'growing into one's own', 'form bonds', amongst other attributes of such a reality can be seen as elements of the four sets of architectural forces. Hence (5.1.2) suggests the mathematical equation that so defines the nature of reality due to the speed of c of the electromagnetic spectrum.

Note that (5.1.2) already implies that Lines 1 – 5 in the Light Matrix of (5.1.1) have been activated, and that the logic of the "magnetic" in the electro-magnetic-ecosystem will automatically precipitate into the material-fabric through the action of Line 6-7 of (5.1.1). This logic is none other than the pre-genetic code as specified by (5.1.2).

Generation of Electromagnetic Spectrum Partial-Singularity 'Wave Archetype' Material-Fabric Pre-Genetic Code

The electromagnetic spectrum contains a range of waves embedded in it. These waves enable many different applications with practical utility apparent in every day life.   So for example there is a region in the electromagnetic spectrum that we call radio waves, and others that we know as microwave, infrared visible light, ultraviolet, x-rays and gamma rays that each make possible many technologies that we use every single day. These ranges essentially code a range of technological-possibility, and therefore the wave-range within the electromagnetic spectrum can be thought of as encoding or of expressing some kind of precipitation or projection or emergence of the property of Knowledge. Or put another way, the property of Knowledge that is found in light, emerges as the wave-range implicit in the electromagnetic spectrum.

The EM spectrum itself, with its vast range of natures from gamma rays through visible light through radio waves, with its implicitness of time-space possibility as suggested by frequency (v) and wavelength (λ), may be thought of as an arrangement of archetypes, and therefore is perhaps a precipitation of system-knowledge, as in Equation 5.1.3:

$$EM_{Spectrum_{Structure}} = Xa + \overline{Yb_{0-n}}$$

$$where \begin{bmatrix} X \in [S_{System_K}] \\ Y \in [S_{System_{Pr}}, S_{System_P}, S_{System_K}, S_{System_N}] \\ a, b \ are \ integers; a > b \end{bmatrix}$$

Eq 5.1.3: Structure of EM Spectrum

Note that (5.1.3) already implies that Lines 1 – 5 in the Light Matrix of (5.1.1) have been activated, and that the

logic of the "wave-archetype" in the electro-magnetic-wavearchetype-masspotential ecosystem will automatically precipitate into the material-fabric through the action of Line 6-7 of (5.1.1). This logic is none other than the pre-genetic code as specified by (5.1.3).

What this also implies is that the significance or intent of the different types of waves that exist can also be expressed by this general equation where the X and Y elements will vary. What precisely these elements are will need to be worked out. A few representative equations, Equations 5.1.4 through 5.1.7 follow:

$$Gamma\ Rays_{Intent} = Xa +$$

$$\overline{Yb_{0-n}}\ where\ \begin{bmatrix} X \in [S_{System_K}] \\ Y \in [S_{System_{Pr}}, S_{System_P}, S_{System_K}, S_{System_N}] \\ a, b\ are\ integers; a > b \end{bmatrix}$$

Eq 5.1.4: Gamma Rays Intent

Knowing that some of the applications of Gamma Rays are in sterilizing and radiotherapy, these can be attributed as elements to the architectural sets. The totality of the use can be thought of as the 'intent' of Gamma Rays. An understanding of the uses will allow a full set of the secondary elements to be mapped thus allowing (5.1.4) to be reverse-engineered.

215

Similarly each of the archetypes present in the electromagnetic spectrum can be mapped out. Sample mapping of intent include x-rays, infrared, and microwaves:

$$X - Rays_{Intent} = Xa +$$

$$\overline{Yb_{0-n}} \quad where \quad \begin{bmatrix} X \in [S_{System_K}] \\ Y \in [S_{System_{Pr}}, S_{System_P}, S_{System_K}, S_{System_N}] \\ a, b \ are \ integers; a > b \end{bmatrix}$$

Eq 5.1.5: X-Rays Intent

$$Infrared_{Intent} = Xa +$$

$$\overline{Yb_{0-n}} \quad where \quad \begin{bmatrix} X \in [S_{System_K}] \\ Y \in [S_{System_{Pr}}, S_{System_P}, S_{System_K}, S_{System_N}] \\ a, b \ are \ integers; a > b \end{bmatrix}$$

Eq 5.1.6: Infrared Intent

$$Microwaves_{Intent} = Xa +$$

$$\overline{Yb_{0-n}} \quad where \quad \begin{bmatrix} X \in [S_{System_K}] \\ Y \in [S_{System_{Pr}}, S_{System_P}, S_{System_K}, S_{System_N}] \\ a, b \ are \ integers; a > b \end{bmatrix}$$

Eq 5.1.7: Microwaves Intent

Generation of Electromagnetic Spectrum Partial-Singularity 'Electro' Material-Fabric Pre-Genetic Code

The range of different wave-types or wavelengths implicit in the electromagnetic spectrum also moves with different frequencies. Since the speed of light is a constant, and it is known that speed is the product of frequency and wavelength, the greater the wavelength the lower will be the frequency of the wave-type. Conversely the less the wavelength the higher will be the frequency of the wave-type. And energy or power of a wave-type will depend on its frequency.

Hence, gamma rays that have a lower wavelength will have a higher frequency, and higher energy associated with it. Radio waves on the other hand that have a higher wavelength will have a lower frequency and therefore lower energy associated with it. So implicit in the electromagnetic spectrum is a gradient of energy. But it is also known that the penetration power is dependent on frequency. Therefore, the higher the frequency or energy, the higher will be the power of the wave-type. Another way to say this is that the property of Power in Light emerges as the energy-gradient in the electromagnetic spectrum. This Power aspect seems to have been captured by the "electro" in electromagnetic.

The energy-gradient implicit in the EM spectrum suggests the power and energy with which knowledge moves and is perhaps a precipitation of system-power, as in Equation 5.1.8:

$$EM\_Spectrum_{Energy} = Xa +$$

$$\overline{Yb_{0-n}} \quad where \quad \begin{bmatrix} X \in [S_{System_P}] \\ Y \in [S_{System_{Pr}}, S_{System_P}, S_{System_K}, S_{System_N}] \\ a, b \ are \ integers; a > b \end{bmatrix}$$

Eq 5.1.8: EM Spectrum Energy

Note that (5.1.8) already implies that Lines 1 – 5 in the Light Matrix of (5.1.1) have been activated, and that the logic of the "electro" in the electro-magnetic-wavearchetype-masspotential ecosystem will automatically precipitate into the material-fabric through the action of Line 6-7 of (5.1.1). This logic is none other than the pre-genetic code as specified by (5.1.8).

This electromagnetic energy, as suggested, is directly proportional to the frequency, of which there is an infinite range as predicted by Maxwell's equations. Different frequencies have different penetration profiles

(HyperPhysics, 2016) and it may be suggested that the nature of the energy is also a precipitation of a range of to be determined meta-functions as suggested in some sample Equations 5.1.9 through 5.1.13. Hence:

$Gamma\ Rays\ _{Nature\ of\ Energy}\ =$

$$Xa + $$
$$\overline{Yb_{0-n}}\ where\ \begin{bmatrix} X \in [S_{System_P}] \\ Y \in [S_{System_{Pr}}, S_{System_P}, S_{System_K}, S_{System_N}] \\ a, b\ are\ integers; a > b \end{bmatrix}$$

Eq 5.1.9: Gamma Rays Nature of Energy

$X - Rays_{Nature\ of\ Energy}\ =$

$$Xa + $$
$$\overline{Yb_{0-n}}\ where\ \begin{bmatrix} X \in [S_{System_P}] \\ Y \in [S_{System_{Pr}}, S_{System_P}, S_{System_K}, S_{System_N}] \\ a, b\ are\ integers; a > b \end{bmatrix}$$

Eq 5.1.10: X-Rays Nature of Energy

$Ultraviolet\ _{Nature\ of\ Energy}\ =$

$$Xa + $$
$$\overline{Yb_{0-n}}\ where\ \begin{bmatrix} X \in [S_{System_P}] \\ Y \in [S_{System_{Pr}}, S_{System_P}, S_{System_K}, S_{System_N}] \\ a, b\ are\ integers; a > b \end{bmatrix}$$

Eq 5.1.11: Ultraviolet Nature of Energy

$White\ Light_{Nature\ of\ Energy}\ =$

$$Xa + $$
$$\overline{Yb_{0-n}}\ where\ \begin{bmatrix} X \in [S_{System_P}] \\ Y \in [S_{System_{Pr}}, S_{System_P}, S_{System_K}, S_{System_N}] \\ a, b\ are\ integers; a > b \end{bmatrix}$$

Eq 5.1.12: White Light Nature of Energy

$$AM\ Radio\ Waves_{Nature\ of\ Energy} =$$

$$Xa + \overline{Yb_{0-n}}\ where\ \left[\begin{array}{c} X \in [S_{System_P}] \\ Y \in [S_{System_{Pr}}, S_{System_P}, S_{System_K}, S_{System_N}] \\ a, b\ are\ integers; a > b \end{array} \right]$$

Eq 5.1.13: AM Radio Waves Nature of Energy

Generation of Electromagnetic Spectrum Partial-Singularity 'Mass Potential' Material-Fabric Pre-Genetic Code

If there is a large range of frequencies implicit in the electromagnetic spectrum then there is also the possibility of different types of masses implicit in the electromagnetic spectrum. Frequency determines energy, and mass and energy are related through Einstein's famous MC-squared equation. So pushing a little further it is not just that mass and energy are related, but a different kind of frequency or wave-type potentially allows a different type of mass to emerge. So the possibility of different types of mass seems to be related to the property of Presence in Light. In other words the property of Presence emerges as the possibility of different types of masses as suggested by the range of mass-possibilities that can emerge from the electromagnetic spectrum.

Mass can be thought of as a container at U within which all possibility happens. In other words it can be thought of as a precipitation or emergence of system-presence as depicted in Equation 5.1.14:

$$EM_{Spectrum_{Mass_{Possibility}}} = Xa + \overline{Yb_{0-n}}$$

$$where \begin{bmatrix} X \in [S_{System_{Pr}}] \\ Y \in [S_{System_{Pr}}, S_{System_P}, S_{System_K}, S_{System_N}] \\ a, b \ are \ integers; a > b \end{bmatrix}$$

Eq 5.1.14: EM Spectrum Mass Possibility

Note that (5.1.14) already implies that Lines 1 – 5 in the Light Matrix of (5.1.1) have been activated, and that the logic of the "masspotential" in the electro-magnetic-wavearchetype-masspotential ecosystem will automatically precipitate into the material-fabric through the action of Line 6-7 of (5.1.1). This logic is none other than the pre-genetic code as specified by (5.1.14).

Further, if the frequencies are infinite, then the possibility of the 'types' of masses or matter is also infinite. The mystery of 'Dark Matter' suggested by scientists to be 27% of our universe, as opposed to 5% of visible matter (NASA-darkmatter, 2016) may have some relation to this. This aspect is also explored in the book Cosmology of Light (Malik, 2018b).

Hypothetically the Equations 5.1.15 through 5.1.19 depict a range of mass possibilities:

$$Gamma \ Rays \ _{Mass_{Possibility}} =$$

$$Xa +$$

$$\overline{Yb_{0-n}} \quad where \begin{bmatrix} X \in [S_{System_{Pr}}] \\ Y \in [S_{System_{Pr}}, S_{System_P}, S_{System_K}, S_{System_N}] \\ a, b \ are \ integers; a > b \end{bmatrix}$$

Eq 5.1.15: Gamma Rays Mass Possibility

$$Ultraviolet \ _{Mass_{Possibility}} =$$

$$Xa +$$

$$\overline{Yb_{0-n}} \quad where \begin{bmatrix} X \in [S_{System_{Pr}}] \\ Y \in [S_{System_{Pr}}, S_{System_P}, S_{System_K}, S_{System_N}] \\ a, b \ are \ integers; a > b \end{bmatrix}$$

Eq 5.1.16: Ultraviolet Mass Possibility

$$Blue \ Light \ _{Mass_{Possibility}} =$$

$$Xa +$$

$$\overline{Yb_{0-n}} \quad where \begin{bmatrix} X \in [S_{System_{Pr}}] \\ Y \in [S_{System_{Pr}}, S_{System_P}, S_{System_K}, S_{System_N}] \\ a, b \ are \ integers; a > b \end{bmatrix}$$

Eq 5.1.17: Blue Light Mass Possibility

$$Microwaves \ _{Mass_{Possibility}} =$$

$$Xa +$$

$$\overline{Yb_{0-n}} \quad where \begin{bmatrix} X \in [S_{System_{Pr}}] \\ Y \in [S_{System_{Pr}}, S_{System_P}, S_{System_K}, S_{System_N}] \\ a, b \ are \ integers; a > b \end{bmatrix}$$

Eq 5.1.18: Microwaves Mass Possibility

$$FM \ Radio \ Waves_{Mass_{Possibility}} =$$

$$Xa +$$

$$\overline{Yb_{0-n}} \quad where \begin{bmatrix} X \in [S_{System_{Pr}}] \\ Y \in [S_{System_{Pr}}, S_{System_P}, S_{System_K}, S_{System_N}] \\ a, b \ are \ integers; a > b \end{bmatrix}$$

Eq 5.1.19: FM Radio Waves Mass Possibility

Summary of Electromagnetic Spectrum Partial-Singularity Material-Fabric Pre-Genetic Code

Summarizing, after the electromagnetic spectrum stage iterations of the Light-Space-Time Emergence equation are complete, the following code-segments will have been generated as specified by Equation 5.1.20, Active Electromagnetic Spectrum Partial-Singularity Material-Fabric Pre-Genetic Code Segments:

$$Light - Space - Time\ Emergence_{EM\ Spectrum} =$$

$$
\left\|
\begin{bmatrix}
c_\infty : [Pr, Po, K, H] \\
\left(\downarrow R_{C_K} = f\left(R_{C_\infty}\right) \right) \\
c_K : [S_{Pr}, S_{Po}, S_K, S_H] \\
\left(\downarrow R_{C_N} = f\left(R_{C_K}\right) \right) \\
c_N : f(S_{Pr} \, x \, S_{Po} \, x \, S_K \, x \, S_H) \\
\left(\downarrow R_{C_U} = f\left(R_{C_N}\right) \right) \\
c_U : [P, V, M, C] \\
\Uparrow \\
c_{0:[D,W,I,C]}
\end{bmatrix}_{Light}
\begin{bmatrix}
M_3 \rightarrow System_X \\
(\uparrow F \rightarrow I) \\
M_2 \rightarrow S_{System_X} \\
(\uparrow Sig \rightarrow F) \\
M_1 \rightarrow Sig_x \\
(\uparrow > P_x) \\
U \rightarrow x_{Space-Time-Energy-Gravity}
\end{bmatrix}_{Space}
\right.
$$

$$
\begin{bmatrix}
M_3 : -\infty \le t \le \infty \\
\downarrow \\
M_2 : 0 \ge t > \infty \\
\downarrow \\
M_1 : 0 > t > \infty \\
\downarrow \\
U \rightarrow \begin{array}{l} t \le E_{Cell}; TC: M_3 \rightarrow U \\ t \sim E_{Human}; TC: U \rightarrow M_3 \end{array}
\end{bmatrix}_{Time}
\qquad TC \rightarrow x_{EM\,Spectrum}
\qquad \langle x_U | x_T \rangle
$$

$\Rightarrow$

$\langle Space - Time - Energy - Gravity \; Material - Fabric \; Pre$
$- Genetic \; Code \rangle +$

$$
\left(
\begin{array}{c}
\sum EM\_Spectrum_{Speed} = Xa + \overline{Yb_{0-n}} \\
X \in [S_{System_N}] \\
where \begin{bmatrix} Y \in [S_{System_{Pr}}, S_{System_P}, S_{System_K}, S_{System_N}] \\ a, b \; are \; integers; a > b \end{bmatrix} \\
\sum EM\_Spectrum_{Str\square\square cture} = Xa + \overline{Yb_{0-n}} \\
X \in [S_{System_K}] \\
where \begin{bmatrix} Y \in [S_{System_{Pr}}, S_{System_P}, S_{System_K}, S_{System_N}] \\ a, b \; are \; integers; a > b \end{bmatrix} \\
\sum EM\_Spectrum_{Energy} = Xa + \overline{Yb_{0-n}} \\
X \in [S_{System_P}] \\
where \begin{bmatrix} Y \in [S_{System_{Pr}}, S_{System_P}, S_{System_K}, S_{System_N}] \\ a, b \; are \; integers; a > b \end{bmatrix} \\
\sum EM_{Spectrum}_{Mass_{Possibility}} = Xa + \overline{Yb_{0-n}} \\
X \in [S_{System_{Pr}}] \\
where \begin{bmatrix} Y \in [S_{System_{Pr}}, S_{System_P}, S_{System_K}, S_{System_N}] \\ a, b \; are \; integers; a > b \end{bmatrix}
\end{array}
\right)
$$

Eq. 5.1.20, Active Electromagnetic Spectrum Partial-Singularity Material-Fabric Pre-Genetic Code Segments

The '$\sum x$' signifies all the possibilities for the 'x' equation-segment and depicts the growing biography by which the singular light-based edifice expresses its materialization, and at this stage, through the electromagnetic spectrum partial-singularity.

SECTION 6: GENERATION OF PARTIAL-SINGULARITIES IN THE SURFACING OF MATTER

So we find that the four underlying properties of Light that we call Harmony, Knowledge, Power, and Presence are of the essence of the speed with which the electromagnetic spectrum moves, the wave-range within the electromagnetic spectrum, the energy-gradient within the electromagnetic spectrum, and the mass-possibilities due to the electromagnetic spectrum, respectively. Further the description – electromagnetic – seems to have captured the Power-Harmony aspects implicit in light. In reality the electro-magnetic spectrum can likely be more completely described as electro-magnetic-wavearchetype-masspotential spectrum.

But further, we will find that layers of matter – quantum particles, which include bosons, and atoms – are also structured or emerge along the same property-lines or property-families of Light.

There is similarly a continuous process of computation that involves the quantum-realms and quantization to create the realities of quantum particles, including bosons, and atoms, to generate matter-based partial-singularities with its associated material-fabric, pre-genetic code.

However, there is an apparent difference in the way matter materializes, in contrast to the electromagnetic spectrum. Chapter 6.1 explores this in greater detail. Chapters, 6.2, 6.3, and 6.4, then, describe a process of computation by which the quantum particles, bosons, and atoms partial-singularities and their associated material-fabric, pre-genetic code is generated respectively.

Chapter 6.1: Containerization of Matter in Partial-Singularities

So far the action of the Light-Space-Time Emergence has generated pre-genetic code. After the Big Bang this pre-genetic code is housed in the material-fabric, which has a universal action on all constructs arising in the universe. Previous chapters explored such actions at the macro space-time-energy-gravity level, and the electromagnetic spectrum level. The emergence of matter however precipitates a phenomenon of distinct material containerization. It is interesting to speculate as to why this may be, especially in reference to the math already developed previously. What is it about the pre-genetic code that in the case of the electromagnetic spectrum maintains its action more as a wave or a field, and what is it about matter, to be discussed in more detail in this section, that causes it to containerize? In this chapter it will be proposed that it is an essential action of the space-time-energy-gravity quadrumvirate that curves on itself so as to speak, to thereby allow such individual containers to be generated. Further, this containerizing action can occur at multiple scale, as will also be explored in this chapter.

Universality Versus Localization of Space, Time, Energy, and Gravity

In Chapter 3.5, A Deeper Look at Quantization of Space, Time, Energy, and Gravity, (3.5.3) proposed a model for the structure of space. This is reproduced here for convenience:

$$Space_{Structure} = \sum_{i,j=1}^{\to \infty} h_{UK}\left(X_i a + \overline{Y_j b_{0-n}}\right)$$

$$where: \begin{bmatrix} X_i \in [S_{System_K}] \\ Y_j \in [S_{System_{Pr}}, S_{System_P}, S_{System_K}, S_{System_N}] \\ a, b, i, j \ are \ integers; a > b \end{bmatrix}$$

Summarizing, space, consisting of a vast array of seeds derived from the properties of Light, is itself an expression of Light's property of Knowledge. It is this essential action of a vast number of seeds that gives space its apparent never-ending-ness. In fact as previously discussed, time, energy and gravity can also be interpreted in terms of these 'seeds'. Equations 6.1.1-4 restate Space, Time, Energy, and Gravity, in terms of this relationship to seeds. Hence, Equation 6.1.1, Relation of Seeds to Space:

$$Space = f(\#seeds)$$
$$= f(Light's \ property \ of \ Knowledge)$$

Eq 6.1.1: Relation of Seeds to Space

Time, bringing forth the meaning contained in the seeds, regardless of circumstance, and even being opposed by circumstance, can be thought of as Light's property of Power. Hence, Equation 6.1.2, Relation of Seeds to Time:

$$Time = f(maturity \ of \ seeds) =$$
$$f(Light's \ property \ of \ Power)$$

228

Eq 6.1.2: Relation of Seeds to Time

Matter or Energy itself, being a container in which space and time can allow deeper properties of Light to become materially tangible, must be an expression of Light's property of Presence. Hence, Equation 6.1.3, Relation of Seeds to Energy / Matter:

$$Matter = f(materialization\ of\ seeds)$$
$$= f(Light's\ property\ of\ Presence)$$

Eq 6.1.3: Relation of Seeds to Energy/Matter

But it is also known from Einstein's General Theory of Relativity that gravity is associated with mass and space, in that it is none other than a mass's instruction telling space how to curve, and again is nothing else that space's instruction telling mass how to move through it (Wheeler, 2000). As such, where mass and space exist, there gravity has to exist as well. Hence it may be inferred that gravity is none other than an expression of Light's property of Harmony, which fixes the collective relationship between object and object. Hence, Equation 6.1.4, Relation of Seeds to Gravity:

$$Gravity = f(cohesion\ of\ seeds)$$
$$= f(Light's\ property\ of\ Harmony)$$

Eq 6.1.4: Relation of Seeds to Gravity

If the number of seeds is large, then the collective action of space, time, energy, and gravity is going to create an apparent universality and the reality of 'never-ending-ness'. By contrast if the number of seeds is smaller, then the action of space, time, energy, and gravity will be relatively localized, and create a more apparent reality of containerization.

Such containerization is likely spurred by the action of quantization already captured by (3.5.2), (3.5.4), (3.5.6), and (3.5.7) reproduced here for convenience.

Hence (3.5.2):

$$Space_{quantization} = h_{UK}\left(Xa + \overline{Yb_{0-n}}\right)$$

$$where: \begin{bmatrix} X \in [S_{System_K}] \\ Y \in [S_{System_{Pr}}, S_{System_P}, S_{System_K}, S_{System_N}] \\ a, b \ are \ integers; a > b \end{bmatrix}$$

(3.5.4):

$$Time_{quantization} = h_{UP}\left(Xa + \overline{Yb_{0-n}}\right)$$

$$where: \begin{bmatrix} X \in [S_{System_P}] \\ Y \in [S_{System_{Pr}}, S_{System_P}, S_{System_K}, S_{System_N}] \\ a, b \ are \ integers; a > b \end{bmatrix}$$

(3.5.6):

$$Energy_{quantization} = h_{UPr}\left(Xa + \overline{Yb_{0-n}}\right)$$

$$where: \begin{bmatrix} X \in [S_{System_{Pr}}] \\ Y \in [S_{System_{Pr}}, S_{System_P}, S_{System_K}, S_{System_N}] \\ a, b \ are \ integers; a > b \end{bmatrix}$$

(3.5.7):

$$Gravity_{quantization} = h_{UH}\left(Xa + \overline{Yb_{0-n}}\right)$$

$$where: \begin{bmatrix} X \in [S_{System_N}] \\ Y \in [S_{System_{Pr}}, S_{System_P}, S_{System_K}, S_{System_N}] \\ a, b \ are \ integers; a > b \end{bmatrix}$$

Universality of Electromagnetic Spectrum

As discussed in Chapter 5, the range of frequencies in the electromagnetic spectrum is infinite. This has been predicted by Maxwell's equations. (5.1.8), the equation for the infinite energy-gradient, (5.1.13) the equation that relates the infinite range of wavelengths to archetypes, and (5.1.14) that relates the infinite range of mass-potential, each essentially capture the infinite seed-range held by the electromagnetic spectrum. The original forms of the equations follow for convenience.

Hence, (5.1.8):

$$EM_{Spectrum_{Energy}} = Xa + \overline{Yb_{0-n}}$$

$$where \begin{bmatrix} X \in [S_{System_P}] \\ Y \in [S_{System_{Pr}}, S_{System_P}, S_{System_K}, S_{System_N}] \\ a, b \ are \ integers; a > b \end{bmatrix}$$

(5.1.13):

$$EM_{Spectrum_{Structure}} = Xa + \overline{Yb_{0-n}}$$

$$where \begin{bmatrix} X \in [S_{System_K}] \\ Y \in [S_{System_{Pr}}, S_{System_P}, S_{System_K}, S_{System_N}] \\ a, b \ are \ integers; a > b \end{bmatrix}$$

(5.1.14):

$$EM_{Spectrum_{Mass_{Possibility}}} = Xa + \overline{Yb_{0-n}}$$

$$where \begin{bmatrix} X \in [S_{System_{Pr}}] \\ Y \in [S_{System_{Pr}}, S_{System_P}, S_{System_K}, S_{System_N}] \\ a, b \ are \ integers; a > b \end{bmatrix}$$

Recall that the infinite frequency-wavelength range becomes possible because of the constancy of the speed of light at c, which creates time-matter reality that marks this universe.

It is because of the infinite seed-range that the action of the electromagnetic spectrum is apparently universalized. Hence when quantization does take place as per the light-space-time emergence equation, (3.1.3), and the potential effects of levels of light on pre-genetic and genetic information equation, (3.6.5), because there are a larger range of seeds the effect of the material-fabric and subsequently on materialization is distributed and spread-out.

Material Localization at Multiple Scale

Since quanta hold within themselves the essence of what

must be projected from properties or function in higher-velocity light to lower-velocity more-material light, and since the four-fold properties are a representation of the inherent oneness of Light, it is reasonable to expect that such four-foldness will continue to have a bias to uphold that oneness in its more material manifestation. In other words, it is reasonable to expect that space-quanta, time-quanta, energy-quanta, and gravity-quanta related to a single particle will operate as a composite fourfold quantum. This means that four quantum fields operate together or as one composite field in the emergence and possibilities represented by a particle and that perhaps is the reason for this apparently intense material localization.

The following subsections explore the notion of intense material localization at different scale – specifically at the atomic-particle, unit-space, big-planet, expanding universe, black hole, and cosmic bounce levels. In the possible iterations of the Light-Space-Time Emergence Equation (3.1.3) the generated code can apply to organizations of increasing size, including those just specified, as suggested by the discussion in Section 4.

Quanta at Atomic-Particle Level

Figure 6.1.1 below, Composite-Quantum at Atomic-Particle Level, depicts a composite-quantum along the space, time, gravity, and energy dimensions. Essentially the graphic is illustrating that all things being equal there is a balanced accumulation of anterior possibility or information along the space, time, gravity, and energy dimensions that plays a primary role in orchestrating the possibility contained in the atomic-particle. Recall that the space dimension contains information to do with unique seeds, the time dimension information to do with the maturity of seeds, the gravity dimension information to do with collectivities of seeds, and the energy dimension information to do with the materialization of seeds. Hence:

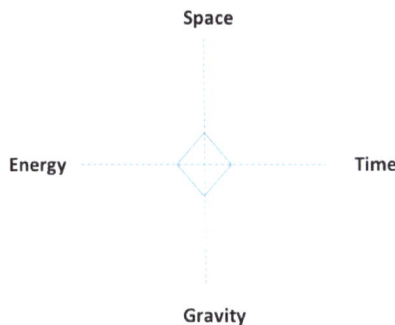

Figure 6.1.1: Composite-Quantum at Atomic-Particle Level

Quanta at Unit-Space Level

The small rhombus in Figure 6.1.2 represents a quantum at the atomic-particle level, while the larger rhombus is an extrapolation of that in some unit space. All things being equal it is the same fourfold accumulation of anterior possibility or information captured by the quantum that will govern particle behavior at the unit-space level. Hence:

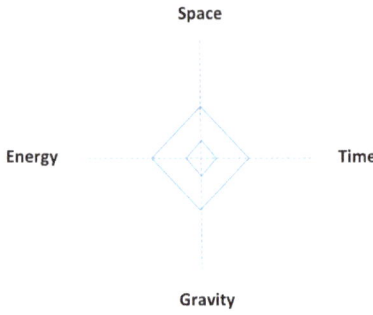

Figure 6.1.2: Composite-Quantum at Unit-Space Level

Further, if Q_A represents a composite-quantum at the atomic-particle level, then Q_{US} will be a summation of such quanta as modeled by Equation 6.1.5:

$$Q_{US} = \Sigma\, Q_A$$

Eq 6.1.5 Composite-Quantum and Unit-Space Level

Quanta at Level of Big-Planet

In Figure 6.1.3, Composite-Quantum at Level of Big-Planet, the smaller rhombus represents a quantum at the unit-space level, while the larger kite-quadrilateral represents a quantum at the Big-Planet level:

234

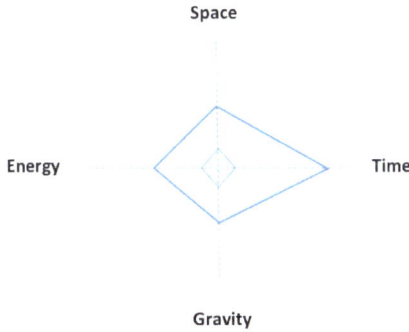

Figure 6.1.3: Composite-Quantum at Level of Big-Planet

The kite-quadrilateral is elongated along the time dimension implying that the larger collectivity of seeds represented by a space such as a Big-Planet will take longer to reach maturity, relative to a seed contained in a unit-space. This time-elongation is also modeled by the '$mod\ (Time_{quantization})$' component of Equation 6.1.6:

$$Q_{BP} = \sum Q_{US} \times mod(Time_{quantization})$$

Eq 6.1.6 Composite-Quantum at the Level of Big-Planet

Note that in (6.1.6), Q_{BP} represents a composite-quantum at the level of a Big-Planet, while Q_{US} represents a composite-quantum at the unit-space level as modeled by (6.1.5). Equation (6.1.6) illustrates the notion of composite-quantum as creating the field within which a Big-Planet can materialize.

The time-elongation captures the underlying dynamic of many more finite steps to maturity for the vaster collectivity of seeds, that therefore manifests as time being experienced more rapidly: hence time speeds-up.

Quanta in Expanding Universe

235

Figure 6.1.4, Composite-Quantum in an Expanding Universe, depicts the relative relationship between the smaller unit-space composite-quantum rhombus and the scaled up expanding-universe composite-quantum rhombus, in which it is estimated that the space, time, gravity, and energy components all go through some related increase:

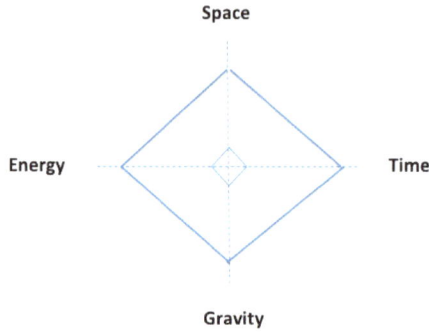

Figure 6.1.4: Composite-Quantum in an Expanding Universe

Equation 6.1.7 models such expanding-universe composite-quantum:

$$Q_{EU} = \Sigma\, Q_{US}$$

Eq 6.1.7 Composite-Quantum in an Expanding Universe

Quanta at Black Hole Level

Figure 6.1.5, Composite-Quantum at the Black Hole Level, illustrates the relatively different quadrilateral, as compared with previously considered composite-quantum in Figures (6.1.1-4):

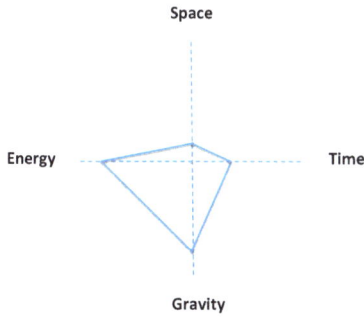

Figure 6.1.5: Composite-Quantum at Black Hole Level

This composite-quantum is hypothesized as being the result of a large number of seeds previously spread in space, coming together to be reformulated as a smaller set or a new seed. The new seed(s) have a different intent than the spread out un-mashed seeds. They will have a very different and relatively shortened maturity-dynamic as depicted along the time-dimension due to a normalization or rationalization of seeds. Further, they will have a very different collectivity-relationship between other seeds outside of the core black-hole formation, as represented by the gravity-dimension. Finally, they will similarly also have a very different energy-materialization dynamic as represented by the energy-dimension, and due to the complexification of the normalized seeds.

Centered on the black hole, space-time-gravity-energy is going to proceed differently than may have existed prior to the formation of the black hole. Such modification is modeled by Equation (6.1.8), where as can be seen there is a summation of previous seeds along each dimension, captured by $\Sigma\, Seed_i$, and a holding of the new essence, captured by h_{Ui} with 'i' being 'K', 'P', 'H' and 'Pr' respectively, which together create the new black hole (BH) composite-quantum, Q_{BH}:

$$Q_{BH} = \begin{bmatrix} mod(Space_{quantization}) \to h_{UK}\,\Sigma\, Seed_K \\ mod(Time_{quantization}) \to h_{UP}\,\Sigma\, Seed_P \\ mod(Gravity_{quantization}) \to h_{UH}\,\Sigma\, Seed_H \\ mod(Energy_{quantization}) \to h_{UPr}\,\Sigma\, Seed_{Pr} \end{bmatrix}$$

Eq 6.1.8 Composite-Quantum at Black-Hole Level

Note that consistent with General Relativity, the dynamic of gravity intensifies, space is compressed, time slows down, and energy-potential increases.

Quanta of Cosmic Bounce

Figure 6.1.6, Composite-Quantum at Cosmic Bounce, depicts the composite-quantum at a possible Cosmic Bounce event. The smaller quadrilateral represents the black hole composite-quantum, while the larger quadrilateral is a scaled up Cosmic Bounce version of Q_{BH}:

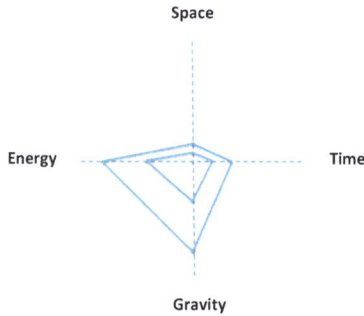

Figure 6.1.6: Composite-Quantum at Cosmic Bounce

The notion of scaling up, is captured by e^M where M is some very large number, in the Cosmic Bounce composite-quantum as modeled by Equation (6.1.9):

$$Q_{CB} = Q_{BH}e^M$$

Eq 6.1.9 Composite-Quantum at Cosmic Bounce

Hence, this brief seed-based analyses suggests that even as organizations scale, there will still be a containerization of matter albeit, possibly, with different quadrumvirate balance. Such containerization is a key dynamic in any partial-singularity and expresses the fundamental urge to maintain material-oneness in accordance with the reality of fourfold oneness of the essential properties of Light in its native state.

The discussion in this chapter also perhaps provides a useful perspective on the dynamics of cosmic singularities such as black-holes, from the perspective of the seed-based analyses foundational to a model based on multiple layers of light.

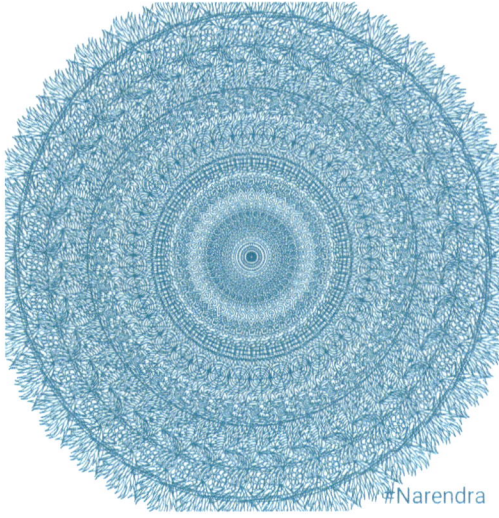
#Narendra

Chapter 6.2: Creation of Quantum Particle Partial-Singularity

The electromagnetic spectrum or perhaps more accurately, the electro-magnetic-wavearchetype-masspotential spectrum, describes one of the first layers or translation of light into pre-genetic material manifestation.

As already explored the finite speed of such spectrums allows the build-up of energies or quanta, and this in turn may express itself as a series or as an array of quantum particles. The vast number of quantum particles that have so far been discovered has in turn yielded what is known as the Standard Model. This model is made up of what is called Quarks, Leptons, Bosons, and a Higgs-Boson (Cottingham, 2007).

But if we look at these four fundamental quantum categories of particles from a property-based or "functional" viewpoint, a different chapter in the exploration of light emerges. All of matter then can be seen as implicit in Light and in fact due to a process of computation by which the implicit codification in Light becomes explicit. The implicit codification becomes explicit by generating material-fabric pre-genetic quantum particle code-segments. These code-segments are added to the existing pre-genetic code-segments and also become binding on future emergences in the material universe. In other words partial-singularities with increasing spheres of influence are progressively created. All previous code-segments create law in a newly emergent partial-singularity hence reinforcing the composite and unified nature of partial-singularities.

As in the previous Section, the Light-Space-Time Emergence equation (3.1.3) can be used to model emergence as it proceeds to this relatively more complex form of fourfold manifestation. (3.6.5), the Potential Effect of Levels of Light on Genetic-Type Information equation, basically sheds light on specific types of mutation that can occur and therefore on the nature of the code-segment that is being generated by (3.1.3). As discussed before, (3.6.5) shows a particular working of (3.1.3) but is wholly contained within it.

This chapter elaborates the action of (3.1.3) and (3.6.5) in the creation of the code-segments that define the quantum particle ecosystem logic that emerges in the quantum particle partial-singularity.

Light's Emergence as Quantum Particles

As discussed previously the Light-Space-Time Emergence equation (3.1.3) being iterative, can be used to model emergence as it proceeds from simpler four-fold to more complex four-fold manifestations. But further, as implied by (3.6.5), the Potential Effect of Levels of Light on Genetic-Type Information equation, any process of organization has the possibility of altering the material-fabric or fabric of existence so long as the bases involved are driven primarily by a meta-level.

As suggested by Equation 6.2.1, Generation of Quantum Particle Partial-Singularity, the architecture and details of quantum particles can be seen to be the result of the application of the Light, Space, and Time matrices as will be elaborated:

$Light - Space - Time\ Emergence_{Quantum\ Particles} =$

$$\left\| \begin{bmatrix} c_\infty: [Pr, Po, K, H] \\ \left(\downarrow R_{C_K} = f\left(R_{C_\infty}\right)\right) \\ c_K: [S_{Pr}, S_{Po}, S_K, S_H] \\ \left(\downarrow R_{C_N} = f\left(R_{C_K}\right)\right) \\ c_N: f\left(S_{Pr} \times S_{Po} \times S_K \times S_H\right) \\ \left(\downarrow R_{C_U} = f\left(R_{C_N}\right)\right) \\ c_U: [P, V, M, C] \\ \Uparrow \\ c_{0:[D,W,I,C]} \end{bmatrix}_{Light} \begin{bmatrix} M_3 \rightarrow System_X \\ (\uparrow F \rightarrow I) \\ M_2 \rightarrow S_{System_X} \\ (\uparrow Sig \rightarrow F) \\ M_1 \rightarrow Sig_X \\ (\uparrow > P_x) \\ U \rightarrow x_{EM\ Spectrum} \end{bmatrix}_{Space} \\ \left[U \rightarrow \begin{matrix} M_3: -\infty \leq t \leq \infty \\ \downarrow \\ M_2: 0 \geq t > \infty \\ \downarrow \\ M_1: 0 > t > \infty \\ \downarrow \\ t \leq E_{Cell}; TC: M_3 \rightarrow U \\ t \sim E_{Human}; TC: U \rightarrow M_3 \end{matrix} \right]_{Time} \quad TC \rightarrow x_{Quantum\ Particles} \right\| \langle x_U | x_T \rangle$$

$\Rightarrow$

$\langle Space - Time - Energy - Gravity\ Material - Fabric\ Pre$
$\qquad\qquad - Genetic\ Code \rangle +$

$(Electromagnetic\ Spectrum\ Material - Fabric\ Pre - Genetic\ Code) +$

$Quantum\ Particles\ Material - Fabric\ Pre - Genetic\ Code$

Eq 6.2.1: *Generation of Quantum Particle Partial-Singularity*

Starting with the Light-Matrix, the top left-hand matrix in (6.2.1), the first line from the top, $C_\infty: [Pr, Po, K, H]$, specifies the fundamental architecture of quantum particles. In this rendering, and as elaborated in subsequent subsections, quarks are an emergence of Light's property of Knowledge, leptons are an emergence

243

of Light's property of Power, bosons are an emergence of Light's property of Harmony, and the Higgs-boson is an emergence of Light's property of Presence. The fundamental architecture of these aspects hence, is an emergence of the properties of Light at ∞.

Line 3 in the Light-Matrix, C_K: $[S_{Pr}, S_{Po}, S_K, S_H]$, elaborates the sets for Presence, Power, Knowledge, and Harmony, each containing multiple elements. For example, as will be explored in greater detail in the section on quarks, various elements derived from the four sets define the behavior of quarks and could be functions such as 'composite-arrangements', 'specifying attributes' amongst others, hence collectively describing quarks' way of being. Specifically, Line 5, C_N: $f(S_{Pr} \times S_{Po} \times S_K \times S_H)$, suggests that unique seeds are created from a combination of such elements from all four sets, with a particular element leading or having more weight, that in effect creates the distinctness possible at the level of quantum particles.

Line 6, ($\downarrow R_{C_U} = f(R_{C_N})$), specifies quantization between the layer where the seeds are formed, and the physical layer, and as explored in Chapter 3.5 and 3.6, will result in Line 7, C_U: $[P, V, M, C]$, hence changing the material-fabric of existence. Note that as in the process describing the generation of the code-segments for space-time-energy-gravity quantization at the time of the Big Bang, the generation of the FBLEE (four-base logic-encoding ecosystem) code-segment as specified by (4.1.5) and the MF (material-fabric) code segment as specified by (4.1.6) are combined together into one equation so that the FBLEE process is apparently transparent.

The possibilities represented by Lines 1 through 5 hence concretize through the quantization represented by Line 6 to become the quantum particles with its physical (related to Presence), vital (related to Power), mental (related to Knowledge), and connection (related to

Harmony) aspects now existing in material reality typified by Light moving at c. Note that just as Line 6 represents a process of quantization relating the layer of reality created by Light traveling at c with the antecedent layers, so too Lines 2 and 4 as previously discussed, also represent quantization of a more subtle kind that ultimately plays a critical part in allowing the material-fabric to express infinite diversity.

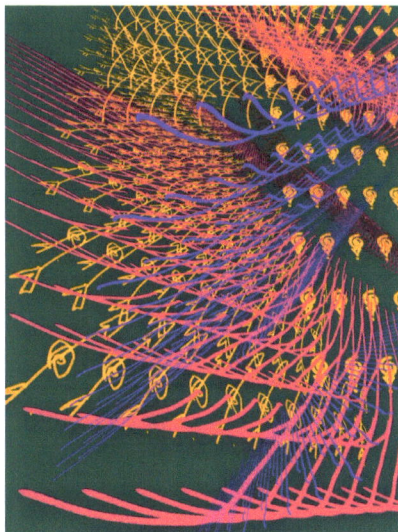

Typically it is the process as captured by the Space-Matrix that will determine if Line 6 is activated. Specifically patterns at the untransformed layer, U, will need to be overcome, as specified by the second-line from the bottom of the Space-Matrix: $(\uparrow > P_x)$. But as specified by the bottom-line of the Time-Matrix, reproduced below, it is only with the advent of the human-system that the automaticity of the action of meta-levels is reversed:

$$U \rightarrow \begin{array}{l} t \le E_{Cell}; \text{TC:} M_3 \rightarrow U \\ t \sim E_{Human}; \text{TC:} U \rightarrow M_3 \end{array}$$

Hence in the case of the quantum particle system, which in this emergence is a pre-human system, the fact that patterns do not need to be overcome means that quantization happens automatically.

Even though quantum particles are a pre-human system and therefore the action of meta-levels are modeled as being automatic, it is nonetheless useful to review (3.6.5),

245

Potential Effect of Levels of Light on Genetic-Type Information, to understand how the relationship with different forms of mutation has been specified:

$$Potential\ Effect\ of\ Levels\ of\ Light\ on\ Genetic$$
$$-\ Type\ Information\ =$$

$$\begin{bmatrix} STATIC\ \langle|[L][S][T]TC \to x_T|_{\langle x_U|x_T\rangle}\rangle \\ \times \\ \left((Y > U{:}\,Z_Q) \lor (Y \leq U{:}\,Z_F) \lor (Y = U{:}\,Z_R)\right) \\ \ni \\ \left(\begin{array}{c} Z \in \mathbb{U}\ (Space, Time, Energy, Gravity) \\ Q{:}\ Quantization;\ F{:}\ Fragmentation;\ R{:}\ Random \end{array}\right) \end{bmatrix} \to h \ni$$

$$h \in \left(\begin{array}{c} [Q]{:}\ Constructive\ zone, \\ [Q]{:}\ Constructive\ zone \land\ Constructive\ mutation, \\ [F]{:}\ Destructive\ mutation, \\ [R]{:}\ Random\ mutation \end{array}\right)$$

Line 1 from the top in the matrix is simply a static form of (3.6.1) the Simplified Light-Space-Time Emergence equation. The static form is designated by 'STATIC', and implies that fundamental operations true of (3.6.1) are being highlighted in (3.6.5). In other words (3.6.1) already has all the operations highlighted in (3.6.5) in it, but by 'freezing' it by making it static, the essential dynamics leading to possible mutations at the genetic level can more clearly be highlighted.

Line 1 is then subjected ($\times$) to a determination of the dominant levels of light that may be active, designated by Line 2, ' $\left((Y > U{:}\,Z_Q) \lor (Y \leq U{:}\,Z_F) \lor (Y = U{:}\,Z_R)\right)$ '. Unpacking this, 'Y > U' implies meta-levels are active and as a result it is possible that Z_Q is going to take place (the subscript 'Q' implies quantum-level action). This also implies activation and potential change of FBLEE. The call from below, as it were, may invoke some function that already exists in the subtle-libraries 'above', so that some already existing function may influence FBLEE through ∞ -entanglement, K-entanglement, or N-entanglement. This may be thought of as a key-and-lock mechanism, where a deep enough visceral urge from below acting as the key, opens an entangled lock to alter FBLEE as per the visceral urge. '$Y \leq U$' implies that only the untransformed levels are active, and therefore also the sub-level where the speed of light is 0 is active and as a result Z_F is going to take place (the subscript 'F' implies 'fragmentation'). '$Y = U$' implies that all levels are active and as a result Z_R is going to take place (the subscript 'R' implies 'random').

Line 3 elaborates the significance of Z_Q, Z_F, and Z_R. Hence Z is the union of potential quantum-operations of space, time, energy, and gravity, designated by '$Z \in \mathbb{U} \, (Space, Time, Energy, Gravity)$'. But the nature of the operations, as suggested in the previous paragraph, is designated   by '$Q: Quantization; F: Fragmentation; R: Random$'. Z_Q, then, implies that the full quantization originating from updated four-base logic-encoding ecosystems (FBLEE) can take place, and will result in lasting material change at the Genetic-Type level. Z_F implies that the essential set will be fragmented and that only libraries at the level of local cellular-level DNA or precipitated material-fabric logic can potentially be altered. Z_R also precludes full quantization, and that some partial local-library constructive or destructive mutation may take place.

The '$\rightarrow h \ni h \in$' segment resolves the outcome of the operations implied by Lines 1 – 3, suggesting that the outcome will be 'h' such that ($\ni$) 'h' is an element ($\in$) of the set specified by the members '[Q]: *Constructive zone*', '[Q}: *Constructive zone AND Constructive mutation*', '[F}: *Destructive mutation*', and '{R}: *Random mutation*'. '[Q]: *Constructive zone*' implies that the in-built buffer has been crossed and that access to the deeper four-base logic-encoding ecosystem (FBLEE) has been granted. '[Q}' in this segment implies that there is the possibility that full-quantization as specified by Line 3 of the previous matrix can take place. Access to this zone is a prerequisite for constructive mutation to occur, as designated by the element ' [*Q*}: *Constructive zone* ∩ *Constructive mutation* ', which implies that full-quantization is going to take place and will result in material change. The '[F]' specifies the relationship between 'Fragmentation' in Line 3 of the previous matrix and destructive mutation. The '[R]' specifies the relationship between 'Random' in Line 3 of the previous matrix and random mutation.

But as just summarized in the Time-Matrix in (6.2.1) Y is by definition greater than U and hence quantization is automatic. In terms of quantum particles such quantization implies that wholeness becomes fully active through specific space, time, energy, and gravity quantization to create an holistic "ecosystem" with its own "quantum particles logic" as it were. The wholeness has now precipitated into the material-fabric and is available to be consciously and unconsciously tapped into. This "logic" or more specifically pre-genetic code is elaborated in the following sub-sections.

Generation of Quantum Particle Partial-Singularity 'Quark' Material-Fabric Pre-Genetic Code

It can be seen first that the nucleus of an atom is made up of a combination of quarks. Specifically, a proton is composed of two "up" quarks and one "down" quark. Quarks have unusual names – up, down, charm, strange, top, bottom, with each subsequent pair belonging to a different "generation" of quarks. A neutron is composed of two "down" quarks and one "up" quark. Protons and neutrons together make up the nucleus of an atom. But we also know that the number of protons in the nucleus specifies the Atomic Number of an atom. Atomic number in turn uniquely identifies the element from the periodic table. Hence, an atomic number of 47, for example, specifies that the element is Silver. In other words it can be suggested that the unique properties of an element, the knowledge of what it is and how it will behave in the universe, is related to the quark. It may be suggested that quarks, therefore, are associated with the precipitation or emergence of Light's property of Knowledge in the quantum world.

Hence, it could be that the signature or code-segment for the family of quarks, as in Equation 6.2.2, is:

$$Sig_{quarks} = Xa + \overline{Yb_{0-n}} \quad where \begin{bmatrix} X \in [S_{System_K}] \\ Y \in [S_{System_{Pr}}, S_{System_P}, S_{System_K}, S_{System_N}] \\ a, b \ are \ integers; a > b \end{bmatrix}$$

249

Eq 6.2.2: Generalized Signature or Code-Segment for Quarks

As can be seen the primary element X is derived from the set of knowledge, S_{System_K}. Various elements, derived from the four sets would define the behavior of quarks and could be functions such as 'composite-arrangements', 'specifying attributes' amongst others, hence collectively describing quarks' way of being.

Note that (6.2.2) already implies that Lines 1 – 5 in the Light Matrix (6.2.1) have been activated, and that the logic of the quark-ecosystem will automatically precipitate into the material-fabric through the action of Line 6-7 of (6.2.1). This logic is none other than the pre-genetic code as specified by (6.2.2).

Generation of Quantum Particle Partial-Singularity 'Lepton' Material-Fabric Pre-Genetic Code

When considering leptons it is useful to know that unlike quarks that only exist in composite arrangements with other quarks, leptons are solitary, point-like particles without internal structure (Arabatzis, 2006). The best-known lepton is the electron. So the electron may be considered as a surrogate for the lepton class. The electron appears to be associated with the flow of energy and power. Further they appear to be the adventurers easily leaving the atom they are a part of. They also form locks or bonds with other atoms through the force of attraction and repulsion. In some sense they seem to be a representation or precipitation or emergence of Light's property of Power.

The signature for the family of leptons, as in Equation 6.2.3, is:

$$Sig_{leptons} = Xa +$$
$$\overline{Yb_{0-n}} \quad where \left[\begin{array}{c} X \in [S_{System_P}] \\ Y \in [S_{System_{Pr}}, S_{System_P}, S_{System_K}, S_{System_N}] \\ a, b \ are \ integers; a > b \end{array} \right]$$

250

Eq 6.2.3: Generalized Signature or Code-Segment for Leptons

As can be seen the primary element X is derived from the set of power, S_{System_P}. Various elements, derived from the four sets would define the behavior of leptons and could be functions such as 'adventurer', 'solitary', 'create attraction', amongst others, hence collectively describing leptons' way of being.

Note that (6.2.3) already implies that Lines 1 – 5 in the Light Matrix (6.2.1) have been activated, and that the logic of the lepton-ecosystem will automatically precipitate into the material-fabric through the action of Line 6-7 of (6.2.1). This logic is none other than the pre-genetic code as specified by (6.2.3).

Generation of Quantum Particle Partial-Singularity 'Boson' Material-Fabric Pre-Genetic Code

The bosons on the other hand are thought of as force-carriers. They are what allow all known matter particles to interact. The three fundamental bosons in this category are the photon, the W and Z bosons, and the gluon. The carrier particle of the electromagnetic force or spectrum is the photon. The carrier particle of the strong nuclear force that holds quarks together is the gluon. The carrier particle for the weak interactions, responsible for the decay of massive quarks and leptons into lighter quarks and leptons, are the W and Z bosons.

Bosons can be thought of as the precipitation of what creates relationship and harmony at the quantum level. Hence they can be thought of as the precipitation or emergence of Light's property of Harmony at the quantum level.

The signature for the family of gauge bosons, as in Equation 6.2.4, is:

$$Sig_{bosons} = Xa +$$

$$\overline{Yb_{0-n}} \quad where \quad \begin{bmatrix} X \in [S_{System_N}] \\ Y \in [S_{System_{Pr}}, S_{System_p}, S_{System_K}, S_{System_N}] \\ a, b \ are \ integers; a > b \end{bmatrix}$$

Eq 6.2.4: Generalized Signature or Code-Segment for Bosons

As can be seen the primary element X is derived from the set of nurturing, S_{System_N}. Various elements, derived from the four sets would define the behavior of bosons and could be functions such as 'creating interaction', 'holding together', amongst others, hence collectively describing bosons' way of being.

Note that (6.2.4) already implies that Lines 1 – 5 in the Light Matrix (6.2.1) have been activated, and that the logic of the boson-ecosystem will automatically precipitate into the material-fabric through the action of Line 6-7 of (6.2.1). This logic is none other than the pre-genetic code as specified by (6.2.4).

Equations in this form give a clue as to how to think about the fundamental particles. While the default approach is always to think about physical qualities, equations in the preceding format allow us to think about detailed functions or qualities associated with particles and give insight into how the process of emergence or complexification of fourfold functionality takes place.

Generation of Quantum Particle Partial-Singularity 'Higgs-Boson' Material-Fabric Pre-Genetic Code

This leaves the other discovered fundamental particle the Higgs-Boson. In ordinary matter, most of the mass is contained in atoms, and the majority of the mass of an atom resides in the nucleus, made of protons and neutrons. Protons and neutrons are each made of three quarks. But it is the quarks that get their mass by interacting with what is known as the Higgs field (Olive, 2014). Hence the Higgs-Boson can be thought of as the mass-giver. In other words it is what gives presence to the quarks and it can be thought of as the precipitation or emergence of Light's property of Presence at the quantum level. Just as there are multiple particles in each of the other 'families' it is likely that there will be multiple particles in the Higgs-Boson family. Recent research at CERN indicates that the Higgs-Boson may have a cousin.

The signature for the Higgs-boson and any other similar particle, as in Equation 6.2.5, is:

$$Sig_{Higgs-boson} = Xa + \overline{Yb_{0-n}}$$

$$where \begin{bmatrix} X \in [S_{System_{Pr}}] \\ Y \in [S_{System_{Pr}}, S_{System_P}, S_{System_K}, S_{System_N}] \\ a, b \ are \ integers; a > b \end{bmatrix}$$

Eq 6.2.5: Generalized Signature or Code-Segment for Higgs-Boson

As can be seen the primary element X is derived from the set of presence, $S_{System_{Pr}}$. Various elements, derived from the four sets would define the behavior of Higgs-bosons and could be a function such as 'creating mass', amongst others, hence collectively describing Higgs-bosons' way of being.

Note that (6.2.5) already implies that Lines 1 – 5 in the Light Matrix (6.2.1) have been activated, and that the logic of the Higgs-boson-ecosystem will automatically precipitate into the material-fabric through the action of Line 6-7 of (6.2.1). This logic is none other than the pre-genetic code as specified by (6.2.5).

Combining all the preceding particle equations it is possible to create a generalized particle equation, as in Equation 6.2.6:

$$Sig_{particle} = Xa +$$
$$\overline{Yb_{0-n}} \quad where \quad \begin{bmatrix} X \in [S_{System_{Pr}}, S_{System_P}, S_{System_K}, S_{System_N}] \\ Y \in [S_{System_{Pr}}, S_{System_P}, S_{System_K}, S_{System_N}] \\ a, b \ are \ integers; a > b \end{bmatrix}$$

Eq 6.2.6: Generalized Signature of Particle

So we see that again the underlying properties of light - Knowledge, Power, Harmony, and Presence - emerge as quarks, leptons, bosons, and Higgs-bosons respectively.

Summary of Quantum Particle Partial-Singularity Material-Fabric Pre-Genetic Code

Summarizing, after the quantum particle stage iterations of the Light-Space-Time Emergence equation are complete, the following code-segments will have been generated as specified by Equation 6.2.7, Active Quantum Particle Partial-Singularity Material-Fabric Pre-Genetic Code Segments:

$$Light - Space - Time\ Emergence_{Quantum\ Particle} =$$

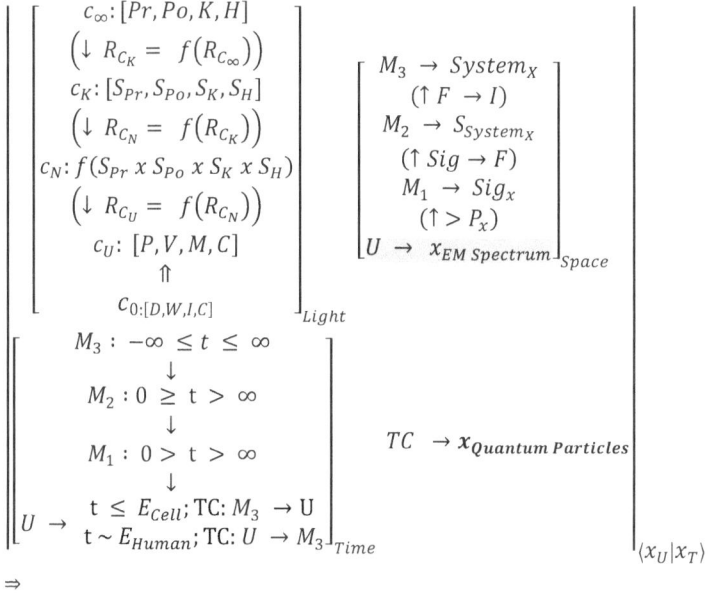

$$\left\|\begin{array}{c}\left[\begin{array}{c}c_\infty\colon [Pr, Po, K, H] \\ \left(\downarrow R_{C_K} = f(R_{C_\infty})\right) \\ c_K\colon [S_{Pr}, S_{Po}, S_K, S_H] \\ \left(\downarrow R_{C_N} = f(R_{C_K})\right) \\ c_N\colon f(S_{Pr}\ x\ S_{Po}\ x\ S_K\ x\ S_H) \\ \left(\downarrow R_{C_U} = f(R_{C_N})\right) \\ c_U\colon [P, V, M, C] \\ \Uparrow \\ c_{0\colon[D,W,I,C]}\end{array}\right]_{Light} \quad \left[\begin{array}{c}M_3\ \rightarrow\ System_X \\ (\uparrow F \rightarrow I) \\ M_2\ \rightarrow\ S_{System_X} \\ (\uparrow Sig \rightarrow F) \\ M_1\ \rightarrow\ Sig_x \\ (\uparrow > P_x) \\ U\ \rightarrow\ x_{EM\ Spectrum}\end{array}\right]_{Space} \\[4em] \left[U \rightarrow \begin{array}{c}M_3\colon\ -\infty\ \leq t\ \leq\ \infty \\ \downarrow \\ M_2\colon 0\ \geq t >\ \infty \\ \downarrow \\ M_1\colon\ 0 > t >\ \infty \\ \downarrow \\ t\ \leq\ E_{Cell}; \text{TC}\colon M_3\ \rightarrow \text{U} \\ t \sim E_{Human}; \text{TC}\colon U\ \rightarrow M_3\end{array}\right]_{Time} \quad TC\ \rightarrow x_{Quantum\ Particles}\end{array}\right\| \langle x_U | x_T \rangle$$

$$\Rightarrow$$

$$\langle Space - Time - Energy - Gravity\ Material - Fabric\ Pre \\ - Genetic\ Code\rangle +$$

$$(Electromagnetic\ Spectrum\ Material - Fabric\ Pre - Genetic\ Code) +$$

$$\left(\begin{array}{c} \sum Sig_{bosons} = Xa + \overline{Yb_{0-n}} \\ \left[\begin{array}{c} X \in [S_{System_N}] \\ where \quad Y \in [S_{System_{Pr}}, S_{System_P}, S_{System_K}, S_{System_N}] \\ a, b \; are \; integers; a > b \end{array} \right] \\ \sum Sig_{quarks} = Xa + \overline{Yb_{0-n}} \\ \left[\begin{array}{c} X \in [S_{System_K}] \\ where \quad Y \in [S_{System_{Pr}}, S_{System_P}, S_{System_K}, S_{System_N}] \\ a, b \; are \; integers; a > b \end{array} \right] \\ \sum Sig_{leptons} = Xa + \overline{Yb_{0-n}} \\ \left[\begin{array}{c} X \in [S_{System_P}] \\ where \quad Y \in [S_{System_{Pr}}, S_{System_P}, S_{System_K}, S_{System_N}] \\ a, b \; are \; integers; a > b \end{array} \right] \\ \sum Sig_{Higgs-boson} = Xa + \overline{Yb_{0-n}} \\ \left[\begin{array}{c} X \in [S_{System_{Pr}}] \\ where \quad Y \in [S_{System_{Pr}}, S_{System_P}, S_{System_K}, S_{System_N}] \\ a, b \; are \; integers; a > b \end{array} \right] \end{array} \right)$$

Eq. 6.2.7, Active Quantum Particle Partial-Singularity Material-Fabric Pre-Genetic Code Segments

The '$\sum x$' signifies all the possibilities for the 'x' equation-segment and depicts the growing biography by which the singular light-based edifice expresses its materialization, and at this stage, through the quantum particle partial-singularity.

Chapter 6.3: Generation of Boson Partial-Singularity

The bosons as mentioned can be thought of as force-carriers and allow all known matter particles to interact.

But when we look at bosons in more detail there are three fundamental bosons – the photon, the W and Z bosons, the gluon - and one hypothetical boson – the graviton.

This chapter looks at the on-going computation involving a process of quantization by which the very logic of bosons precipitates into the material-fabric hence generating the boson partial-singularity.

As in the previous Section, the Light-Space-Time Emergence equation (3.1.3) can be used to model emergence as it proceeds to this relatively more complex form of fourfold manifestation. (3.6.5), the Potential Effect of Levels of Light on Genetic-Type Information equation, basically sheds light on specific types of mutation that can occur and therefore on the nature of the code-segment that is being generated by (3.1.3). As discussed before, (3.6.5) shows a particular working of (3.1.3) but is wholly contained within it.

This chapter elaborates the action of (3.1.3) and (3.6.5) in the creation of the code-segments that define the boson ecosystem logic. While this development may precede the generation of quantum particle code, for simplicity it will be assumed that this happens at the same time.

Light's Emergence as Bosons

As discussed previously the Light-Space-Time Emergence equation (3.1.3) being iterative, can be used to model emergence as it proceeds from simpler four-fold to more complex four-fold manifestations. Hence (3.1.3) has already been applied to suggest the emergence of the space-time-energy-gravity quadrumvirate, the electromagnetic spectrum, and quantum particles in

257

general. Here it will be applied to suggest the emergence of a sub-class of quantum particles, Bosons. Hence as in (6.2.1) the starting point will be assumed to be the electromagnetic spectrum.

As suggested by Equation 6.3.1, Generation of Boson Partial-Singularity, the architecture and details of bosons can be seen to be the result of the application of the Light, Space, and Time matrices as will be elaborated:

$$Light - Space - Time\ Emergence_{Bosons} =$$

$$
\left\| \left\|
\begin{array}{c}
c_\infty: [Pr, Po, K, H] \\
\left(\downarrow R_{C_K} = f(R_{C_\infty}) \right) \\
c_K: [S_{Pr}, S_{Po}, S_K, S_H] \\
\left(\downarrow R_{C_N} = f(R_{C_K}) \right) \\
c_N: f(S_{Pr}\ x\ S_{Po}\ x\ S_K\ x\ S_H) \\
\left(\downarrow R_{C_U} = f(R_{C_N}) \right) \\
c_U: [P, V, M, C] \\
\Uparrow \\
c_{0:[D,W,I,C]}
\end{array}
\right]_{Light}
\left[
\begin{array}{c}
M_3\ \to\ System_X \\
(\uparrow F\ \to\ I) \\
M_2\ \to\ S_{System_X} \\
(\uparrow Sig\ \to\ F) \\
M_1\ \to\ Sig_X \\
(\uparrow > P_x) \\
U\ \to\ x_{EM\ Spectrum}
\end{array}
\right]_{Space}
\right.
$$

$$
\left.
\left[U\ \to
\begin{array}{c}
M_3 :\ -\infty \le t \le \infty \\
\downarrow \\
M_2 : 0 \ge t > \infty \\
\downarrow \\
M_1 : 0 > t > \infty \\
\downarrow \\
t \le E_{Cell}; TC: M_3 \to U \\
t \sim E_{Human}; TC: U \to M_3
\end{array}
\right]_{Time}
\quad TC\ \to\ x_{Bosons}
\right\|_{\langle x_U | x_T \rangle}
$$

$$\Rightarrow$$

$$\langle Space - Time - Energy - Gravity\ Material - Fabric\ Pre$$
$$- Genetic\ Code \rangle +$$

$$(Electromagnetic\ Spectrum\ Material - Fabric\ Pre - Genetic\ Code) +$$

$$FBLEEE < \cdots > + Boson\ Material - Fabric\ Pre - Genetic\ Code$$

Eq 6.3.1: Generation of Boson Partial-Singularity

Starting with the Light-Matrix, the top left-hand matrix in (6.3.1), the first line from the top, $C_\infty: [Pr, Po, K, H]$, specifies the fundamental architecture of bosons. As explored previously in this chapter, Gluons are an emergence of Light's property of Knowledge, W and Z bosons are an emergence of Light's property of Power, Photons are an emergence of Light's property of Presence, and the Graviton is an emergence of Light's property of Harmony. The fundamental architecture of these aspects hence, is an emergence of the properties of Light at ∞.

Line 3 in the Light-Matrix,

$C_K: [S_{Pr}, S_{Po}, S_K, S_H]$,
elaborates the sets for Presence, Power, Knowledge, and Harmony, each containing multiple elements. For example, as will be explored in the section on photons, various elements derived from the four sets define the behavior of photons and could be functions such as 'pervasiveness', 'multiple wave handler', amongst others, hence collectively describing photons' way of being. Specifically, Line 5, $C_N: f(S_{Pr} \times S_{Po} \times S_K \times S_H)$, suggests that unique seeds are created from a combination of such elements from all four sets, with a particular element leading or having more weight, that in effect creates the distinctness possible at the level of bosons.

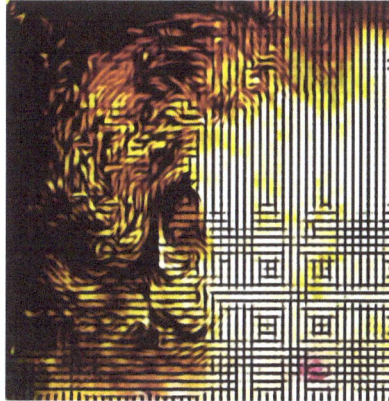

Line 6, $(\downarrow R_{C_U} = f(R_{C_N}))$, specifies quantization between the layer where the seeds are formed, and the physical layer, and as explored in Chapter 3.5 and 3.6, will result in Line 7, $C_U: [P, V, M, C]$, hence changing the material-fabric of existence. Note that as in the process describing the generation of the code-segments for space-time-energy-gravity quantization at the time of the Big Bang, the generation of the FBLEE (four-base logic-encoding ecosystem) code-segment as specified by (4.1.5) and the MF (material-fabric) code segment as specified by (4.1.6) are combined together into one equation so that the FBLEE process is apparently transparent. Note also that in (6.2.1), following the '$\Rightarrow$' the FBLEEE <...> code-segment implies that four-base logic-encoding entanglement likely with the generation of the other quantum-particle FBLEE code-segments as specified by (6.3.1).

The possibilities represented by Lines 1 through 5 hence concretize through the quantization represented by Line 6 to become the bosons with its physical (related to Presence), vital (related to Power), mental (related to Knowledge), and connection (related to Harmony) aspects now existing in material reality typified by Light moving at c. Note that just as Line 6 represents a process of quantization relating the layer of reality created by Light traveling at c with the antecedent layers, so too Lines 2 and 4 as previously discussed, also represent quantization of a more subtle kind that ultimately plays a critical part in allowing the material-fabric to express infinite diversity.

Typically it is the process as captured by the Space-Matrix that will determine if Line 6 is activated. Specifically patterns at the untransformed layer, U, will need to be overcome, as specified by the second-line from the bottom of the Space-Matrix: ($\uparrow > P_x$). But as specified by the bottom-line of the Time-Matrix, reproduced below, it is only with the advent of the human-system that the automaticity of the action of meta-levels is reversed:

$$U \rightarrow \begin{array}{l} t \leq E_{Cell}; \text{TC: } M_3 \rightarrow U \\ t \sim E_{Human}; \text{TC: } U \rightarrow M_3 \end{array}$$

260

Hence in the case of bosons, which in this emergence is a pre-human system, the fact that patterns do not need to be overcome means that quantization happens automatically.

Even though bosons are a pre-human system and therefore the action of meta-levels are modeled as being automatic, it is useful to review (3.6.5), Potential Effect of Levels of Light on Genetic-Type Information, to understand how the relationship with different forms of mutation has been specified:

$$
\begin{array}{c}
Potential\ Effect\ of\ Levels\ of\ Light\ on\ Genetic \\
-\ Type\ Information\ = \\
\begin{bmatrix}
STATIC\ \langle |[L][S][T]TC \rightarrow x_T|_{\langle x_U|x_T\rangle}\rangle \\
\times \\
\left((Y > U\colon Z_Q) \vee (Y \leq U\colon Z_F) \vee (Y = U\colon Z_R) \right) \\
\ni \\
\left(\begin{array}{c} Z \in \mathbb{U}\ (Space, Time, Energy, Gravity) \\ Q\colon Quantization;\ F\colon Fragmentation;\ R\colon Random \end{array} \right)
\end{bmatrix} \rightarrow h \ni \\
h \in \left(\begin{array}{c} [Q]\colon Constructive\ zone, \\ [Q]\colon Constructive\ zone \wedge Constructive\ mutation, \\ [F]\colon Destructive\ mutation, \\ [R]\colon Random\ mutation \end{array} \right)
\end{array}
$$

Line 1 from the top in the matrix is simply a static form of (3.6.1) the Simplified Light-Space-Time Emergence equation. The static form is designated by 'STATIC', and implies that fundamental operations true of (3.6.1) are being highlighted in (3.6.5). In other words (3.6.1) already has all the operations highlighted in (3.6.5) in it, but by 'freezing' it by making it static, the essential dynamics leading to possible mutations at the genetic level can more clearly be highlighted.

Line 1 is then subjected ($\times$) to a determination of the dominant levels of light that may be active, designated by Line 2, ' $\left((Y > U\colon Z_Q) \vee (Y \leq U\colon Z_F) \vee (Y = U\colon Z_R) \right)$ '. Unpacking this, '$Y > U$' implies meta-levels are active and as a result it is possible that Z_Q is going to take place (the subscript 'Q' implies quantum-level action). This also implies activation and potential change of FBLEE. The call from below, as it were, may invoke some function that already exists in the subtle-libraries 'above', so that some already existing function may influence FBLEE through ∞ -entanglement, K-entanglement, or N-entanglement. This may be thought of as a key-and-lock mechanism, where a deep enough visceral urge from below acting as the key, opens an entangled lock to alter FBLEE as per the visceral urge. 'Y $\leq$ U' implies that only the untransformed levels are active, and therefore also the sub-level where the speed of light is 0 is active and as a result Z_F is going to take place (the subscript 'F' implies 'fragmentation'). 'Y $=$ U' implies that all levels are active and as a result Z_R is going to take place (the subscript 'R' implies 'random').

Line 3 elaborates the significance of Z_Q, Z_F, and Z_R. Hence Z is the union of potential quantum-operations of space, time, energy, and gravity, designated by '$Z \in \mathbb{U}$ ($Space, Time, Energy, Gravity$)'. But the nature of the operations, as suggested in the previous paragraph, is designated   by '$Q: Quantization; F: Fragmentation; R: Random$'. Z_Q, then, implies that the full quantization originating from updated four-base logic-encoding ecosystems (FBLEE) can take place, and will result in lasting material change at the Genetic-Type level. Z_F implies that the essential set will be fragmented and that only libraries at the level of local cellular-level DNA or precipitated material-fabric logic can potentially be altered. Z_R also precludes full quantization, and that some partial local-library constructive or destructive mutation may take place.

The '$\rightarrow h \ni h \in$' segment resolves the outcome of the operations implied by Lines 1 – 3, suggesting that the outcome will be 'h' such that ($\ni$) 'h' is an element ($\in$) of the set specified by the members '[Q]: $Constructive\ zone$',

'[Q}: *Constructive zone AND Constructive mutation*', '[F};
Destructive mutation', and '{R}: *Random mutation*'. '[Q]:
Constructive zone' implies that the in-built buffer has been
crossed and that access to the deeper four-base logic-
encoding ecosystem (FBLEE) has been granted. '[Q}' in
this segment implies that there is the possibility that full-
quantization as specified by Line 3 of the previous matrix
can take place. Access to this zone is a prerequisite for
constructive mutation to occur, as designated by the
element   '   [Q}: *Constructive zone* ∩
Constructive mutation ', which implies that full-
quantization is going to take place and will result in
material change. The '[F]' specifies the relationship
between 'Fragmentation' in Line 3 of the previous matrix
and destructive mutation. The '[R]' specifies the
relationship between 'Random' in Line 3 of the previous
matrix and random mutation.

But as just summarized in the Time-Matrix in (6.3.1) Y is
by definition greater than U and hence quantization is
automatic. In terms of bosons such quantization implies
that wholeness becomes fully active through specific
space, time, energy, and gravity quantization to create an
holistic "ecosystem" with its own "boson logic" as it
were. The wholeness has now precipitated into the
material-fabric and is available to be consciously and
unconsciously tapped into. This "logic" or more
specifically pre-genetic code is elaborated in the
following sub-sections.

Generation of Boson Partial-Singularity 'Photon' Material-Fabric Pre-Genetic Code

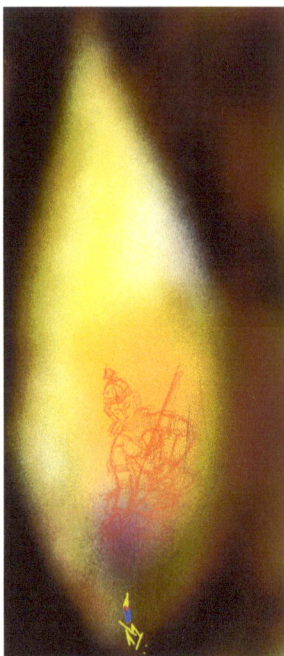

The photon is the carrier particle of the electromagnetic force. But in the scheme of things the electromagnetic force pervades everything and as explored in the section on the electromagnetic spectrum appears to be foundational to this reality we are in. So it could be said that it is related to Presence or is an emergence of Light's property of Presence at the level of quantum force-carriers.

Hence, an equation for the photon, Equation 6.3.2 could be:

$$Sig_{photon} = Xa + \overline{Yb_{0-n}} \ where \ \left[\begin{array}{c} X \in [S_{System_{Pr}}] \\ Y \in [S_{System_{Pr}}, S_{System_{P}}, S_{System_{K}}, S_{System_{N}}] \\ a, b \ are \ integers; a > b \end{array} \right]$$

Eq 6.3.2: Signature of Photon

While the primary element X, would continue to be the same as X for bosons, as in Equation 6.1.4, there will be additional secondary elements Y that will further qualify the attributes or functionality of photons. As an example such elements could be 'pervasiveness', 'multiple wave handler', amongst others.

Note that (6.3.2) already implies that Lines 1 – 5 in the Light Matrix (6.3.1) have been activated, and that the logic of the photon-ecosystem will automatically precipitate

264

into the material-fabric through the action of Line 6-7 of (6.3.1). This logic is none other than the pre-genetic code as specified by (6.3.2).

Generation of Boson Partial-Singularity 'Gluon' Material-Fabric Pre-Genetic Code

The gluon is the carrier particle of what is known as the strong nuclear force and holds quarks together in their inherently composite arrangements. But we have posited that quarks are related to or are an emergence of Knowledge at the quantum-particle level. Hence it must be that the gluon is an emergence of Light's property of Knowledge at the level of quantum force-carriers.

Hence, Equation 6.3.3 is an equation for the gluon:

$$Sig_{gluon} = Xa +$$
$$\overline{Yb_{0-n}} \quad where \left[\begin{array}{c} X \in [S_{System_K}] \\ Y \in [S_{System_{Pr}}, S_{System_P}, S_{System_K}, S_{System_N}] \\ a, b \ are \ integers; a > b \end{array} \right]$$

Eq 6.3.3: Signature of Gluon

The primary element X would continue to be the same as in the equation for the boson. There are likely additional secondary elements Y such as 'concentrated', 'intense connection', amongst others that would collectively determine the behavior of gluons.

Note that (6.3.3) already implies that Lines 1 – 5 in the Light Matrix (6.3.1) have been activated, and that the logic of the gluon-ecosystem will automatically precipitate into the material-fabric through the action of Line 6-7 of (6.3.1). This logic is none other than the pre-genetic code as specified by (6.3.3).

Generation of Boson Partial-Singularity 'W and Z Bosons' Material-Fabric Pre-Genetic Code

The W and Z bosons are the carrier particle for the weak interactions, responsible for the decay of massive quarks and leptons into lighter quarks and leptons. But this usually is accompanied by the release of energy and power and so it must be that W and Z bosons are an emergence of Light's property of Power at the level of quantum force-carriers.

Hence, Equation 6.3.4, for W and Z bosons is as follows:

$$Sig_{W,Z\ bosons} = Xa +$$
$$\overline{Yb_{0-n}} \quad where \quad \begin{bmatrix} X \in [S_{System_P}] \\ Y \in [S_{System_{Pr}}, S_{System_P}, S_{System_K}, S_{System_N}] \\ a, b\ are\ integers; a > b \end{bmatrix}$$

Eq 6.3.4: Signature of W and Z Bosons

The primary element X would continue to be the same as in the equation for the boson. There are likely additional secondary elements Y such as 'weak link', 'release of energy', amongst others that would collectively determine the behavior of W and Z bosons.

Note that (6.3.4) already implies that Lines 1 – 5 in the Light Matrix (6.3.1) have been activated, and that the logic of the W and Z-bosons-ecosystem will automatically precipitate into the material-fabric through the action of Line 6-7 of (6.3.1). This logic is none other than the pre-genetic code as specified by (6.3.4).

Generation of Boson Partial-Singularity 'Graviton' Material-Fabric Pre-Genetic Code

This leaves the hypothetical graviton that is thought to be the carrier particle for the force of gravity. But gravity, it is thought, is what holds astronomical objects together in relationship. Hence it must be that gravitons are an emergence of Light's property of Harmony at the level of quantum force-carriers.

Hence, Equation 6.3.5, for the hypothetical graviton is as follows:

$$Sig_{graviton} = Xa +$$

$$\overline{Yb_{0-n}} \quad where \quad \begin{bmatrix} X \in [S_{System_N}] \\ Y \in [S_{System_{Pr}}, S_{System_P}, S_{System_K}, S_{System_N}] \\ a, b \ are \ integers; a > b \end{bmatrix}$$

Eq 6.3.5: Signature of Graviton

The primary element X would continue to be the same as in the equation for the boson. There are likely additional secondary elements Y such as 'mass curving space', 'space defining object movement', amongst others that would collectively determine the behavior of gravitons.

Note that (6.3.5) already implies that Lines 1 – 5 in the Light Matrix (6.3.1) have been activated, and that the logic of the graviton-ecosystem will automatically precipitate into the material-fabric through the action of Line 6-7 of (6.3.1). This logic is none other than the pre-genetic code as specified by (6.3.5).

So we see that the underlying properties of light - Presence, Knowledge, Power, and Harmony - emerge as photons, gluons, W and Z bosons, and the hypothetical graviton at the quantum force-carrier level.

Summary of Boson Partial-Singularity Material-Fabric Pre-Genetic Code

Summarizing, after the boson stage iteration of the Light-Space-Time Emergence equation is complete, the following code-segments will have been generated as specified by Equation 6.3.6, Active Boson Partial-Singularity Material-Fabric Pre-Genetic Code Segments:

$$Light - Space - Time \ Emergence_{Boson} =$$

$$\left[\begin{array}{c} \begin{bmatrix} c_\infty : [Pr, Po, K, H] \\ \left(\downarrow R_{C_K} = f\left(R_{C_\infty}\right) \right) \\ c_K : [S_{Pr}, S_{Po}, S_K, S_H] \\ \left(\downarrow R_{C_N} = f\left(R_{C_K}\right) \right) \\ c_N : f(S_{Pr} \times S_{Po} \times S_K \times S_H) \\ \left(\downarrow R_{C_U} = f\left(R_{C_N}\right) \right) \\ c_U : [P, V, M, C] \\ \Uparrow \\ c_{0:[D,W,I,C]} \end{bmatrix}_{Light} \quad \begin{bmatrix} M_3 \rightarrow System_X \\ (\uparrow F \rightarrow I) \\ M_2 \rightarrow S_{System_X} \\ (\uparrow Sig \rightarrow F) \\ M_1 \rightarrow Sig_x \\ (\uparrow > P_x) \\ U \rightarrow x_{EM \; Spectrum} \end{bmatrix}_{Space} \\[2em] \begin{bmatrix} M_3 : -\infty \le t \le \infty \\ \downarrow \\ M_2 : 0 \ge t > \infty \\ \downarrow \\ M_1 : 0 > t > \infty \\ \downarrow \\ U \rightarrow \begin{array}{l} t \le E_{Cell}; TC: M_3 \rightarrow U \\ t \sim E_{Human}; TC: U \rightarrow M_3 \end{array} \end{bmatrix}_{Time} \quad TC \rightarrow x_{Bosons} \end{array}\right]_{\langle x_U | x_T \rangle}$$

$\Rightarrow$

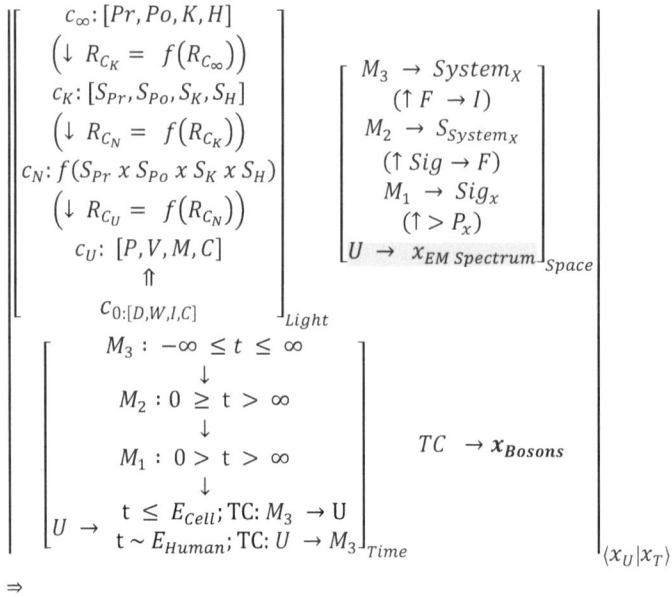

$\langle Space - Time - Energy - Gravity \; Material - Fabric \; Pre - Genetic \; Code \rangle +$

$(ELectromagnetic \; Spectrum \; Material - Fabric \; Pre - Genetic \; Code)$
$+$

$FBLEEE < \cdots > +$

$$\left(\begin{array}{c} Sig_{graviton} = Xa + \overline{Yb_{0-n}} \\[4pt] where \begin{bmatrix} X \in [S_{System_N}] \\ Y \in [S_{System_{Pr}}, S_{System_P}, S_{System_K}, S_{System_N}] \\ a, b \text{ are integers}; a > b \end{bmatrix} \\[16pt] Sig_{gluon} = Xa + \overline{Yb_{0-n}} \\[4pt] where \begin{bmatrix} X \in [S_{System_K}] \\ Y \in [S_{System_{Pr}}, S_{System_P}, S_{System_K}, S_{System_N}] \\ a, b \text{ are integers}; a > b \end{bmatrix} \\[16pt] Sig_{W,Z\ bosons} = Xa + \overline{Yb_{0-n}} \\[4pt] where \begin{bmatrix} X \in [S_{System_P}] \\ Y \in [\square\square_{System_{Pr}}, S_{System_P}, S_{System_K}, S_{System_N}] \\ a, b \text{ are integers}; a > b \end{bmatrix} \\[16pt] Sig_{photon} = Xa + \overline{Yb_{0-n}} \\[4pt] where \begin{bmatrix} X \in [S_{System_{Pr}}] \\ Y \in [S_{System_{Pr}}, S_{System_P}, S_{System_K}, S_{System_N}] \\ a, b \text{ are integers}; a > b \end{bmatrix} \end{array}\right)$$

Eq. 6.3.6, Active Boson Partial-Singularity Material-Fabric Pre-Genetic Code Segments

The '$\sum x$' signifies all the possibilities for the 'x' equation-segment and depicts the growing biography by which the singular light-based edifice expresses its materialization, and at this stage, through the boson partial-singularity.

Chapter 6.4: Generation of Atom Partial-Singularity

Quantum particles in turn create atoms, and all known atoms can be classified into one of four groups: The s-Group, p-Group, d-Group, and f-Group. But what are these groups and are they too emergences of Light as it continues its journey of crystalizing the possibility or potentiality or fourfold functionality within it?

As in the previous chapters, the Light-Space-Time Emergence equation (3.1.3) can be used to model emergence as it proceeds to this relatively more complex form of fourfold manifestation. (3.6.5), the Potential Effect of Levels of Light on Genetic-Type Information equation, basically sheds light on specific types of mutation that can occur and therefore on the nature of the code-segment that is being generated by (3.1.3). As discussed before, (3.6.5) shows a particular working of (3.1.3) but is wholly contained within it.

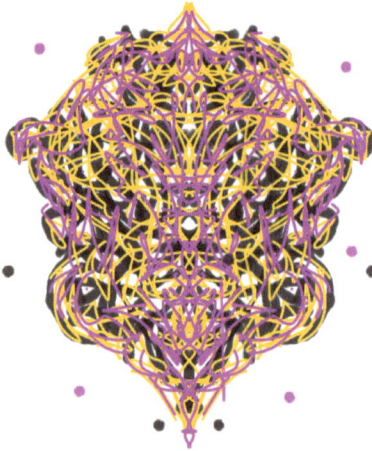

This chapter elaborates the action of (3.1.3) and (3.6.5) in the creation of the code-segments that define the logic of the atom ecosystem thereby generating the atom partial-singularity.

Light's Emergence as Atoms

As discussed previously the Light-Space-Time Emergence equation (3.1.3) being iterative, can be used to model emergence as it proceeds from simpler four-fold to more complex four-fold manifestations. Hence (3.1.3) has already been applied to suggest the emergence of the space-time-energy-gravity quadrumvirate, the electromagnetic spectrum, quantum particles in general, and bosons as a further instance of a particular kind of quantum particle. Here it will be applied to suggest the emergence of Atoms. But further, as implied by (3.6.5), the Potential Effect of Levels of Light on Genetic-Type Information equation, any process of organization, such as the architecture and cohesiveness of atoms, has the possibility of altering the material-fabric so long as the bases involved are driven primarily by a meta-level.

As suggested by Equation 6.4.1 (same as 4.2.2), Generation of Atom Partial-Singularity, the architecture and details of atoms can be seen to be the result of the application of the Light, Space, and Time matrices as will be elaborated:

$Light - Space - Time\ Emergence_{Atoms} =$

$$\left[\begin{array}{l} \begin{bmatrix} c_\infty: [Pr, Po, K, H] \\ \left(\downarrow R_{C_K} = f(R_{C_\infty})\right) \\ c_K: [S_{Pr}, S_{Po}, S_K, S_H] \\ \left(\downarrow R_{C_N} = f(R_{C_K})\right) \\ c_N: f(S_{Pr} \times S_{Po} \times S_K \times S_H) \\ \left(\downarrow R_{C_U} = f(R_{C_N})\right) \\ c_U: [P, V, M, C] \\ \Uparrow \\ c_{0:[D,W,I,C]} \end{bmatrix}_{Light} \quad \begin{bmatrix} M_3 \to System_X \\ (\uparrow F \to I) \\ M_2 \to S_{System_X} \\ (\uparrow Sig \to F) \\ M_1 \to Sig_X \\ (\uparrow > P_x) \\ U \to x_{Quantum\ Particles} \end{bmatrix}_{Space} \\ \\ U \to \begin{bmatrix} M_3: -\infty \le t \le \infty \\ \downarrow \\ M_2: 0 \ge t > \infty \\ \downarrow \\ M_1: 0 > t > \infty \\ \downarrow \\ t \le E_{Cell};\ TC: M_3 \to U \\ t \sim E_{Human};\ TC: U \to M_3 \end{bmatrix}_{Time} \quad TC \to x_{Atoms} \end{array}\right]_{\langle x_U | x_T \rangle}$$

$\Rightarrow$

$\langle Space - Time - Energy - Gravity\ Material - Fabric\ Pre - Genetic\ Code \rangle +$

$(Electromagnetic\ Spectrum\ Material - Fabric\ Pre - Genetic\ Code) +$

$(Quantum\ Particles\ Material - Fabric\ Pre - Genetic\ Code) +$

$Atoms\ Material - Fabric\ Pre - Genetic\ Code$

Eq 6.4.1: Generation of Atom Partial-Singularity

Starting with the Light-Matrix, the top left-hand matrix in (6.4.1), the first line from the top, $c_\infty: [Pr, Po, K, H]$, specifies the fundamental architecture of atoms. As will be explored in this chapter, p-Group atoms are an emergence of Light's property of Knowledge, s-Group atoms are an emergence of Light's property of Power, d-Group atoms are an emergence of Light's property of Presence, and the f-Group atoms are an emergence of Light's property of Harmony. The fundamental architecture of these aspects hence, is an emergence of the properties of Light at ∞.

Line 3 in the Light-Matrix, C_K: $[S_{Pr}, S_{Po}, S_K, S_H]$, elaborates the sets for Presence, Power, Knowledge, and Harmony, each containing multiple elements. For example, as explored shortly in the section on the s-Group, various elements derived from the four sets define the behavior of s-Group atoms and could be functions such as 'power', 'energy', 'adventure', and 'courage', amongst others, hence collectively describing s-Groups atoms' way of being. Specifically, Line 5, C_N: $f(S_{Pr} \times S_{Po} \times S_K \times S_H)$, suggests that unique seeds are created from a combination of such elements from all four sets, with a particular element leading or having more weight, that in effect creates the distinctness possible at the level of atoms.

Line 6, $(\downarrow R_{C_U} = f(R_{C_N}))$, specifies quantization between the layer where the seeds are formed, and the physical layer, and as explored in Chapter 3.5 and 3.6, will result in Line 7, C_U: $[P, V, M, C]$, hence changing the material-fabric of existence. Note that as in the process describing the generation of the code-segments for space-time-energy-gravity quantization at the time of the Big Bang, the generation of the FBLEE (four-base logic-encoding ecosystem) code-segment as specified by (4.1.5) and the MF (material-fabric) code segment as specified by (4.1.6) are combined together into one equation so that the FBLEE process is apparently transparent.

The possibilities represented by Lines 1 through 5 hence concretize through the quantization represented by Line 6 to become the atoms with its physical (related to Presence), vital (related to Power), mental (related to Knowledge), and connection (related to Harmony) aspects now existing in material reality typified by Light moving at c. Note that just as Line 6 represents a process of quantization relating the layer of reality created by

Light traveling at c with the antecedent layers, so too Lines 2 and 4 as previously discussed, also represent quantization of a more subtle kind that ultimately plays a critical part in allowing the material-fabric to express infinite diversity.

Typically it is the process as captured by the Space-Matrix that will determine if Line 6 is activated. Specifically patterns at the untransformed layer, U, will need to be overcome, as specified by the second-line from the bottom of the Space-Matrix: ($\uparrow > P_x$). But as specified by the bottom-line of the Time-Matrix, reproduced below, it is only with the advent of the human-system that the automaticity of the action of meta-levels is reversed:

$$U \to \begin{matrix} t \leq E_{Cell}; \text{TC}: M_3 \to U \\ t \sim E_{Human}; \text{TC}: U \to M_3 \end{matrix}$$

Hence in the case of atoms, which in this emergence is a pre-human system, the fact that patterns do not need to be overcome means that quantization happens automatically.

Even though atoms are a pre-human system and therefore the action of meta-levels are modeled as being automatic, it is nonetheless useful to review (3.6.5), Potential Effect of Levels of Light on Genetic-Type Information, to understand how the relationship with different forms of mutation has been specified:

$$Potential\ Effect\ of\ Levels\ of\ Light\ on\ Genetic$$
$$-\ Type\ Information =$$
$$\begin{bmatrix} STATIC\ \langle |[L][S][T]TC \to x_T|_{\langle x_U | x_T \rangle} \rangle \\ \times \\ \left((Y > U: Z_Q) \vee (Y \leq U: Z_F) \vee (Y = U: Z_R) \right) \\ \ni \\ \left(\begin{matrix} Z \in \mathbb{U}\ (Space, Time, Energy, Gravity) \\ Q: Quantization; F: Fragmentation; R: Random \end{matrix} \right) \end{bmatrix} \to h \ni$$

274

$$h \in \begin{pmatrix} [Q]: \textit{Constructive zone,} \\ [Q]: \textit{Constructive zone} \wedge \textit{Constructive mutation,} \\ [F]: \textit{Destructive mutation,} \\ [R]: \textit{Random mutation} \end{pmatrix}$$

Line 1 from the top in the matrix is simply a static form of (3.6.1) the Simplified Light-Space-Time Emergence equation. The static form is designated by 'STATIC', and implies that fundamental operations true of (3.6.1) are being highlighted in (3.6.5). In other words (3.6.1) already has all the operations highlighted in (3.6.5) in it, but by 'freezing' it by making it static, the essential dynamics leading to possible mutations at the genetic level can more clearly be highlighted.

Line 1 is then subjected ($\times$) to a determination of the dominant levels of light that may be active, designated by Line 2, ' $\Big((Y > U: Z_Q) \vee (Y \leq U: Z_F) \vee (Y = U: Z_R) \Big)$ '. Unpacking this, 'Y > U' implies meta-levels are active and as a result it is possible that Z_Q is going to take place (the subscript 'Q' implies quantum-level action). This also implies activation and potential change of FBLEE. The call from below, as it were, may invoke some function that already exists in the subtle-libraries 'above', so that some already existing function may influence FBLEE through ∞ -entanglement, K-entanglement, or N-entanglement. This may be thought of as a key-and-lock mechanism, where a deep enough visceral urge from below acting as the key, opens an entangled lock to alter FBLEE as per the visceral urge. 'Y $\leq$ U' implies that only the untransformed levels are active, and therefore also the sub-level where the speed of light is 0 is active and as a result Z_F is going to take place (the subscript 'F' implies 'fragmentation'). 'Y = U' implies that all levels are active and as a result Z_R is going to take place (the subscript 'R' implies 'random').

Line 3 elaborates the significance of Z_Q, Z_F, and Z_R. Hence Z is the union of potential quantum-operations of

275

space, time, energy, and gravity, designated by '$Z \in \mathbb{U} \, (Space, Time, Energy, Gravity)$'. But the nature of the operations, as suggested in the previous paragraph, is designated by 'Q: $Quantization$; F: $Fragmentation$; R: $Random$'. Z_Q, then, implies that the full quantization originating from updated four-base logic-encoding ecosystems (FBLEE) can take place, and will result in lasting material change at the Genetic-Type level. Z_F implies that the essential set will be fragmented and that only libraries at the level of local cellular-level DNA or precipitated material-fabric logic can potentially be altered. Z_R also precludes full quantization, and that some partial local-library constructive or destructive mutation may take place.

The '$\rightarrow h \ni h \in$' segment resolves the outcome of the operations implied by Lines 1 – 3, suggesting that the outcome will be 'h' such that ($\ni$) 'h' is an element ($\in$) of the set specified by the members '[Q]: $Constructive\ zone$', '[Q]: $Constructive\ zone\ AND\ Constructive\ mutation$', '[F]; $Destructive\ mutation$', and '{R}: $Random\ mutation$'. '[Q]: $Constructive\ zone$' implies that the in-built buffer has been crossed and that access to the deeper four-base logic-encoding ecosystem (FBLEE) has been granted. '[Q]' in this segment implies that there is the possibility that full-quantization as specified by Line 3 of the previous matrix can take place. Access to this zone is a prerequisite for constructive mutation to occur, as designated by the element '[Q]: $Constructive\ zone \cap Constructive\ mutation$', which implies that full-quantization is going to take place and will result in material change. The '[F]' specifies the relationship between 'Fragmentation' in Line 3 of the previous matrix and destructive mutation. The '[R]' specifies the relationship between 'Random' in Line 3 of the previous matrix and random mutation.

But as just summarized in the Time-Matrix in (6.4.1) Y is by definition greater than U and hence quantization is

276

automatic. In terms of atoms such quantization implies that wholeness becomes fully active through specific space, time, energy, and gravity quantization to create an holistic "ecosystem" with its own "atoms logic" as it were. The wholeness has now precipitated into the material-fabric and is available to be consciously and unconsciously tapped into. This "logic" or more specifically pre-genetic code is elaborated in the following sub-sections and suggests the increasing sphere of influence of the atom partial-singularity that now also contains light-engendered laws for all atoms.

Generation of Atom Partial-Singularity 's-Group' Material-Fabric Pre-Genetic Code

The s-Group consists primarily of what are known as

alkali metals and alkali earth metals. These alkali metals are known to easily lose electrons and form what is known as positive ions. When they lose electrons energy is gained, but when the electrons are taken up by other atoms in proximity there is a lot of energy released. Some have referred to these groups as "violent worlds" (Tweed, 2003), and it has been pointed out that stars shine because they are transmuting vast amount of hydrogen into helium, both of which are s-Group elements. So one gets the sense that the s-Group may be an emergence of Light's property of Power.

Stars and suns are also known to be the crucibles where all the different kinds of atoms are created. So these furnaces of power by virtue of their heat and high pressure are able to force electrons and protons and neutrons to come together to create all the different types of atoms known in the universe.

But philosophically what are the s-Group atoms or elements? These are atoms where there is an equal likelihood of an electron being anywhere in a symmetrical sphere around the nucleus. The other groups are all similarly defined by likelihoods of electrons being within a possible pattern around the nucleus. The patterns that distinguishes s-Group atoms is a sphere, and since all other patterns can be thought of as occurring within some sphere, in some sense this is like an imprint or precipitation or emergence that allows other kinds of emergences to surface within it. So the elements that are part of the s-Group may, in a sense, be thought of as the adventurers with courage who venture into a brave new world to create some foundation by which all other element-creations can follow.

The fact that hydrogen and helium are known to constitute 98% of the Universe (Heiserman, 1991) relative to other elements therefore makes sense in this view, especially since hydrogen and helium provide the fuel with which the star-furnaces manufacture all other elements.

So s-Group elements seem to embody functions such as power, energy, adventure, courage, and can be thought of as an emergence of the property of Light to do with Power.

Hence, a series of equations linked to S_{System_P} as the prime set can be suggested, starting with the s-Group mapping, as in Equation 6.4.2:

$$Element\ _{s-Group}\ =\ Xa\ +$$

$$\overline{Yb_{0-n}}\ \ where\ \left[\begin{matrix} X\ \in [S_{System_P}] \\ Y\ \in [S_{System_{Pr}}, S_{System_P}, S_{System_K}, S_{System_N}] \\ a,b\ are\ integers;\ a > b \end{matrix}\right]$$

Eq 6.4.2: s-Group Element

Note that (6.4.2) already implies that Lines 1 – 5 in the Light Matrix (6.4.1) have been activated, and that the logic of the s-Group-ecosystem will automatically precipitate into the material-fabric through the action of Line 6-7 of (6.4.1). This logic is none other than the pre-genetic code as specified by (6.4.2).

Further, the equivalent mapping between traditional element groupings and S_{System_P} as in Equations 6.4.3 and 6.4.4 can be specified:

$$Element\ _{Alkali\ metal}\ =$$

$$Xa\ +$$

$$\overline{Yb_{0-n}}\ \ where\ \left[\begin{matrix} X\ \in [S_{System_P}] \\ Y\ \in [S_{System_{Pr}}, S_{System_P}, S_{System_K}, S_{System_N}] \\ a,b\ are\ integers;\ a > b \end{matrix}\right]$$

Eq 6.4.3: Alkali Metal Element

$$Element\ _{Alkali\ earth\ metal}\ =$$

$$Xa\ +$$

$$\overline{Yb_{0-n}}\ \ where\ \left[\begin{matrix} X\ \in [S_{System_P}] \\ Y\ \in [S_{System_{Pr}}, S_{System_P}, S_{System_K}, S_{System_N}] \\ a,b\ are\ integers;\ a > b \end{matrix}\right]$$

Eq 6.4.4: Alkali Earth Metal Element

Further, as a representative element belonging to the Alkali Metal group the equation for Lithium (Li), as in Equation 6.4.5, would be:

$$Element_{Lithium} = Xa +$$

$$\overline{Yb_{0-n}} \quad where \begin{bmatrix} X \in [S_{System_P}] \\ Y \in [S_{System_{Pr}}, S_{System_P}, S_{System_K}, S_{System_N}] \\ a, b \ are \ integers; a > b \end{bmatrix}$$

Eq 6.4.5: Lithium

Note that the primary element X in all these cases may be a function or attribute along the lines of 'expresses power'. The secondary elements Y in all these would have some elements being the same, and then would have specific differences as one gets into sub-classes and the individual elements themselves.

Generation of Atom Partial-Singularity 'p-Group' Material-Fabric Pre-Genetic Code

Atoms or elements belonging to the p-Group are those with the likelihood of electrons occurring equally on either side of the nucleus, like a dumbbell of sorts.

There are some very significant elements in this group that are part of the metal, metalloid, non-metal, halogen, and noble gas sub-groupings. Carbon, Nitrogen, Oxygen, and Silicon are some of the sample elements. Looking at the types of elements present in this group it is as though all the element possibilities have been represented within it. It is perhaps that the possibility of ideas behind all elements has emerged in this group and one can hypothesize that this group may be a reflection of the property of Knowledge, forming archetypes from which all other elements are created.

Philosophically, the one spherical shell or probability cloud (s) becoming two shells like a dumbbell (p) signifies the creation of an essential polarity within a unit space. So in a sense the one becoming two is the first instance of variability in space. The two, spaced in 3-dimensions around the nucleus in a sense create six switches becoming an attractor or allowing a vaster number of different kinds of elements to surface. So there is a sense of the 'idea' of the element that comes into focus.

But further, the essential elements that allow both thinking and virtual thinking machines to come into being, are also contained within this group. Carbon is the basis of DNA and of all life. The fact that Silicon, directly below it in the periodic table and therefore sharing essential qualities, is considered the basis of all virtual thinking machines is therefore perhaps significant and may reinforce the notion that the p-Group is a precipitation of Light's property of Knowledge.

Hence, a series of equations linked to S_{System_K} as the prime set can be suggested, starting with the p-Group mapping, as in Equation 6.4.6:

$$Element_{p-Group} = Xa + \overline{Yb_{0-n}} \quad where \begin{bmatrix} X \in [S_{System_K}] \\ Y \in [S_{System_{Pr}}, S_{System_P}, S_{System_K}, S_{System_N}] \\ a, b \ are \ integers; a > b \end{bmatrix}$$

Eq 6.4.6: p-Group Element

Note that (6.4.6) already implies that Lines 1 – 5 in the Light Matrix (6.4.1) have been activated, and that the logic of the p-Group-ecosystem will automatically precipitate into the material-fabric through the action of Line 6-7 of (6.4.1). This logic is none other than the pre-genetic code as specified by (6.4.6).

Further, the equivalent mapping between traditional element groupings and S_{System_K} as in Equations 6.4.7 through 6.4.12 can also be specified:

$$Element_{Metal} = Xa + \overline{Yb_{0-n}}$$

$$where \begin{bmatrix} X \in [S_{System_K}] \\ Y \in [S_{System_{Pr}}, S_{System_P}, S_{System_K}, S_{System_N}] \\ a, b \ are \ integers; a > b \end{bmatrix}$$

Eq 6.4.7: Metal Element

$$Element_{Metalloid} = Xa + \overline{Yb_{0-n}} \quad where \begin{bmatrix} X \in [S_{System_K}] \\ Y \in [S_{System_{Pr}}, S_{System_P}, S_{System_K}, S_{System_N}] \\ a, b \ are \ integers; a > b \end{bmatrix}$$

Eq 6.4.8: Metalloid Element

$$Element_{Non-Metal} = Xa +$$

$$\overline{Yb_{0-n}} \quad where \quad \begin{bmatrix} X \in [S_{System_K}] \\ Y \in [S_{System_{Pr}}, S_{System_P}, S_{System_K}, S_{System_N}] \\ a, b \ are \ integers; a > b \end{bmatrix}$$

Eq 6.4.9: Non-Metal Element

$$Element_{Halogen} = Xa +$$

$$\overline{Yb_{0-n}} \quad where \quad \begin{bmatrix} X \in [S_{System_K}] \\ Y \in [S_{System_{Pr}}, S_{System_P}, S_{System_K}, S_{System_N}] \\ a, b \ are \ integers; a > b \end{bmatrix}$$

Eq 6.4.10: Halogen Element

$$Element_{Noble\ Gas} = Xa +$$

$$\overline{Yb_{0-n}} \quad where \quad \begin{bmatrix} X \in [S_{System_K}] \\ Y \in [S_{System_{Pr}}, S_{System_P}, S_{System_K}, S_{System_N}] \\ a, b \ are \ integers; a > b \end{bmatrix}$$

Eq 6.4.11: Noble Gas Element

Further, as a representative element belonging to the Non-Metal group the equation for Carbon (C), as in Equation 6.4.12, would be:

$$Element_{Carbon} = Xa +$$

$$\overline{Yb_{0-n}} \quad where \quad \begin{bmatrix} X \in [S_{System_K}] \\ Y \in [S_{System_{Pr}}, S_{System_P}, S_{System_K}, S_{System_N}] \\ a, b \ are \ integers; a > b \end{bmatrix}$$

Eq 6.4.12: Carbon

Note that the primary element X in all these cases may be a function or attribute along the lines of 'knowledge'. The secondary elements Y in all these would have some elements being the same, and then would have specific differences as one gets into sub-classes and the individual elements themselves.

Generation of Atom Partial-Singularity 'd-Group' Material-Fabric Pre-Genetic Code

The d-Group comprises the Transition Metals. These metals are generally hard and strong, exhibit corrosive resistance, and can be thought of as workhorse elements. Many industrial and well-known elements sit in this group: Titanium, Chromium, Manganese, Iron, Cobalt, Nickel, Copper, Zinc, Silver, Platinum, and Gold, amongst others.

The elements and atoms in the d-Group are equally likely to show up in four possible lobes or probability-spaces around the nucleus. Four lobes occurring in multiple possible planes around the nucleus will likely create a space of stability, since there is a possibility of four lobes creating the four vertices of a tetrahedron (Fuller, 1982) that has been positioned as one of the most stable shapes in the universe.

Much of the constructed world around us is created from these elements. Further, most of the series in the group easily lose one or more electrons thereby easily combining with other atoms to form a vast array of compounds. Also, looking more broadly at the function of these elements, it can be seen that these metals exist for service, to help bring about perfection in the constructed world, to help much of the machinery in which they are used, and to assist the processes dependent on them to be completed with diligence. Hence, these transition metals appear to be an emergence of Light's property of Presence.

Therefore, a series of equations linked to $S_{System_{Pr}}$ as the prime set can be suggested, starting with the d-Group mapping, as in Equation 6.4.13:

$$Element_{d-Group} = Xa +$$

$$\overline{Yb_{0-n}} \quad where \left[\begin{array}{c} X \in [S_{System_{Pr}}] \\ Y \in [S_{System_{Pr}}, S_{System_P}, S_{System_K}, S_{System_N}] \\ a, b \ are \ integers; a > b \end{array} \right]$$

Eq 6.4.13: d-Group Element

Note that (6.4.13) already implies that Lines 1 – 5 in the Light Matrix (6.4.1) have been activated, and that the logic of the d-Group-ecosystem will automatically precipitate into the material-fabric through the action of Line 6-7 of (6.4.1). This logic is none other than the pre-genetic code as specified by (6.4.13).

Further, the equivalent mapping between traditional element groupings and $S_{System_{Pr}}$, as in Equation 6.4.14, can also be specified:

$$Element_{Transition\ Metal} =$$

$$Xa +$$

$$\overline{Yb_{0-n}} \quad where \left[\begin{array}{c} X \in [S_{System_{Pr}}] \\ Y \in [S_{System_{Pr}}, S_{System_P}, S_{System_K}, S_{System_N}] \\ a, b \ are \ integers; a > b \end{array} \right]$$

Eq 6.4.14: Transition Metal Element

Further, as a representative element belonging to the Transition-Metal group the equation for Gold (Au), as in Equation 6.4.15, would be:

$$Element_{Gold} = Xa +$$

$$\overline{Yb_{0-n}} \quad where \left[\begin{array}{c} X \in [S_{System_{Pr}}] \\ Y \in [S_{System_{Pr}}, S_{System_P}, S_{System_K}, S_{System_N}] \\ a, b \ are \ integers; a > b \end{array} \right]$$

Eq 6.4.15: Gold

Note that the primary element X in all these cases may be a function or attribute along the lines of 'work horse'. The secondary elements Y in all these would have some elements being the same, and then would have specific differences as one gets into sub-classes and the individual elements themselves.

Generation of Atom Partial-Singularity 'f-Group' Material-Fabric Pre-Genetic Code

The f-Group comprises of the Lanthanides and Actinides. Philosophically, elements in the f-Group consist of 6 lobes around the nucleus within which an electron may be found. 6 lobes will exist in multiple planes around the nucleus and suggests the notion of extended relationship and collectivity: the attempt to build larger and larger bonds within a small space. Considering this it is likely that the f-Group is an emergence of Light's property of Harmony.

Thinking about Lanthanides, some interesting facts may reinforce this notion. First, the spin of electrons in a lanthanides' outer shell is aligned, creating a strong magnetic field. The notion of creating a strong magnetic field seems to be consistent with the notion of engendering a collectivity through the ordered attraction and repulsion of elements. Second, these elements curiously occur together in nature often in the same ores (Gray, 2009) and are chemically interchangeable also suggesting the notion of forming a tight intra-group collectivity.

Actinides on the other hand are inherently radioactive. This implies that these elements have inherently crossed a threshold of stability and have the urge, over their own half-lives, to decompose into other elements. This natural urge may suggest some boundary conditions on the notion of collectivity and nurturing, giving additional insight into the nature of collectivity and nurturing. Further, the entire actinide group, as opposed to the lanthanide group that is inherently stable, is unstable. It is curious that both these should be part of the f-Group, and they must provide insight into boundary conditions into the notion of collectivity in elements.

Hence, a series of equations linked to S_{System_N} as the prime set can be suggested, starting with the f-Group mapping, as in Equation 6.4.16:

$$Element_{f-Group} = Xa +$$
$$\overline{Yb_{0-n}} \ \ where \ \begin{bmatrix} X \in [S_{System_N}] \\ Y \in [S_{System_{Pr}}, S_{System_P}, S_{System_K}, S_{System_N}] \\ a, b \ are \ integers; a > b \end{bmatrix}$$

Eq 6.4.16: f-Group Element

Note that (6.4.16) already implies that Lines 1 – 5 in the Light Matrix (6.4.1) have been activated, and that the logic of the f-Group-ecosystem will automatically precipitate into the material-fabric through the action of Line 6-7 of (6.4.1). This logic is none other than the pre-genetic code as specified by (6.4.16).

Further, the equivalent mapping between traditional element groupings and S_{System_N}, as in Equations 6.4.17 and 6.4.18, can also be specified:

$$Element\,_{Lanthanide} = Xa +$$

$$\overline{Yb_{0-n}} \;\; where \begin{bmatrix} X \in [S_{System_N}] \\ Y \in [S_{System_{Pr}}, S_{System_P}, S_{System_K}, S_{system_N}] \\ a, b\ are\ integers; a > b \end{bmatrix}$$

Eq 6.4.17: Lanthanide Element

$$Element\,_{Actinide} = Xa + \overline{Yb_{0-n}}$$

$$where \begin{bmatrix} X \in [S_{System_N}] \\ Y \in [S_{System_{Pr}}, S_{System_P}, S_{System_K}, S_{System_N}] \\ a, b\ are\ integers; a > b \end{bmatrix}$$

Eq 6.4.18: Actinide Element

Further, as a representative element belonging to the Lanthanide group the equation for Lanthanum (La), as in Equation 6.3.19, would be:

$$Element\,_{Lanthanum} = Xa + \overline{Yb_{0-n}}$$

$$where \begin{bmatrix} X \in [S_{System_N}] \\ Y \in [S_{System_{Pr}}, S_{System_P}, S_{System_K}, S_{System_N}] \\ a, b\ are\ integers; a > b \end{bmatrix}$$

Eq 6.4.19: Lanthanum

Note that the primary element X in all these cases may be a function or attribute along the lines of 'experiments in collectivity'. The secondary elements Y in all these would have some elements being the same, and then would have specific differences as one gets into sub-classes and the individual elements themselves.

Summary of Atom Partial-Singularity Material-Fabric Pre-Genetic Code

Summarizing, after the atoms stage iterations of the Light-Space-Time Emergence equation are complete, the following code-segments will have been generated as specified by Equation 6.4.20, Active Atom Partial-Singularity Material-Fabric Pre-Genetic Code Segments:

$Light - Space - Time \ Emergence_{Atoms} =$

$$\left\| \begin{bmatrix} \begin{array}{c} c_\infty : [Pr, Po, K, H] \\ \left(\downarrow R_{C_K} = f\left(R_{C_\infty}\right) \right) \\ c_K : [S_{Pr}, S_{Po}, S_K, S_H] \\ \left(\downarrow R_{C_N} = f\left(R_{C_K}\right) \right) \\ c_N : f(S_{Pr} \times S_{Po} \times S_K \times S_H) \\ \left(\downarrow R_{C_U} = f\left(R_{C_N}\right) \right) \\ c_U : [P, V, M, C] \\ \Uparrow \\ c_{0:[D,W,I,C]} \end{array} \end{bmatrix}_{Light} \quad \begin{bmatrix} M_3 \rightarrow System_X \\ (\uparrow F \rightarrow I) \\ M_2 \rightarrow S_{System_X} \\ (\uparrow Sig \rightarrow F) \\ M_1 \rightarrow Sig_x \\ (\uparrow > P_x) \\ U \rightarrow x_{Quantum\ Particles} \end{bmatrix}_{Space} \right.$$

$$\begin{bmatrix} M_3 : -\infty \leq t \leq \infty \\ \downarrow \\ M_2 : 0 \geq t > \infty \\ \downarrow \\ M_1 : 0 > t > \infty \\ \downarrow \\ U \rightarrow \begin{array}{c} t \leq E_{Cell}; TC: M_3 \rightarrow U \\ t \sim E_{Human}; TC: U \rightarrow M_3 \end{array} \end{bmatrix}_{Time} \quad TC \rightarrow x_{Atoms} \left. \right\|_{\langle x_U | x_T \rangle}$$

$\Rightarrow$

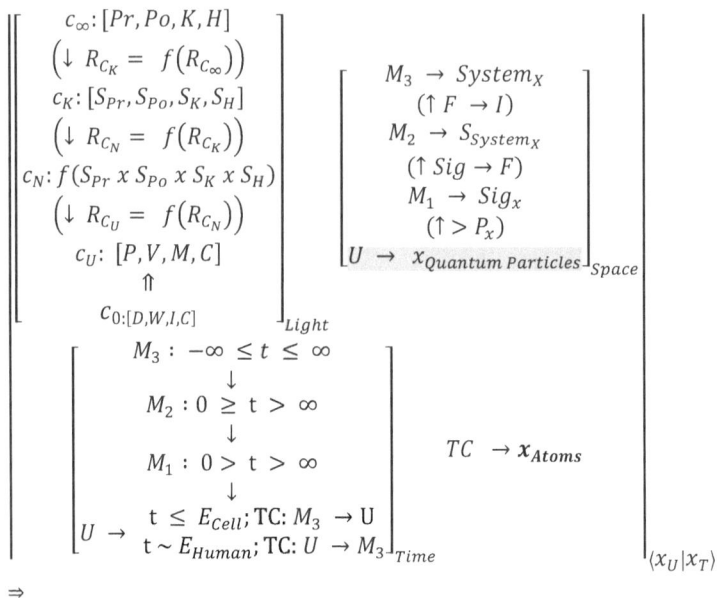

$\langle Space - Time - Energy - Gravity\ Material - Fabric\ Pre$
$\qquad - Genetic\ Code \rangle +$

$(Electromagnetic\ Spectrum\ Material - Fabric\ Pre - Genetic\ Code) +$

$(Quantum\ Particles\ Material - Fabric\ Pre - Genetic\ Code) +$

$$\left(\begin{array}{c} \sum Element_{f-Group} = Xa + \overline{Yb_{0-n}} \\ where \left[\begin{array}{c} X \in [S_{System_N}] \\ Y \in [S_{System_{Pr}}, S_{System_P}, S_{System_K}, S_{System_N}] \\ a, b\ are\ integers; a > b \end{array}\right] \\ \sum Element_{p-Group} = Xa + \overline{Yb_{0-n}} \\ where \left[\begin{array}{c} X \in [S_{System_K}] \\ Y \in [S_{System_{Pr}}, S_{System_P}, S_{System_K}, S_{System_N}] \\ a, b\ are\ integers; a > b \end{array}\right] \\ \sum Element_{s-Group} = Xa + \overline{Yb_{0-n}} \\ where \left[\begin{array}{c} X \in [S_{System_P}] \\ Y \in [S_{System_{Pr}}, S_{System_P}, S_{System_K}, S_{System_N}] \\ a, b\ are\ integers; a > b \end{array}\right] \\ \sum Element_{d-Group} = Xa + \overline{Yb_{0-n}} \\ where \left[\begin{array}{c} X \in [S_{System_{Pr}}] \\ Y \in [S_{System_{Pr}}, S_{System_P}, S_{System_K}, S_{System_N}] \\ a, b\ are\ integers; a > b \end{array}\right] \end{array}\right)$$

Eq. 6.4.20, Active Atom Partial-Singularity Material-Fabric Pre-Genetic Code Segments

The '$\sum x$' signifies all the possibilities for the 'x' equation-segment and depicts the growing biography by which the singular light-based edifice expresses its materialization, and at this stage, through the atom partial-singularity.

SECTION 7: GENERATION OF PARTIAL-SINGULARITIES IN THE SURFACING OF LIFE

This section explores the quantum-level computation and the generation of genetic code that is associated with the

emergence of life through the cell, complex human attributes such as thoughts and feelings, and uniqueness of individuality. As suggested genetic code articulates a biography, and in this case, the emerging biography depicting an increase in light-based singularities' sphere of influence through the emergence of additional light-generated "laws".

So far the action of the Light-Space-Time Emergence has generated pre-genetic code. After the Big Bang this pre-genetic code is housed in the material-fabric, which has a universal action on all constructs arising in the universe. The emergence of matter, as discussed in Section 6, precipitates a phenomenon of material containerization. This was hypothesized as being due to an essential action of the space-time-energy-gravity quadrumvirate being contained within a smaller "space" as a result of the number of driver seeds.

In this section it is assumed that just as the space-time-energy-gravity quadrumvirate has an impact on every material emergence, so too will the action of the electromagnetic spectrum have an action on every emergence of life. Some dynamics of this possibility are explored in Chapter 7.1.

Chapter 7.2 will explore the generation of the genetic code responsible for the emergence and complexification of four-foldness through the primary molecular plans of nucleic acids, proteins, lipids, and polysaccharides.

Chapter 7.3 will relate key human attributes of sensations, urges, desires, wills, feelings, emotions, and thought to the continuing journey of fourfold complexification and suggest that genetic code is also generated to support these attributes.

Chapter 7.4 will relate truer individuality to the fourfold properties implicit in Light and suggest the generation of the genetic code tied to this.

Chapter 7.1: Effect of Electromagnetic Spectrum on Life

As explored in Section 5, while there is a significant volume of material-fabric pre-genetic code generated for the electromagnetic spectrum, this chapter will focus on some aspects of the visible-light portion of the electromagnetic spectrum that is known to have a direct impact on life.

Research indicates that in the visible light spectrum specified by light with a wavelength between 380 nanometers to 760 nanometers, each color has specific physical and psychological properties (Martel, 2018). For example, colors in the spectrum that range from red to yellow, are known as warm colors and generally are known to be stimulating, fortifying, and energizing. Colors in the spectrum that range from green to violet are known as cool colors and generally are known to be calming, sedative, and analgesic.

This suggests that there is a specific set of functions associated with each color that conceivably uses that color as a vehicle to express or concretize itself. It further suggests that there are segments of electromagnetic spectrum related code that may be fundamental to all of life and are likely deeply embedded in or influence all of life. This is evident in that colors are known to generate specific responses in human bodies. The equilibrium of the autonomic nervous system, for example, which is related to the equilibrium of the sympathetic and parasympathetic nervous system, can be influenced by the application of warm and cool colors (Spitler, 2011).

Research into generic properties associated with colors (Van Obberghen, 2007; Deppe, 2013) suggests the specific functions related to the wave-archetype or structure, the energy-profile, and the mass-potential for each of the colors.

The speed or magnetic profile of each color is assumed to be that of c as derived in Chapter 5.1. Hence as in (5.1.2) all colors will be assumed to have the magnetic properties associated with light moving at c:

$$EM\_Spectrum_{Speed} = Xa +$$

$$\overline{Yb_{0-n}} \quad where \begin{bmatrix} X \in [S_{System_N}] \\ Y \in [S_{System_{Pr}}, S_{System_P}, S_{System_K}, S_{System_N}] \\ a, b \text{ are integers}; a > b \end{bmatrix}$$

Red

Red is associated with the wavelength 635 to 760 nanometers.

In its association with life the energy-profile of 'red' is suggested to engender qualities to do with 'tonifying', 'warming', 'antianemia', 'antimigraine', amongst possible others. Assuming that the primary or x-element has to do with tonifying, then various y-elements in combination can create the other engendered qualities. These relationships may be depicted in Equation 7.1.1, Red Energy, in the following manner:

$$Red_{Energy} =' tonifying'.a + \overline{Yb_{0-n}}$$

$$where \begin{bmatrix} tonifying \in [S_{System_P}] \\ Y \in [S_{System_{Pr}}, S_{System_P}, S_{System_K}, S_{System_N}] \\ a, b \text{ are integers}; a > b \end{bmatrix}$$

Eq 7.1.1: Red Energy

The wave-archetype or intent has to do with a primary or x-element associated with the function 'passion'. Other y-elements in combination will engender intent such as 'extroversion'. Hence, these relationships may be depicted by Equation 7.1.2, Red Intent:

$$Red_{Intent} =' passion'.a + \overline{Yb_{0-n}}$$

$$where \begin{bmatrix} passion \in [S_{System_K}] \\ Y \in [S_{System_{Pr}}, S_{System_P}, S_{System_K}, S_{System_N}] \\ a, b \ are \ integers; a > b \end{bmatrix}$$

Eq 7.1.2: Red Intent

The mass-potential or body-part most linked to red is suggested to be the 'reproductive system', the 'urogenital system', and 'system to do with blood pressure', amongst

others. So there is likely a build up of mass in such a manner that facilitates the creation of these systems. An assumption can be made that there is a 'red-mass' x-element and perhaps other y-elements that yield the different systems when built up. Equation 7.1.3, Reproductive System – Red Mass Possibility, suggests how this can be depicted:

$$Reproductive \ System - Red_{Mass_{Possibility}} =' red - mass'.a + \overline{Yb_{0-n}}$$

$$where \begin{bmatrix} 'red - mass' \in [S_{System_{Pr}}] \\ Y \in [S_{System_{Pr}}, S_{System_P}, S_{System_K}, S_{System_N}] \\ a, b \ are \ integers; a > b \end{bmatrix}$$

Eq 7.1.3: Reproductive System - Red Mass Possibility

Note that (7.1.1 – 3) are proposed to be part of the material-fabric. As such they are of a pre-genetic type but also inform or even help form the cell-based genetic code.

Orange

Orange is associated with the wavelength 590 to 635 nanometers.

In its association with life the energy-profile of 'orange' is suggested to engender qualities to do with 'invigorating' and 'adrenal stimulant' amongst possible others. Assuming that the primary or x-element has to do with 'invigorating', then various y-elements in combination can create the other engendered qualities. These relationships may be depicted in Equation 7.1.4, Orange Energy, in the following manner:

$$Orange_{Energy} =' invigorating'.a + \overline{Yb_{0-n}}$$

$$where \begin{bmatrix} invigorating \in [S_{System_P}] \\ Y \in [S_{System_{Pr}}, S_{System_P}, S_{System_K}, S_{System_N}] \\ a, b \ are \ integers; a > b \end{bmatrix}$$

Eq 7.1.4: Orange Energy

The wave-archetype or intent has to do with a primary or x-element associated with the function 'joy'. Other y-elements in combination will engender intent such as 'antidepressant' and 'sociability'. Hence, these relationships may be depicted by Equation 7.1.5, Orange Intent:

$$Orange_{Intent} =' joy'.a + \overline{Yb_{0-n}}$$

$$where \begin{bmatrix} joy \in [S_{System_K}] \\ Y \in [S_{System_{Pr}}, S_{System_P}, S_{System_K}, S_{System_N}] \\ a, b \ are \ integers; a > b \end{bmatrix}$$

Eq 7.1.5: Orange Intent

The mass-potential or body-part most linked to orange is suggested to be the 'adrenal glands', 'vasomotricity', and 'musculoskeletal system', amongst possible others. So there is likely a build up of mass in such a manner that facilitates the creation of these systems. An assumption can be made that there is a 'orange-mass' x-element and perhaps other y-elements that yield the different systems when built up. Equation 7.1.6, Adrenal Gland - Orange Mass Possibility, suggests how this can be depicted:

$$Adrenal\ Glands - Orange_{Mass_{Possibility}} =' orange - mass'.a + \overline{Yb_{0-n}}$$

$$where \begin{bmatrix} 'orange - mass' \in [S_{System_{Pr}}] \\ Y \in [S_{System_{Pr}}, S_{System_P}, S_{System_K}, S_{System_N}] \\ a, b\ are\ integers; a > b \end{bmatrix}$$

Eq 7.1.6: Adrenal Glands - Orange Mass Possibility

Note that (7.1.4 – 6) are proposed to be part of the material-fabric. As such they are of a pre-genetic type but also inform or even help form the cell-based genetic code.

Yellow

Yellow is associated with the wavelength 570 to 590 nanometers.

In its association with life the energy-profile of 'yellow' is suggested to engender qualities to do with 'digestive stimulant', and 'lymphatic stimulant' amongst possible others. Assuming that the primary or x-element has to do with 'stimulant', then various y-elements in combination can create the other engendered qualities. These relationships may be depicted in Equation 7.1.7, Yellow Energy, in the following manner:

$$Yellow_{Energy} =' stimulant'.a + \overline{Yb_{0-n}}$$

$$where \begin{bmatrix} stimulant \in [S_{System_P}] \\ Y \in [S_{System_{Pr}}, S_{System_P}, S_{System_K}, S_{System_N}] \\ a, b \ are \ integers; a > b \end{bmatrix}$$

Eq 7.1.7: Yellow Energy

The wave-archetype or intent has to do with a primary or x-element associated with the function 'optimism'. then various y-elements in combination can create the other engendered qualities. Hence, these relationships may be depicted by Equation 7.1.8, Yellow Intent:

$$Yellow_{Intent} =' optimism'.a + \overline{Yb_{0-n}}$$

$$where \begin{bmatrix} optimism \in [S_{System_K}] \\ Y \in [S_{System_{Pr}}, S_{System_P}, S_{System_K}, S_{System_N}] \\ a, b \ are \ integers; a > b \end{bmatrix}$$

Eq 7.1.8: Yellow Intent

The mass-potential or body-part most linked to yellow is suggested to be the 'digestive system', the 'intestine', the 'spleen', the 'pancreas' amongst possible others. So there is likely a build up of mass in such a manner that facilitates the creation of these systems. An assumption can be made that there is a 'yellow-mass' x-element and perhaps other y-elements that yield the different systems when built up. Equation 7.1.9, Digestive System - Yellow Mass Possibility, suggests how this can be depicted:

$$Digestive\ System - Yellow_{Mass_{Possibility}} =' yellow - mass'.a + \overline{Yb_{0-n}}$$

$$where \left[\begin{array}{c} 'yellow - mass' \in [S_{System_{Pr}}] \\ Y \in [S_{System_{Pr}}, S_{System_P}, S_{System_K}, S_{System_N}] \\ a, b\ are\ integers; a > b \end{array} \right]$$

Eq 7.1.9: Digestive System - Yellow Mass Possibility

Note that (7.1.7 – 9) are proposed to be part of the material-fabric. As such they are of a pre-genetic type but also inform or even help form the cell-based genetic code.

Lemon

Lemon is associated with the wavelength 550 to 570 nanometers.

In its association with life the energy-profile of 'lemon' is suggested to engender qualities to do with 'antiallergy', 'detoxifying', and 'addresses chronic conditions', amongst possible others. Assuming that the primary or x-element has to do with 'detoxifying', then various y-elements in combination can create the other engendered qualities Energy, in the following manner:

$$Red_{Energy} =' detoxifying'.a + \overline{Yb_{0-n}}$$

$$where \left[\begin{array}{c} detoxifying \in [S_{System_P}] \\ Y \in [S_{System_{Pr}}, S_{System_P}, S_{System_K}, S_{System_N}] \\ a, b\ are\ integers; a > b \end{array} \right]$$

Eq 7.1.10: Lemon Energy

The wave-archetype or intent has to do with a primary or x-element associated with the function 'peace'. Other y-elements in combination will engender intent such as 'flexibility' and 'opening', amongst possible others. Hence, these relationships may be depicted by Equation 7.1.11, Lemon Intent:

$$Red_{Intent} =' peace'.a + \overline{Yb_{0-n}}$$

$$where \begin{bmatrix} peace \in [S_{System_K}] \\ Y \in [S_{System_{Pr}}, S_{System_P}, S_{System_K}, S_{System_N}] \\ a, b \ are \ integers; a > b \end{bmatrix}$$

Eq 7.1.11: Lemon Intent

The mass-potential or body-part most linked to lemon is suggested to be 'joints', 'gallbladder', 'diaphragm' and 'vision', amongst possible others. So there is likely a build up of mass in such a manner that facilitates the creation of these systems. An assumption can be made that there is a 'lemon-mass' x-element and perhaps other y-elements that yield the different systems when built up. Equation 7.1.12, Joints – Lemon Mass Possibility, suggests how this can be depicted:

$$Joints - Lemon_{Mass_{Possibility}} =' lemon - mass'.a + \overline{Yb_{0-n}}$$

$$where \begin{bmatrix} 'lemon - mass' \in [S_{System_{Pr}}] \\ Y \in [S_{System_{Pr}}, S_{System_P}, S_{System_K}, S_{System_N}] \\ a, b \ are \ integers; a > b \end{bmatrix}$$

Eq 7.1.12: Joints - Lemon Mass Possibility

Note that (7.1.10 – 12) are proposed to be part of the material-fabric. As such they are of a pre-genetic type but also inform or even help form the cell-based genetic code.

Green

Green is associated with the wavelength 520 to 550 nanometers.

In its association with life the energy-profile of 'green' is suggested to engender qualities to do with 'general balancing', 'neutralizing', and 'anti-infection' amongst possible others. Assuming that the primary or x-element has to do with 'balancing', then various y-elements in combination can create the other engendered qualities. These relationships may be depicted in Equation 7.1.13, Green Energy, in the following manner:

$$Green_{Energy} =' balancing'.a + \overline{Yb_{0-n}}$$

$$where \left[\begin{array}{c} balancing \in [S_{System_P}] \\ Y \in [S_{System_{Pr}}, S_{System_P}, S_{System_K}, S_{System_N}] \\ a, b\ are\ integers; a > b \end{array} \right]$$

Eq 7.1.13: Green Energy

The wave-archetype or intent has to do with a primary or x-element associated with the function 'love'. Other y-elements in combination will engender intent such as 'abundance' and 'adaptability', amongst possible others. Hence, these relationships may be depicted by Equation 7.1.14, Green Intent:

$$Green_{Intent} =' love'.a + \overline{Yb_{0-n}}$$

$$where \left[\begin{array}{c} love \in [S_{System_K}] \\ Y \in [S_{System_{Pr}}, S_{System_P}, S_{System_K}, S_{System_N}] \\ a, b\ are\ integers; a > b \end{array} \right]$$

Eq 7.1.14: Green Intent

The mass-potential or body-part most linked to green is suggested to be the 'liver', 'lungs', 'thymus', and 'immune system', amongst possible others. So there is likely a build up of mass in such a manner that facilitates the creation of these systems. An assumption can be made that there is a 'green-mass' x-element and perhaps other y-elements that yield the different systems when built up. Equation 7.1.15, Liver – Green Mass Possibility, suggests how this can be depicted:

$$Liver - Green_{Mass_{Possibility}} =' green - mass'.a + \overline{Yb_{0-n}}$$

$$where \begin{bmatrix} 'green - mass' \in [S_{System_{Pr}}] \\ Y \in [S_{System_{Pr}}, S_{System_P}, S_{System_K}, S_{System_N}] \\ a, b \ are \ integers; a > b \end{bmatrix}$$

Eq 7.1.15: Liver - Green Mass Possibility

Note that (7.1.13 – 15) are proposed to be part of the material-fabric. As such they are of a pre-genetic type but also inform or even help form the cell-based genetic code.

Cyan

Cyan is associated with the wavelength 490 to 520 nanometers.

In its association with life the energy-profile of 'cyan' is suggested to engender qualities to do with 'cleansing', 'refreshing', and 'addressing acute conditions' amongst possible others. Assuming that the primary or x-element has to do with 'cleansing', then various y-elements in combination can create the other engendered qualities. These relationships may be depicted in Equation 7.1.16, Cyan Energy, in the following manner:

$$Cyan_{Energy} =' cleansing'.a + \overline{Yb_{0-n}}$$

$$where \begin{bmatrix} cleansing \in [S_{System_P}] \\ Y \in [S_{System_{Pr}}, S_{System_P}, S_{System_K}, S_{System_N}] \\ a, b \ are \ integers; a > b \end{bmatrix}$$

Eq 7.1.16: Cyan Energy

The wave-archetype or intent has to do with a primary or x-element associated with the function 'wholeness'. Other y-elements in combination will engender intent such as 'autonomy', and 'fulcrum of emotion vs. thought', amongst possible others. Hence, these relationships may be depicted by Equation 7.1.17, Cyan Intent:

$$Cyan_{Intent} =' \ wholeness'. a + \overline{Yb_{0-n}}$$

$$where \begin{bmatrix} wholeness \in [S_{System_K}] \\ Y \in [S_{System_{Pr}}, S_{System_P}, S_{System_K}, S_{System_N}] \\ a, b \ are \ integers; a > b \end{bmatrix}$$

Eq 7.1.17: Cyan Intent

The mass-potential or body-part most linked to cyan is suggested to be the 'arms and hands', 'parathyroid glands', and 'skin'. amongst others. So there is likely a build up of mass in such a manner that facilitates the creation of these systems. An assumption can be made that there is a 'cyan-mass' x-element and perhaps other y-elements that yield the different systems when built up. Equation 7.1.18, Skin – Cyan Mass Possibility, suggests how this can be depicted:

$$Skin - Cyan_{Mass_{Possibility}} =' \ cyan - mass'. a + \overline{Yb_{0-n}}$$

$$where \begin{bmatrix} 'cyan - mass' \in [S_{System_{Pr}}] \\ Y \in [S_{System_{Pr}}, S_{System_P}, S_{System_K}, S_{System_N}] \\ a, b \ are \ integers; a > b \end{bmatrix}$$

Eq 7.1.18: Skin - Cyan Mass Possibility

Note that (7.1.16 – 18) are proposed to be part of the material-fabric. As such they are of a pre-genetic type but also inform or even help form the cell-based genetic code.

Blue

Blue is associated with the wavelength 460 to 490 nanometers.

In its association with life the energy-profile of 'blue' is suggested to engender qualities to do with 'calming', 'anti-inflammation', and 'antipyretic', amongst possible others. Assuming that the primary or x-element has to do with 'calming', then various y-elements in combination can create the other engendered qualities. These relationships may be depicted in Equation 7.1.19, Blue Energy, in the following manner:

$$Blue_{Energy} =' calming'.a + \overline{Yb_{0-n}}$$

$$where \begin{bmatrix} calming \in [S_{System_P}] \\ Y \in [S_{System_{Pr}}, S_{System_P}, S_{System_K}, S_{System_N}] \\ a, b \ are \ integers; a > b \end{bmatrix}$$

Eq 7.1.19: Blue Energy

The wave-archetype or intent has to do with a primary or x-element associated with the function 'antistress'. Other y-elements in combination will engender intent such as 'introversion' and 'endorphins stimulant', amongst possible others.

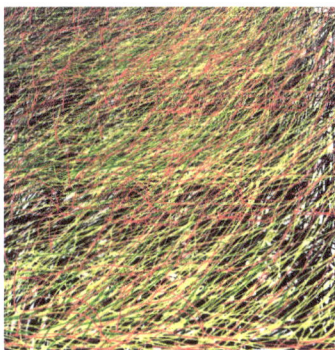

Hence, these relationships may be depicted by Equation 7.1.20, Blue Intent:

$$Blue_{Intent} =' antistress'.a + \overline{Yb_{0-n}}$$

$$where \begin{bmatrix} antistress \in [S_{System_K}] \\ Y \in [S_{System_{Pr}}, S_{System_P}, S_{System_K}, S_{System_N}] \\ a, b \ are \ integers; a > b \end{bmatrix}$$

Eq 7.1.20: Blue Intent

The mass-potential or body-part most linked to blue is suggested to be the 'nose, throat, neck', 'thyroid gland', breathing', 'ears', amongst possible others. So there is likely a build up of mass in such a manner that facilitates the creation of these systems. An assumption can be made that there is a 'blue-mass' x-element and perhaps other y-elements that yield the different systems when built up. Equation 7.1.21, Ears – Blue Mass Possibility, suggests how this can be depicted:

$$Ears - Blue_{Mass_{Possibility}} =' blue - mass'.a + \overline{Yb_{0-n}}$$

$$where \begin{bmatrix} 'blue - mass' \in [S_{System_{Pr}}] \\ Y \in [S_{System_{Pr}}, S_{System_P}, S_{System_K}, S_{System_N}] \\ a, b \ are \ integers; a > b \end{bmatrix}$$

Eq 7.1.21: Ears - Blue Mass Possibility

Note that (7.1.19 – 21) are proposed to be part of the material-fabric. As such they are of a pre-genetic type but also inform or even help form the cell-based genetic code.

Indigo

Indigo is associated with the wavelength 430 to 460 nanometers.

In its association with life the energy-profile of 'indigo' is suggested to engender qualities to do with 'sleep-inducing' and 'muscle relaxant', amongst possible others. Assuming that the primary or x-element has to do with 'relaxant', then various y-elements in combination can create the other engendered qualities. These relationships may be depicted in Equation 7.1.22, Indigo Energy, in the following manner:

$$Indigo_{Energy} =' relaxant'.a + \overline{Yb_{0-n}}$$

$$where \begin{bmatrix} relaxant \in [S_{System_P}] \\ Y \in [S_{System_{Pr}}, S_{System_P}, S_{System_K}, S_{System_N}] \\ a, b\ are\ integers; a > b \end{bmatrix}$$

Eq 7.1.22: Indigo Energy

The wave-archetype or intent has to do with a primary or x-element associated with the function 'intuition'. Other y-elements in combination will engender intent such as 'imagination' and 'anxiolytic', amongst possible others. Hence, these relationships may be depicted by Equation 7.1.23, Indigo Intent:

$$Indigo_{Intent} =' intuition'.a + \overline{Yb_{0-n}}$$

$$where \begin{bmatrix} intuition \in [S_{System_K}] \\ Y \in [S_{System_{Pr}}, S_{System_P}, S_{System_K}, S_{System_N}] \\ a, b\ are\ integers; a > b \end{bmatrix}$$

Eq 7.1.23: Indigo Intent

The mass-potential or body-part most linked to indigo is suggested to be the 'sinuses', 'pituitary', 'forehead and occiput', and 'limbic system', amongst possible others. So there is likely a build up of mass in such a manner that facilitates the creation of these systems. An assumption can be made that there is a 'indigo-mass' x-element and perhaps other y-elements that yield the different systems when built up. Equation 7.1.24, Pituitary – Indigo Mass Possibility, suggests how this can be depicted:

$$Pituitary - Indigo_{Mass_{Possibility}} =' indigo - mass'.a + \overline{Yb_{0-n}}$$

$$where \begin{bmatrix} 'indigo - mass' \in [S_{System_{Pr}}] \\ Y \in [S_{System_{Pr}}, S_{System_P}, S_{System_K}, S_{System_N}] \\ a, b \ are \ integers; a > b \end{bmatrix}$$

Eq 7.1.24: Pituitary - Indigo Mass Possibility

Note that (7.1.22 – 24) are proposed to be part of the material-fabric. As such they are of a pre-genetic type but also inform or even help form the cell-based genetic code.

Violet

Violet is associated with the wavelength 380 to 430 nanometers.

In its association with life the energy-profile of 'violet' is suggested to engender qualities to do with 'immunostimulant', amongst possible others. Assuming that the primary or x-element has to do with 'stimulant', then various y-elements in combination can create the other engendered qualities. These relationships may be depicted in Equation 7.1.25, Violet Energy, in the following manner:

$$Violet_{Energy} =' stimulant'.a + \overline{Yb_{0-n}}$$

$$where \begin{bmatrix} stimulant \in [S_{System_P}] \\ Y \in [S_{System_{Pr}}, S_{System_P}, S_{System_K}, S_{System_N}] \\ a, b \ are \ integers; a > b \end{bmatrix}$$

Eq 7.1.25: Violet Energy

The wave-archetype or intent has to do with a primary or x-element associated with the function 'spirituality'.

Other y-elements in combination will engender intent such as 'trust' and 'dignity', amongst possible others. Hence, these relationships may be depicted by Equation 7.1.26, Violet Intent:

$$Violet_{Intent} =' spirituality'.a + \overline{Yb_{0-n}}$$

$$where \begin{bmatrix} spirituality \in [S_{System_K}] \\ Y \in [S_{System_{Pr}}, S_{System_P}, S_{System_K}, S_{System_N}] \\ a, b \ are \ integers; a > b \end{bmatrix}$$

Eq 7.1.26: Violet Intent

The mass-potential or body-part most linked to violet is suggested to be the 'cerebral cortex', 'pineal gland', and 'memory, amongst possible others. So there is likely a build up of mass in such a manner that facilitates the creation of these systems. An assumption can be made that there is a 'violet-mass' x-element and perhaps other y-elements that yield the different systems when built up. Equation 7.1.27, Cerebral Cortex– Violet Mass Possibility, suggests how this can be depicted:

$$Cerebral \ Cortex - Violet_{Mass_{Possibility}} =' violet - mass'.a + \overline{Yb_{0-n}}$$

$$where \begin{bmatrix} 'violet - mass' \in [S_{System_{Pr}}] \\ Y \in [S_{System_{Pr}}, S_{System_P}, S_{System_K}, S_{System_N}] \\ a, b \ are \ integers; a > b \end{bmatrix}$$

Eq 7.1.27: Cerebral Cortex - Violet Mass Possibility

Note that (7.1.25 – 27) are proposed to be part of the material-fabric. As such they are of a pre-genetic type but also inform or even help form the cell-based genetic code.

Summary Note

This chapter on the effect of the electromagnetic spectrum on life is perhaps a clear example of how the sphere of influence of emerging partial-singularities increases. The electromagnetic spectrum partial-singularity exercises increased yet cohesive complexification of four-foldness in the pre-genetic material-fabric, not marginalizing the comprehensive logic of the space-time-energy-gravity quadrumvirate already in place with the emergence of the Big Bang partial-singularity. Variation in wavelength in the electromagnetic spectrum is the mechanism by which antecedent functional possibility encoded in Light moving at speeds greater than c materializes in an initial pre-material medium at light traveling at c. Such variation in wavelength creates diverse functional possibility in the intent, energetics, and mass-possibility of visible light. As life-related partial-singularities emerge, such pre-existing logic or law is automatically operative so that there is deep consistency with it in the surfacing of more complex tissues and even organs, as suggested by the preceding discussion on mass-

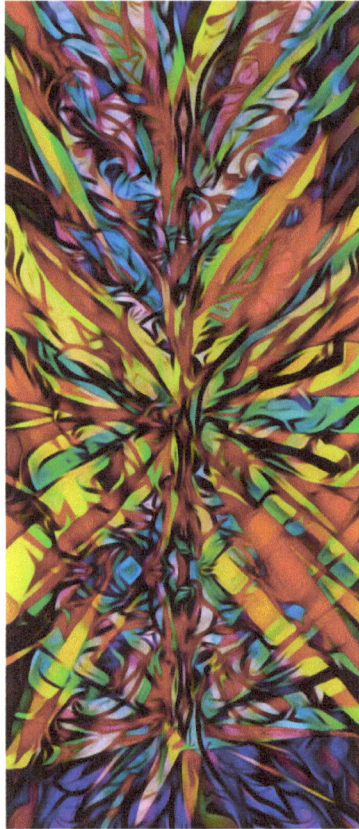

possibility. Cross-over of such fourfold consistency from one type or partial-singularity to another, as in the case from the electromagnetic partial-singularity, to matter-based partial-singularities, to life-based partial-singularities is perhaps the hallmark of light-based singularities indicating that all that emerges in such a manner is of one light-based edifice.

Chapter 7.2: Generation of Living Cell Partial-Singularity

The living cell has a universe of adaptability embedded in it. In 'The Machinery of Life', Goodsell, an Associate Professor of Molecular Biology at the Scripps Research Institute (Goodsell, 2010) suggests that every living thing on Earth uses a similar set of molecules to eat, to breathe, to move, and to reproduce. There are molecular machines that do the myriad things that distinguish living organisms that are identical in all living cells. This nanoscale machinery of cells uses four basic molecular plans with unique chemical personalities: nucleic acids, proteins, lipids, and polysaccharides.

Extrapolating from previous chapters there will likely be a vast number of iterations of the Light-Space-Time Emergence Equation to generate the four basic molecular plans. Equation 4.2.4, Cells Genetic Code, created in Chapter 4.2 that explored the post Big-Bang generation of pre-genetic code, genetic code, and post-genetic code, is reproduced here for convenience. The starting point, x_U, is 'Molecules', and the ending point, x_T, is 'Cells':

$Light - Space - Time\ Emergence_{Cells} =$

$$
\left|\left|\begin{array}{c}
\left[\begin{array}{c}
c_\infty : [Pr, Po, K, H] \\
\left(\downarrow R_{C_K} = f(R_{C_\infty})\right) \\
c_K : [S_{Pr}, S_{Po}, S_K, S_H] \\
\left(\downarrow R_{C_N} = f(R_{C_K})\right) \\
c_N : f(S_{Pr} \times S_{Po} \times S_K \times S_H) \\
\left(\downarrow R_{C_U} = f(R_{C_N})\right) \\
c_U : [P, V, M, C] \\
\Uparrow \\
c_{0:[D,W,I,C]}
\end{array}\right]_{Light}
\end{array}\right.\right.
\quad
\left[\begin{array}{c}
M_3 \rightarrow System_X \\
(\uparrow F \rightarrow I) \\
M_2 \rightarrow S_{System_X} \\
(\uparrow Sig \rightarrow F) \\
M_1 \rightarrow Sig_x \\
(\uparrow > P_x) \\
U \rightarrow x_{Molecules}
\end{array}\right]_{Space}
$$

$$
\left[U \rightarrow \begin{array}{c}
M_3 : -\infty \leq t \leq \infty \\
\downarrow \\
M_2 : 0 \geq t > \infty \\
\downarrow \\
M_1 : 0 > t > \infty \\
\downarrow \\
t \leq E_{Cell}; \text{TC}: M_3 \rightarrow U \\
t \sim E_{Human}; \text{TC}: U \rightarrow M_3
\end{array}\right]_{Time}
\qquad TC \rightarrow x_{Cells}
$$

$$
\left.\left.\vphantom{\begin{array}{c}1\\2\\3\end{array}}\right|\right|_{\langle x_U | x_T \rangle}
$$

$$
\Rightarrow
$$

$\langle Space - Time - Energy - Gravity\ Material - Fabric\ Pre$
$- Genetic\ Code \rangle +$

$\langle Electromagnetic\ Spectrum\ \ Material - Fabric\ Pre - Genetic\ Code \rangle$
$+$

$\langle Quantum - Particle\ Material - Fabric\ Pre - Genetic\ Code \rangle +$

$\langle Atoms\ Material - Fabric\ Pre - Genetic\ Code \rangle +$

$\langle Molecules\ Material - Fabric\ Pre - Genetic\ Code \rangle + LSTE \langle ... \rangle$
$+ Cells\ Genetic\ Code$

There are many likely iterations of this equation between molecules and cells and this in general is depicted by '$LSTE \langle ... \rangle$', where LSTE signifies iterations(s) of the Light-Space-Time Emergence equation. Also note that an assumption is made that all the previous material-fabric code-segments are intimately active at the level of cellular genetic code. This may be through the device of entanglement, or perhaps even by the material-fabric code segments being embedded in cellular genetic code, or perhaps by some protein-structures that exist in each cell that give the cell the capacity to read the pre-genetic code embedded in the material-fabric.

The emergence of a quadrumvirate cell-based structure specified by four molecular plans is significant in that the genetic code is now being housed in a post material-fabric structure. It is likely the collective and simultaneous action of nucleic acids, polysaccharides, lipids, and proteins that facilitates this post material-fabric housing ability. While in this chapter the code generated for each of these molecular plans appears to be separate, as an emergence of properties of Light they should be thought of as happening in such a manner so as to encapsulate the essential unity of the four properties.

Strictly speaking (4.2.4) should reflect the transition between the pre-genetic to genetic housing structure, and this is depicted as 'PGGT <...>', Pre-Genetic to Genetic Transition, in Equation 7.2.1, Generation of Cells Partial-Singularity:

$$Light - Space - Time\ Emergence_{Cells} =$$

$$\left\|\begin{bmatrix} c_\infty : [Pr, Po, K, H] \\ \left(\downarrow R_{C_K} = f\left(R_{C_\infty}\right)\right) \\ c_K : [S_{Pr}, S_{Po}, S_K, S_H] \\ \left(\downarrow R_{C_N} = f\left(R_{C_K}\right)\right) \\ c_N : f(S_{Pr} \times S_{Po} \times S_K \times S_H) \\ \left(\downarrow R_{C_U} = f\left(R_{C_N}\right)\right) \\ c_U : [P, V, M, C] \\ \Uparrow \\ c_{0:[D,W,I,C]} \end{bmatrix}_{Light} \begin{bmatrix} M_3 \to System_X \\ (\uparrow F \to I) \\ M_2 \to S_{System_X} \\ (\uparrow Sig \to F) \\ M_1 \to Sig_x \\ (\uparrow > P_x) \\ U \to x_{Molecules} \end{bmatrix}_{Space} \right. $$

$$\left. \begin{bmatrix} M_3 : -\infty \le t \le \infty \\ \downarrow \\ M_2 : 0 \ge t > \infty \\ \downarrow \\ M_1 : 0 > t > \infty \\ \downarrow \\ U \to \begin{array}{l} t \le E_{Cell}; \text{TC}: M_3 \to U \\ t \sim E_{Human}; \text{TC}: U \to M_3 \end{array} \end{bmatrix}_{Time} \quad TC \to x_{Cells} \right\| \langle x_U | x_T \rangle$$

$$\Rightarrow$$

$\langle Space - Time - Energy - Gravity\ Material - Fabric\ Pre$
$- Genetic\ Code \rangle +$

$\langle Electromagnetic\ Spectrum\ \ Material - Fabric\ Pre - Genetic\ Code \rangle$
$+$

$\langle Quantum - Particle\ \ Material - Fabric\ Pre - Genetic\ Code \rangle +$

$\langle Atoms\ \ Material - Fabric\ Pre - Genetic\ Code \rangle +$

$\langle Molecules\ \ Material - Fabric\ Pre - Genetic\ Code \rangle + LSTE\ \langle \dots \rangle$
$+ PGGT < \cdots > +$

Cells Genetic Code

Eq 7.2.1: Generation of Cells Partial-Singularity

Light's Emergence as Living Cells

315

As discussed previously the Light-Space-Time Emergence equation (3.1.3) being iterative, can be used to model emergence as it proceeds from simpler four-fold to more complex four-fold manifestations. Hence (3.1.3) has already been applied to suggest the emergence of the space-time-energy-gravity quadrumvirate, the electromagnetic spectrum, quantum particles in general, bosons as a further instance of a particular kind of quantum particle, and atoms. Here it will be applied to suggest the emergence of Living Cells as summarized by (7.2.1). But further, as implied by (3.6.5), the Potential Effect of Levels of Light on Genetic-Type Information equation, any process of deeper organization, such as is responsible for the architecture and cohesiveness of living cells, has the possibility of altering the material-fabric or a post material-fabric housing structure so long as the bases involved are driven primarily by a meta-level.

Starting with the Light-Matrix, the top left-hand matrix in (7.2.1), the first line from the top, $C_\infty : [Pr, Po, K, H]$, specifies the fundamental architecture of living cells. As will be explored shortly, Nucleic Acids are an emergence of Light's property of Knowledge, Polysaccharides are an emergence of Light's property of Power, Proteins are an emergence of Light's property of Presence, and Lipids are an emergence of Light's property of Harmony. The fundamental architecture of these aspects hence, is an emergence of the properties of Light at ∞.

Line 3 in the Light-Matrix, C_K: $[S_{Pr}, S_{Po}, S_K, S_H]$, elaborates the sets for Presence, Power, Knowledge, and Harmony, each containing multiple elements. For example, as will be explored in the section on Proteins, various elements derived from the four sets define the behavior of Proteins and could be functions such as 'exist for service', 'to bring about perfection at the level of the cell', 'extreme diligence and perseverance', amongst others, hence collectively describing Proteins' way of being. Specifically, Line 5, $C_{N:}\ f(S_{Pr} \times S_{Po} \times S_K \times S_H)$, suggests that unique seeds are created from a combination of such elements from all four sets, with a particular element leading or having more weight, that in effect creates the distinctness of proteins possible at the level of cells.

Line 6, $(\downarrow R_{C_U} = f(R_{C_N}))$, specifies quantization between the layer where the seeds are formed, and the physical layer, and as explored in Chapter 3.5 and 3.6, will result in Line 7, C_U: $[P, V, M, C]$, hence changing the post material-fabric housing structure. Note that as in the process describing the generation of the code-segments for space-time-energy-gravity quantization at the time of the Big Bang, the FBLEE (four-base logic-encoding ecosystem) process is apparently transparent.

The possibilities represented by Lines 1 through 5 hence concretize through the quantization represented by Line 6 to become or enhance living cells with its subtle physical (related to Presence), vital (related to Power), mental (related to Knowledge), and connection (related to Harmony) aspects now existing in material reality typified by Light moving at c. Note that just as Line 6 represents a process of quantization relating the layer of reality created by Light traveling at c with the antecedent layers, so too Lines 2 and 4 as previously discussed, also represent quantization of a more rarefied kind that ultimately plays a critical part in allowing the material-fabric to express infinite diversity.

Typically it is the process as captured by the Space-Matrix that will determine if Line 6 is activated. Specifically patterns at the untransformed layer, U, will need to be overcome, as specified by the second-line from the bottom of the Space-Matrix: $(\uparrow > P_x)$. But as specified by the bottom-line of the Time-Matrix, reproduced below, it is only with the advent of the human-system that the automaticity of the action of meta-levels is reversed:

$$U \; \to \; \begin{array}{l} t \leq E_{Cell}; TC: M_3 \; \to U \\ t \sim E_{Human}; TC: U \; \to M_3 \end{array}$$

Hence in the case of living cells, which in this emergence is a pre-human system, the fact that patterns do not need to be overcome means that quantization more or less happens automatically.

Even though cells are a pre-human system and therefore the action of meta-levels are modeled as being automatic, it is nonetheless useful to review (3.6.5), Potential Effect of Levels of Light on Genetic-Type Information, to understand how the relationship with different forms of mutation has been specified:

$$Potential\ Effect\ of\ Levels\ of\ Light\ on\ Genetic$$
$$-Type\ Information =$$
$$\left[\begin{array}{c} STATIC\ \langle|[L][S][T]TC \to x_T|_{\langle x_U|x_T\rangle} \\ \times \\ \left((Y > U: Z_Q) \vee (Y \leq U: Z_F) \vee (Y = U: Z_R)\right) \\ \ni \\ \left(\begin{array}{c} Z \in \mathbb{U}\ (Space, Time, Energy, Gravity) \\ Q: Quantization; F: Fragmentation; R: Random \end{array}\right) \end{array}\right] \to h \ni$$

$$h \in \left(\begin{array}{c} [Q]: Constructive\ zone, \\ [Q]: Constructive\ zone \wedge Constructive\ mutation, \\ [F]: Destructive\ mutation, \\ [R]: Random\ mutation \end{array}\right)$$

Line 1 from the top in the matrix is simply a static form of (3.6.1) the Simplified Light-Space-Time Emergence equation. The static form is designated by 'STATIC', and implies that fundamental operations true of (3.6.1) are being highlighted in (3.6.5). In other words (3.6.1) already has all the operations highlighted in (3.6.5) in it, but by 'freezing' it by making it static, the essential dynamics leading to possible mutations at the genetic level can more clearly be highlighted.

Line 1 is then subjected ($\times$) to a determination of the dominant levels of light that may be active, designated by Line 2, '$\left((Y > U: Z_Q) \vee (Y \leq U: Z_F) \vee (Y = U: Z_R) \right)$'. Unpacking this, '$Y > U$' implies meta-levels are active and as a result it is possible that Z_Q is going to take place (the subscript 'Q' implies quantum-level action). This also implies activation and potential change of FBLEE. The call from below, as it were, may invoke some function that already exists in the subtle-libraries 'above', so that some already existing function may influence FBLEE through ∞-entanglement, K-entanglement, or N-entanglement. This may be thought of as a key-and-lock mechanism, where a deep enough visceral urge from below acting as the key, opens an entangled lock to alter FBLEE as per the visceral urge. '$Y \leq U$' implies that only the untransformed levels are active, and therefore also the sub-level where the speed of light is 0 is active and as a result Z_F is going to take place (the subscript 'F' implies 'fragmentation'). '$Y = U$' implies that all levels are active and as a result Z_R is going to take place (the subscript 'R' implies 'random').

Line 3 elaborates the significance of Z_Q, Z_F, and Z_R. Hence Z is the union of potential quantum-operations of space, time, energy, and gravity, designated by '$Z \in \mathbb{U}(Space, Time, Energy, Gravity)$'. But the nature of the operations, as suggested in the previous paragraph, is designated   by

' Q: *Quantization*; F: *Fragmentation*; R: *Random*'. Z_Q, then, implies that the full quantization originating from updated four-base logic-encoding ecosystems (FBLEE) can take place, and will result in lasting material change at the Genetic-Type level. Z_F implies that the essential set will be fragmented and that only libraries at the level of local cellular-level DNA or precipitated material-fabric logic can potentially be altered. Z_R also precludes full quantization, and that some partial local-library constructive or destructive mutation may take place.

The ' $\rightarrow h \ni h \in$ ' segment resolves the outcome of the operations implied by Lines 1 – 3, suggesting that the outcome will be 'h' such that ($\ni$) 'h' is an element ($\in$) of the set specified by the members '[Q]: *Constructive zone*', '[Q}: *Constructive zone AND Constructive mutation*', '[F}; *Destructive mutation*', and '{R}: *Random mutation*'. '[Q]: *Constructive zone*' implies that the in-built buffer has been crossed and that access to the deeper four-base logic-encoding ecosystem (FBLEE) has been granted. '[Q}' in this segment implies that there is the possibility that full-quantization as specified by Line 3 of the previous matrix can take place. Access to this zone is a prerequisite for constructive mutation to occur, as designated by the element ' [Q}: *Constructive zone* $\cap$ *Constructive mutation* ', which implies that full-quantization is going to take place and will result in material change. The '[F]' specifies the relationship between 'Fragmentation' in Line 3 of the previous matrix and destructive mutation. The '[R]' specifies the relationship between 'Random' in Line 3 of the previous matrix and random mutation.

But as just summarized in the Time-Matrix in (7.2.1) Y is by definition greater than U and hence quantization is automatic. In terms of living cells such quantization implies that wholeness becomes fully active through specific space, time, energy, and gravity quantization to create an holistic "ecosystem" with its own "living cell

320

logic" as it were. The wholeness has now precipitated into the genetic code in the cellular structure. Note that such precipitation is captured by the action of 'PGGT $<...>$' that can be assumed to be triggered by Z_Q. This "logic" or more specifically genetic code is elaborated in the following sub-sections and suggests the increasing sphere of influence of the cellular-level partial-singularity that now also contains light-engendered laws for all living cells.

Generation of Cellular-Level Partial-Singularity 'Nucleic Acids' Genetic Code

Nucleic acids basically encode information. They store and transmit the genome, the hereditary information needed to keep the cell alive. They function as the cell's librarians and contain information on how to make proteins and when to make them.

They are hence, the keepers of a cell's knowledge, its wisdom, its ability to make laws, the vehicle to spread knowledge within cells and to the next generation of cells. Being so, one can see that there is similarity with the set for system-knowledge highlighted earlier in Chapter 2.3. Reproducing Equation 2.3.3:

$$S_{System_K} \ni [Wisdom, Law\ Making, Spread\ of\ Knowledge\ ...]$$

Nucleic acids can therefore be thought of as a precipitation of system-knowledge at the cellular level.

Hence, a nucleic acid will have a generalized signature, as in Equation 7.2.2, derived from the system-knowledge family:

$$Sig_{nucleic\ acid} = Xa + \overline{Yb_{0-n}}\ where\ \begin{bmatrix} X \in [S_{System_K}] \\ Y \in [S_{System_{Pr}}, S_{System_P}, S_{System_K}, S_{System_N}] \\ a, b\ are\ integers; a > b \end{bmatrix}$$

Eq 7.2.2: Nucleic Acid

The primary element X could be an attribute or function such as 'keeper of knowledge'. Secondary elements Y could be 'protein laws', 'generational knowledge', amongst others. The collectivity of elements as per the equation would specify the character of a nucleic acid.

Note that (7.2.2) already implies that Lines 1 – 5 in the Light Matrix (7.2.1) have been activated, and that the logic of the nucleic-acid-ecosystem will precipitate into the post material-fabric cell-based genetic structure through the action of Line 6-7 of (7.2.1). This action of precipitating into the cell-based structure as opposed to the material-fabric is designated by 'PGGT <...>' as specified in (7.2.1).

DNA and RNA, two types of nucleic acids, would hence have the equations as specified by Equation 7.2.3 and 7.2.4 respectively.

$$Sig_{DNA} = Xa + \overline{Yb_{0-n}} \quad where \begin{bmatrix} X \in [S_{System_K}] \\ Y \in [S_{System_{Pr}}, S_{System_P}, S_{System_K}, S_{System_N}] \\ a, b \ are \ integers; a > b \end{bmatrix}$$

Eq 7.2.3: DNA

$$Sig_{RNA} = Xa + \overline{Yb_{0-n}} \quad where \begin{bmatrix} X \in [S_{System_K}] \\ Y \in [S_{System_{Pr}}, S_{System_P}, S_{System_K}, S_{System_N}] \\ a, b \ are \ integers; a > b \end{bmatrix}$$

Eq 7.2.4: RNA

The primary element X in (7.2.3) and (7.2.4) would be the same as that for nucleic acids (7.2.2). The secondary elements Y however will be a larger and more specific set with many elements in common with (7.2.2). Note also that the genetic code as specified by the following sub-sections on proteins, polysaccharides, and lipids, would be translated into DNA-format and stored in nucleic acids as specified by (7.2.3).

Generation of Cellular-Level Partial-Singularity 'Proteins' Genetic Code

Proteins are the cells work-horses. Look anywhere in a cell and one will see proteins at work. Proteins are built in thousands of shapes and sizes, each performing a different function. As Goodsell describes, "some are built simply to adopt a defined shape, assembling into rods, nets, hollow spheres, and tubes. Some are molecular motors, using energy to rotate, or flex, or crawl. Many are chemical catalysts that perform chemical reactions atom-by-atom, transferring and transforming chemical groups exactly as needed." With their wide potential for diversity, proteins are constructed to perform most of the everyday tasks of the cells. In fact human cells build around 30,000 different kinds of proteins to execute on the diverse array of cellular level tasks.

Proteins hence, exist for service, to bring about perfection at the level of the cell, are characterized by extreme diligence and perseverance, and so on. Being so, one can see that there is similarity with the set for system-presence highlighted earlier in Chapter 2.3. Reproducing Equation 2.3.1:

$S_{System_{Pr}} \ni [Service, Perfection, Diligence, Perseverance, ...]$

Proteins can therefore be thought of as a precipitation of system-presence at the cellular level.

Hence, a protein could have a generalized signature, as in Equation 7.2.5, derived from the system-presence family:

$$Sig_{protein} = Xa +$$

$$\overline{Yb_{0-n}} \quad where \quad \begin{bmatrix} X \in [S_{System_{Pr}}] \\ Y \in [S_{System_{Pr}}, S_{System_{P}}, S_{System_{K}}, S_{System_{N}}] \\ a, b \ are \ integers; a > b \end{bmatrix}$$

Eq 7.2.5: Protein

This could yield a vast number of functional proteins. In fact it may be possible that the 30,000 or so known proteins created by the human cell could each be specified by a signature equation of this nature. It may be possible to map existing proteins to functionality as suggested by the four sets of molecular plans.

Note that (7.2.5) already implies that Lines 1 – 5 in the Light Matrix (7.2.1) have been activated, and that the logic of the protein-ecosystem will precipitate into the post material-fabric cell-based genetic structure through the action of Line 6-7 of (7.2.1). This action of precipitating into the cell-based structure as opposed to the material-fabric is designated by 'PGGT <…>' as specified in (7.2.1).

Consider Insulin, for example. Insulin regulates the metabolism of carbohydrates, fats and protein by promoting the absorption of, especially, glucose from the blood into fat, liver and skeletal muscle cells. Equation 7.2.6 for Insulin would hence be:

$$Sig_{insulin} = Xa +$$

$$\overline{Yb_{0-n}} \quad where \left[\begin{array}{c} X \in [S_{System_{Pr}}] \\ Y \in [S_{System_{Pr}}, S_{System_P}, S_{System_K}, S_{System_N}] \\ a, b \ are \ integers; a > b \end{array} \right]$$

Eq 7.2.6: Insulin

The primary element X could be an attribute or function such as 'workhorse'. Secondary elements Y could be 'metabolic regulation', 'glucose absorption', 'blood to fat channel', amongst others. The collectivity of elements as per the equation would specify the character of insulin.

Consider Histones as another example. They are the chief protein components of chromatin, acting as spools around which DNA winds, and playing a role in gene regulation. Without histones, the unwound DNA in chromosomes would be very long (a length to width ratio of more than 10 million to 1 in human DNA). Equation 7.2.7 for Histones would hence be:

$$Sig_{histones} = Xa$$

$$+ \overline{Yb_{0-n}} \quad where \left[\begin{array}{c} X \in [S_{System_{Pr}}] \\ Y \in [S_{System_{Pr}}, S_{System_P}, S_{System_K}, S_{System_N}] \\ a, b \ are \ integers; a > b \end{array} \right]$$

Eq 7.2.7: Histones

The secondary element Y would have elements such as 'gene regulation', 'unwound DNA management', amongst others.

Generation of Cellular-Level Partial-Singularity 'Lipids' Genetic Code

Lipids by themselves are tiny molecules, but when grouped together form the largest structures of the cell. When placed in water, lipid molecules aggregate to form huge waterproof sheets. These sheets easily form boundaries at multiple levels and allow concentrated interactions and work to be performed within a cell. Hence, the nucleus and the mitochondria are contained within lipid-defined compartments. Similarly, each cell itself is contained within a lipid-defined boundary.

Lipids are therefore promoters of relationship, of harmony in the cell, of nurturing the cell-level division of labor, of allowing specialization and uniqueness to emerge, hence perhaps of earlier forms of compassion and love, and so on. The notion of such early forms of compassion is consistent with the biologist's perspective that at some point a gene for compassion was developed in pre-human species (Wright, 2009). Being so, one can see that there is similarity with the set for system-nurturing highlighted earlier in Chapter 2.3. Reproducing Equation 2.3.4:

$$S_{System_N} \ni [Love, Compassion, Harmony, Relationship \dots]$$

This function of harmonization suggests that lipids can therefore be thought of as a precipitation of system-nurturing at the cellular level.

Lipids could have a generalized signature, as in Equation 7.2.8, derived from the system-nurturing family:

$$Sig_{lipid} = Xa +$$

$$\overline{Yb_{0-n}} \quad where \begin{bmatrix} X \in [S_{System_N}] \\ Y \in [S_{System_{Pr}}, S_{System_P}, S_{System_K}, S_{System_N}] \\ a, b \ are \ integers; a > b \end{bmatrix}$$

Eq 7.2.8: Lipid

Note that (7.2.8) already implies that Lines $1 - 5$ in the Light Matrix (7.2.1) have been activated, and that the logic of the lipid-ecosystem will precipitate into the post material-fabric cell-based genetic structure through the action of Line 6-7 of (7.2.1). This action of precipitating into the cell-based structure as opposed to the material-fabric is designated by 'PGGT $<...>$' as specified in (7.2.1).

Specific lipids such as monoglycerides and phospholipids could have the following equations:

$$Sig_{monoglyceride} = Xa +$$

$$\overline{Yb_{0-n}} \quad where \left[\begin{array}{c} X \in [S_{System_N}] \\ Y \in [S_{System_{Pr}}, S_{System_P}, S_{System_K}, S_{System_N}] \\ a, b \ are \ integers; a > b \end{array} \right]$$

Eq 7.2.9: Monoglyceride

$$Sig_{phospholipids} = Xa +$$

$$\overline{Yb_{0-n}} \quad where \left[\begin{array}{c} X \in [S_{System_N}] \\ Y \in [S_{System_{Pr}}, S_{System_P}, S_{System_K}, S_{System_N}] \\ a, b \ are \ integers; a > b \end{array} \right]$$

Eq 7.2.10: Phospholipids

The primary element X shared by each of the lipids could be an attribute or function such as 'compartmentalization'. Secondary elements Y could be of the nature of 'work breakdown', 'intra-cell love', amongst others, and would vary with each different kind of lipid.

Generation of Cellular-Level Partial-Singularity 'Polysaccharides' Genetic Code

Polysaccharides are long, often branched chains of sugar molecules. Sugars are covered with hydroxyl groups, which associate to form storage containers. As a result polysaccharides function as the storehouse of cell's energy. In addition polysaccharides are also used to build some of the most durable biological structures. The stiff shell of insects, for example are made of long polysaccharides.

Polysaccharides function to create energy, power, courage, strength thereby readying the cell for adventure, and so on. Being so, one can see that there is similarity with the set for system-power highlighted previously in Chapter 2.3. Reproducing Equation 2.3.2:

$$S_{System_P} \ni [Power, Courage, Adventure, Justice, ...]$$

Providing energy and strength, polysaccharides can be thought of as a precipitation of system-power at the cellular level.

Polysaccharides could have a generalized signature, as in Equation 7.2.11, derived from the system-power family:

$$Sig_{polysaccharide} = Xa + \overline{Yb_{0-n}} \quad where \begin{bmatrix} X \in [S_{System_P}] \\ Y \in [S_{System_{Pr}}, S_{System_P}, S_{System_K}, S_{System_N}] \\ a, b \ are \ integers; a > b \end{bmatrix}$$

Eq 7.2.11: Polysaccharide

Note that (7.2.11) already implies that Lines 1 – 5 in the Light Matrix (7.2.1) have been activated, and that the logic of the polysaccharide-ecosystem will precipitate into the post material-fabric cell-based genetic structure through the action of Line 6-7 of (7.2.1). This action of precipitating into the cell-based structure as opposed to the material-fabric is designated by 'PGGT <...>' as specified in (7.2.1).

Glycogen is an example of a polysaccharide. Glycogen forms an energy reserve that can be quickly mobilized to meet a sudden need for glucose, but one that is less compact and more immediately available as an energy reserve than say triglycerides. Equation 7.2.12 for Glycogen follows:

$$Sig_{glycogen} = Xa +$$

$$\overline{Yb_{0-n}} \quad where \left[\begin{array}{c} X \in [S_{System_P}] \\ Y \in [S_{System_{Pr}}, S_{System_P}, S_{System_K}, S_{System_N}] \\ a, b \ are \ integers; a > b \end{array} \right]$$

Eq 7.1.12: Glycogen

Cellulose is another example of a polysaccharide. Cellulose is a polymer made with repeated glucose units bonded together by beta-linkages. Humans and many animals lack an enzyme to break the beta-linkages, so they do not digest cellulose. Equation 7.2.13 for Cellulose follows:

$$Sig_{cellulose} = Xa +$$

$$\overline{Yb_{0-n}} \quad where \left[\begin{array}{c} X \in [S_{System_P}] \\ Y \in [S_{System_{Pr}}, S_{System_P}, S_{System_K}, S_{System_N}] \\ a, b \ are \ integers; a > b \end{array} \right]$$

Eq 7.2.13: Cellulose

The primary element for the preceding polysaccharides X would be along the lines of 'energy storage'. The secondary elements may vary, with a Y element for Glycogen being 'rapid energy deployment' for example, and a Y element for Cellulose being 'bonded energy', for example.

Summary of Cellular-Level Partial-Singularity Genetic Code

Summarizing, after the cellular stage iterations of the Light-Space-Time Emergence equation are complete, the following code-segments will have been generated as specified by Equation 7.2.14, Active Cellular-Level Partial-Singularity Pre-Genetic and Genetic Code:

$$Light - Space - Time \ Emergence_{Cells} =$$

$$\left\|\begin{bmatrix} \begin{bmatrix} c_\infty : [Pr, Po, K, H] \\ \left(\downarrow R_{C_K} = f\left(R_{C_\infty}\right)\right) \\ c_K : [S_{Pr}, S_{Po}, S_K, S_H] \\ \left(\downarrow R_{C_N} = f\left(R_{C_K}\right)\right) \\ c_N : f(S_{Pr} \, x \, S_{Po} \, x \, S_K \, x \, S_H) \\ \left(\downarrow R_{C_U} = f\left(R_{C_N}\right)\right) \\ c_U : [P, V, M, C] \\ \Uparrow \\ c_{0 : [D, W, I, C]} \end{bmatrix}_{Light} \\ \begin{bmatrix} M_3 : -\infty \le t \le \infty \\ \downarrow \\ M_2 : 0 \ge t > \infty \\ \downarrow \\ M_1 : 0 > t > \infty \\ \downarrow \\ U \to \begin{array}{l} t \le E_{Cell}; TC: M_3 \to U \\ t \sim E_{Human}; TC: U \to M_3 \end{array} \end{bmatrix}_{Time} \quad \begin{array}{c} \begin{bmatrix} M_3 \to System_X \\ (\uparrow F \to I) \\ M_2 \to S_{System_X} \\ (\uparrow Sig \to F) \\ M_1 \to Sig_x \\ (\uparrow > P_x) \\ U \to \quad x_{Molecules} \end{bmatrix}_{Space} \\ \\ TC \to x_{Cells} \end{array} \end{bmatrix}\right\|_{\langle x_U | x_T \rangle}$$

$$\Rightarrow$$

$\langle Space - Time - Energy - Gravity \; Material - Fabric \; Pre$
$\qquad - Genetic \; Code \rangle \; +$

$\langle Electromagnetic \; Spectrum \;\; Material - Fabric \; Pre - Genetic \; Code \rangle$
$\qquad +$

$\langle Quantum - Particle \; Material - Fabric \; Pre - Genetic \; Code \rangle \; +$

$\langle Atoms \;\; Material - Fabric \; Pre - Genetic \;\; Code \rangle \; +$

$\langle Molecules \;\; Material - Fabric \; Pre - Genetic \; Code \rangle + LSTE \; \langle \dots \rangle$
$\qquad + PGGT < \dots > \; +$

$$\left(\begin{array}{c} \sum Sig_{lipid} = Xa + \overline{Yb_{0-n}} \\ where \begin{bmatrix} X \in [S_{System_N}] \\ Y \in [S_{System_{Pr}}, S_{System_P}, S_{System_K}, S_{System_N}] \\ a, b \text{ are integers}; a > b \end{bmatrix} \\ \sum Sig_{nucleic\ acid} = Xa + \overline{Yb_{0-n}} \\ where \begin{bmatrix} X \in [S_{System_K}] \\ Y \in [S_{System_{Pr}}, S_{System_P}, S_{System_K}, S_{System_N}] \\ a, b \text{ are integers}; a > b \end{bmatrix} \\ \sum Sig_{polysaccharide} = Xa + \overline{Yb_{0-n}} \\ where \begin{bmatrix} X \in [S_{System_P}] \\ Y \in [S_{System_{Pr}}, S_{System_P}, S_{System_K}, S_{System_N}] \\ a, b \text{ are integers}; a > b \end{bmatrix} \\ \sum Sig_{protein} = Xa + \overline{Yb_{0-n}} \\ where \begin{bmatrix} X \in [S_{System_{Pr}}] \\ Y \in [S_{System_{Pr}}, S_{System_P}, S_{System_K}, S_{System_N}] \\ a, b \text{ are integers}; a > b \end{bmatrix} \end{array}\right)$$

Eq. 7.2.14, Active Cellular-Level Partial-Singularity Pre-Genetic and Genetic Code

Note that the final code segment following 'PGGT <...>' contains the code for all possible ($\sum x$) lipids, nucleic acids, polysaccharides, and proteins, respectively. As mentioned previously in the sub-section on nucleic acids, this final code segment would also be contained in the nucleic acid segment in some form.

Each equation-segment in (7.2.14) will generate a vast body of "code". As a reminder the language of this code is essentially four-fold, deriving from properties of light, and function-based, and depicts the growing biography by which the singular light-based edifice expresses its materialization, and at this stage, through the cellular-level partial-singularity.

Chapter 7.3: Generation of Fundamental Capacities of Self Partial-Singularity

As human beings we experience sensations, urges and desires and wills, feelings and emotions, and thought.

These are key aspects of our being and becoming and critical aspects of how choice at both the individual and collective levels may be determined. Further, there is known to be a tight relationship between these fundamental capacities of being and the effect on bodies. This implies that there is a deeply embedded relationship between these capacities and the actual functioning of bodies and cells.

This chapter suggests that this deeply embedded relationship is founded on genetic code that is created with the emergence of these capacities. This chapter hence, will go over the quantum-level computation that suggests how light emerges as these capacities, how these precipitate into the post material-fabric genetic housing structure, and what the specific equation-segment function-based code manifests as.

Light's Emergence as Fundamental Capacities of Self

As discussed previously the Light-Space-Time Emergence equation (3.1.3) being iterative, can be used to model emergence as it proceeds from simpler four-fold to more complex four-fold manifestations. Hence (3.1.3) has already been applied to suggest the emergence of the space-time-energy-gravity quadrumvirate, the electromagnetic spectrum, quantum particles in general, bosons as a further instance of a particular kind of quantum particle, atoms, and living cells.

Here, a revised form of it building off the PGGT form of the cells genetic code (7.2.1) will be applied to suggest the emergence of fundamental capacities of self, Equation 7.3.1, Generation of Fundamental Capacities of Self Partial-Singularity. The architecture and details of capacities of self can be seen to be the result of the application of the Light, Space, and Time matrices as will be elaborated. But further, as implied by (3.6.5), the Potential Effect of Levels of Light on Genetic-Type Information equation, any process of deeper organization, such as is responsible for the architecture and cohesiveness of capacities of self, has the possibility of altering the material-fabric or a post material-fabric housing structure so long as the bases involved are driven primarily by a meta-level.

Hence:

$$Light - Space - Time\ Emergence_{Fundamental\ Capacities\ of\ Self} =$$

$$\left|\left|\begin{array}{c} \begin{bmatrix} c_\infty: [Pr, Po, K, H] \\ (\downarrow R_{C_K} = f(R_{C_\infty})) \\ c_K: [S_{Pr}, S_{Po}, S_K, S_H] \\ (\downarrow R_{C_N} = f(R_{C_K})) \\ c_N: f(S_{Pr} \; x \; S_{Po} \; x \; S_K \; x \; S_H) \\ (\downarrow R_{C_U} = f(R_{C_N})) \\ c_U: [P, V, M, C] \\ \Uparrow \\ c_{0:[D,W,I,C]} \end{bmatrix}_{Light} \quad \begin{bmatrix} M_3 \to System_X \\ (\uparrow F \to I) \\ M_2 \to S_{System_X} \\ (\uparrow Sig \to F) \\ M_1 \to Sig_x \\ (\uparrow > P_x) \\ U \to x_{Cells} \end{bmatrix}_{Space} \\ U \to \begin{bmatrix} M_3: -\infty \le t \le \infty \\ \downarrow \\ M_2: 0 \ge t > \infty \\ \downarrow \\ M_1: 0 > t > \infty \\ \downarrow \\ t \le E_{Cell}; TC: M_3 \to U \\ t \sim E_{Human}; TC: U \to M_3 \end{bmatrix}_{Time} \qquad TC \to x_{Capacities\ of\ Self} \end{array}\right|\right|_{\langle x_U | x_T \rangle}$$

$\Rightarrow$

$\langle Space - Time - Energy - Gravity\ Material - Fabric\ Pre - Genetic\ Code \rangle +$

$\langle Electromagnetic\ Spectrum\ \ Material - Fabric\ Pre - Genetic\ Code \rangle +$

$\langle Quantum - Particle\ Material - Fabric\ Pre - Genetic\ Code \rangle +$

$\langle Atoms\ Material - Fabric\ Pre - Genetic\ Code \rangle +$

$\langle Molecules\ Material - Fabric\ Pre - Genetic\ Code \rangle + LSTE \langle ... \rangle + PGGT < \cdots > +$

$Cells\ Genetic\ Code + LSTE < \cdots > + Fundamental\ Capacities\ of\ Self\ Genetic\ Code$

Eq 7.3.1: Generation of Fundamental Capacities of Self Partial-Singularity

Starting with the Light-Matrix, the top left-hand matrix in (7.3.1), the first line from the top, $C_\infty: [Pr, Po, K, H]$, specifies the architecture of the fundamental capacities of self to be introduced in this chapter. Hence, Thoughts are an emergence of Light's property of Knowledge, Urges, Desires, and Wills are an emergence of Light's property of Power, Sensations are an emergence of Light's property of Presence, and Feelings and Emotions are an emergence of Light's property of Harmony. The fundamental architecture of these aspects hence, is an emergence of the properties of Light at ∞.

Line 3 in the Light-Matrix, C_K: $[S_{Pr}, S_{Po}, S_K, S_H]$, elaborates the sets for Presence, Power, Knowledge, and Harmony, each containing multiple elements. For example, as will be explored in the section on Sensations, various elements derived from the four sets define the behavior of Sensations and could be functions such as 'tangible', 'take notice of', amongst others, hence collectively describing Sensations' way of being. Specifically, Line 5, $C_{N:} f(S_{Pr} \times S_{Po} \times S_K \times S_H)$, suggests that unique seeds are created from a combination of such elements from all four sets, with a particular element leading or having more weight, that in effect creates the distinctness possible at the level of fundamental capacities of self.

Line 6, ($\downarrow R_{C_U} = f(R_{C_N})$), specifies quantization between the layer where the seeds are formed, and the physical layer, and as explored in Chapter 3.5 and 3.6, will result in Line 7, C_U: $[P, V, M, C]$, hence changing the post material-fabric housing structure. Note that as in the process describing the generation of the code-segments for space-time-energy-gravity quantization at the time of the Big Bang, the FBLEE (four-base logic-encoding ecosystem) process is apparently transparent.

The possibilities represented by Lines 1 through 5 hence concretize through the quantization represented by Line 6 to become or further enhance the capacities of self with its subtle physical (related to Presence), vital (related to Power), mental (related to Knowledge), and connection (related to Harmony) aspects now existing in material reality typified by Light moving at c. Note that just as Line 6 represents a process of quantization relating the layer of reality created by Light traveling at c with the antecedent layers, so too Lines 2 and 4 as previously discussed, also represent quantization of a more subtle kind that ultimately plays a critical part in allowing the material-fabric to express infinite diversity.

Typically it is the process as captured by the Space-Matrix that will determine if Line 6 is activated. Specifically patterns at the untransformed layer, U, will need to be overcome, as specified by the second-line from the bottom of the Space-Matrix: $(\uparrow > P_x)$. But as specified by the bottom-line of the Time-Matrix, reproduced below, it is only with the advent of the human-system that the automaticity of the action of meta-levels is reversed:

$$U \rightarrow \begin{array}{l} t \leq E_{Cell}; \text{TC: } M_3 \rightarrow U \\ t \sim E_{Human}; \text{TC: } U \rightarrow M_3 \end{array}$$

Hence in the case of sensations, which in this emergence is largely a post-human system, the fact that patterns do need to be overcome means that quantization requires effort to happen. Given this, it is useful to review Equation 3.6.5, Potential Effect of Levels of Light on Genetic-Type Information:

$$Potential\ Effect\ of\ Levels\ of\ Light\ on\ Genetic$$
$$-\ Type\ Information =$$

$$\begin{bmatrix} STATIC\ \langle |[L][S][T]TC \rightarrow x_T|_{\langle x_U | x_T \rangle} \rangle \\ \times \\ \left((Y > U: Z_Q)\ \vee\ (Y \leq U: Z_F)\ \vee\ (Y = U: Z_R) \right) \\ \ni \\ \left(\begin{array}{c} Z \in \mathbb{U}\ (Space, Time, Energy, Gravity) \\ Q: Quantization; F: Fragmentation; R:\ Random \end{array} \right) \end{bmatrix} \rightarrow h \ni$$

$$h \in \left(\begin{array}{c} [Q]: Constructive\ zone, \\ [Q]: Constructive\ zone\ \wedge\ Constructive\ mutation, \\ [F]: Destructive\ mutation, \\ [R]: Random\ mutation \end{array} \right)$$

Line 1 from the top in the matrix is simply a static form of (3.6.1) the Simplified Light-Space-Time Emergence equation. The static form is designated by 'STATIC', and implies that fundamental operations true of (3.6.1) are being highlighted in (3.6.5). In other words (3.6.1) already has all the operations highlighted in (3.6.5) in it, but by 'freezing' it by making it static, the essential dynamics leading to possible mutations at the genetic level can more clearly be highlighted.

Line 1 is then subjected (×) to a determination of the dominant levels of light that may be active, designated by Line 2, ' $\left((Y > U: Z_Q) \vee (Y \leq U: Z_F) \vee (Y = U: Z_R) \right)$ '. Unpacking this, '$Y > U$' implies meta-levels are active and as a result it is possible that Z_Q is going to take place (the subscript 'Q' implies quantum-level action). This also implies activation and potential change of FBLEE. The call from below, as it were, may invoke some function that already exists in the subtle-libraries 'above', so that some already existing function may influence FBLEE through ∞-entanglement, K-entanglement, or N-entanglement. This may be thought of as a key-and-lock mechanism, where a deep enough visceral urge from below acting as the key, opens an entangled lock to alter FBLEE as per the visceral urge. '$Y \leq U$' implies that only the untransformed levels are active, and therefore also the sub-level where the speed of light is 0 is active and as a result Z_F is going to take place (the subscript 'F' implies 'fragmentation'). '$Y = U$' implies that all levels are active and as a result Z_R is going to take place (the subscript 'R' implies 'random').

Line 3 elaborates the significance of Z_Q, Z_F, and Z_R. Hence Z is the union of potential quantum-operations of space, time, energy, and gravity, designated by '$Z \in \mathbb{U}\,(Space, Time, Energy, Gravity)$'. But the nature of the operations, as suggested in the previous paragraph, is designated   by

' Q: *Quantization*; F: *Fragmentation*; R: *Random* '. Z_Q , then, implies that the full quantization originating from updated four-base logic-encoding ecosystems (FBLEE) can take place, and will result in lasting material change at the genetic level. Z_F implies that the essential set will be fragmented and that only libraries at the level of local cellular-level DNA or can potentially be altered. Z_R also precludes full quantization, and that some partial cellular-level DNA constructive or destructive mutation may take place.

The ' $\rightarrow h \ni h \in$' segment resolves the outcome of the operations implied by Lines 1 – 3, suggesting that the outcome will be 'h' such that ($\ni$) 'h' is an element ($\in$) of the set specified by the members '[Q]: *Constructive zone*', '[Q]: *Constructive zone AND Constructive mutation*', '[F]: *Destructive mutation*', and '{R}: *Random mutation*'. '[Q]: *Constructive zone*' implies that the in-built buffer has been crossed and that access to the deeper four-base logic-encoding ecosystem (FBLEE) has been granted. '[Q]' in this segment implies that there is the possibility that full-quantization as specified by Line 3 of the previous matrix can take place. Access to this zone is a prerequisite for constructive mutation to occur, as designated by the element '[Q]: *Constructive zone* ∩ *Constructive mutation* ', which implies that full-quantization is going to take place and will result in material change. The '[F]' specifies the relationship between 'Fragmentation' in Line 3 of the previous matrix and destructive mutation. The '[R]' specifies the relationship between 'Random' in Line 3 of the previous matrix and random mutation.

But as just summarized in the Time-Matrix in (7.3.1) Y is by definition not greater than U and hence quantization is not automatic. In terms of fundamental capacities of self, such quantization implies that increasing wholeness can become fully active through specific space, time, energy, and gravity quantization to create an holistic "ecosystem" with its own "fundamental capacities of self logic" as it were. Possible precipitation into the post material-fabric housing structure is captured by the action of 'PGGT <...>' that can be assumed to be triggered by Z_Q. This "logic" or more specifically genetic code is elaborated in the following sub-sections and suggests the increasing sphere of influence of the fundamental capacities of self partial-singularity that now also contains light-engendered laws for all selves.

Generation of Fundamental Capacities of Self Partial-Singularity 'Sensations' Genetic Code

Sensations are those things we experience with our senses. We see things, hear things, and smell things, taste things, can touch things. This ability to enter into relationship with objects through sensation is nothing other than a result of the emergence of Light's property of Presence. We become present to Presence through the device of sensation. Sensation can be thought of as the means by which this property of Light – Presence - molds or ingrains itself in us as human beings. Its potentiality, all which is contained in this aspect of Light, becomes available to us through the power of sensation. Hence an equation, Equation 7.3.2, will generally represent the family of sensations. Some elements that it would comprise of may be 'tangible', 'take notice of', amongst others.

$$Sig_{sensation} = Xa +$$

$$\overline{Yb_{0-n}} \ where \ \begin{bmatrix} X \in [S_{System_{Pr}}] \\ Y \in [S_{System_{Pr}}, S_{System_P}, S_{System_K}, S_{System_N}] \\ a, b \ are \ integers; a > b \end{bmatrix}$$

Eq 7.3.2: Sensation

There could also be equations for hearing, seeing, tasting, touching, and smelling.

But there is also a deeper experience of sensation that is possible. When we see things, for instance, what are we seeing? Is it just the surface rendering of the play of matter, or do we see that the fullness of Light is still there, with all its potentiality and possibility, in the smallest thing we look at? Do we see that the whole universe and more is present in all its fullness in the least thing that we easily ignore, or belittle, or loathe? When we touch things is it the seeming concreteness of the play of the particles or atoms or chains of molecules that we touch? Or is it the Love and Light and the vastness of all that IS that allows itself to be as a small corner that we touch so as to make infinity be felt by something so finite?

Such a deeper contact offered through sensation suggests a subset of (7.3.2) with secondary elements perhaps described as 'fullness of Light', 'contacting infinity', amongst others, thus also yielding an equation form (7.3.3):

$$Sig_{deeper-sensation} = Xa +$$

$$\overline{Yb_{0-n}} \quad where \quad \begin{bmatrix} X \in [S_{System_{Pr}}] \\ Y \in [S_{System_{Pr}}, S_{System_P}, S_{System_K}, S_{System_N}] \\ a, b \; are \; integers; a > b \end{bmatrix}$$

340

Eq 7.3.3: Deeper Sensation

Note that (7.3.2 and 7.3.3) already implies that Lines 1 – 5 in the Light Matrix (7.3.1) have been activated, and that the logic of the sensations- ecosystem will precipitate into the post material-fabric cell-based genetic structure through the action of Line 6-7 of (7.3.1). This action of precipitating into the cell-based structure as opposed to the material-fabric is designated by 'PGGT <...>' as specified in (7.3.1).

Generation of Fundamental Capacities of Self Partial-Singularity 'Urges, Desires & Wills' Genetic Code

Urges and desires and wills are similarly a play of the emergence of Light's property of Power. In the mystery of focus, the vastness of Light has projected itself in us into an apparent smallness that is in reality everything that is. And this smallness is trying through urge and desire and will to connect viscerally or even intentionally to other smallnesses that similarly are nothing other than the fullness of Light projected into a small smorgasbord of selected function. So the urge or desire for food, or companionship, or of possession, or of climbing a peak, is nothing other than Light's compressed property of Power, trying to reach more of the fullness that it is through a fulfillment of the urge or desire or will that it masquerades as. Hence urges can be represented as Equation 7.3.4:

$$Sig_{urges} = Xa + \overline{Yb_{0-n}} \ where \begin{bmatrix} X \in [S_{System_P}] \\ Y \in [S_{System_{Pr}}, S_{System_P}, S_{System_K}, S_{System_N}] \\ a, b \ are \ integers; a > b \end{bmatrix}$$

341

Elements may be of the type of 'grasp', 'possess', 'deeply connect',
amongst
others.

Note that
(7.3.4)
already
implies
that Lines
1 – 5 in the Light Matrix (7.3.1) have been activated, and that the logic of the urges-ecosystem will precipitate into the post material-fabric cell-based genetic structure through the action of Line 6-7 of (7.3.1). This action of precipitating into the cell-based structure as opposed to the material-fabric is designated by 'PGGT <...>' as specified in (7.3.1).

Generation of Fundamental Capacities of Self Partial-Singularity 'Feelings & Emotion' Genetic Code

Feelings and emotions are a play of the emergence of Light's property of Harmony or Nurturing. Its instrument is the Heart and it generates an array of emotions that are an indication or active radar of whether we are moving toward or away from a reality of harmony, whether based on our small self or some larger Self of Light. Gradually, by navigating with these emotions and feelings we can get to a state where we always feel positive emotions which basically means we have more truly

342

entered into relationship with some larger continent of Light. An equation for feelings is as represented by Equation 7.3.5:

$$Sig_{feelings} = Xa +$$

$$\overline{Yb_{0-n}} \ \ where \ \begin{bmatrix} X \in [S_{System_N}] \\ Y \in [S_{System_{Pr}}, S_{System_P}, S_{System_K}, S_{System_N}] \\ a, b \ are \ integers; a > b \end{bmatrix}$$

Eq 7.3.5: Feelings

Note that (7.3.5) already implies that Lines 1 – 5 in the Light Matrix (7.3.1) have been activated, and that the logic of the feelings-ecosystem will precipitate into the post material-fabric cell-based genetic structure through the action of Line 6-7 of (7.3.1). This action of precipitating into the cell-based structure as opposed to the material-fabric is designated by 'PGGT <…>' as specified in (7.3.1).

Generation of Fundamental Capacities of Self Partial-Singularity 'Thought' Genetic Code

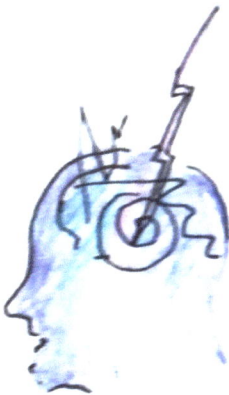

Thoughts are a play of the emergence of Light's property of Knowledge. Through the thought we can become greater or conceptualize things greater or begun to enter into relationship with some things other than our small self. Thought allows us to connect to more "othernesses" or even the oneness of the reality of Light. An equation for thoughts is the following:

$$Sig_{thoughts} = Xa +$$

$$\overline{Yb_{0-n}} \quad where \quad \begin{bmatrix} X \in [S_{System_K}] \\ Y \in [S_{System_{Pr}}, S_{System_P}, S_{System_K}, S_{System_N}] \\ a, b \ are \ integers; a > b \end{bmatrix}$$

Eq 7.3.6: Thoughts

Note that (7.3.5) already implies that Lines 1 – 5 in the Light Matrix (7.3.1) have been activated, and that the logic of the thoughts-ecosystem will precipitate into the post material-fabric cell-based genetic structure through the action of Line 6-7 of (7.3.1). This action of precipitating into the cell-based structure as opposed to the material-fabric is designated by 'PGGT <...>' as specified in (7.3.1).

Summary of Fundamental Capacities of Self Partial-Singularity Genetic Code

Summarizing, after the fundamental capacities of self

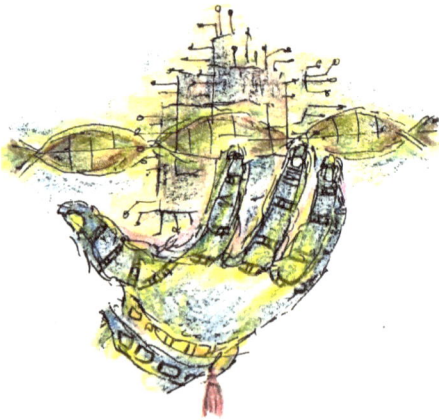

iterations of the Light-Space-Time Emergence equation are complete, the following code-segments will have been generated as specified by Equation 7.3.7, Active Fundamental Capacities of Self Partial-Singularity Pre-Genetic and Genetic Code:

$$Light - Space - Time \ Emergence_{Fundamental \ Capacities \ of \ Self} =$$

344

$$\left| \begin{array}{l} \begin{bmatrix} \begin{array}{c} c_\infty : [Pr, Po, K, H] \\ \left(\downarrow R_{C_K} = f\left(R_{C_\infty}\right) \right) \\ c_K : [S_{Pr}, S_{Po}, S_K, S_H] \\ \left(\downarrow R_{C_N} = f\left(R_{C_K}\right) \right) \\ c_N : f(S_{Pr} \times S_{Po} \times S_K \times S_H) \\ \left(\downarrow R_{C_U} = f\left(R_{C_N}\right) \right) \\ c_U : [P, V, M, C] \\ \Uparrow \\ c_{0:[D,W,I,C]} \end{array} \end{bmatrix}_{Light} \begin{bmatrix} M_3 \to System_X \\ (\uparrow F \to I) \\ M_2 \to S_{System_X} \\ (\uparrow Sig \to F) \\ M_1 \to Sig_x \\ (\uparrow > P_x) \\ U \to x_{Cells} \end{bmatrix}_{Space} \\ \begin{bmatrix} M_3 : -\infty \le t \le \infty \\ \downarrow \\ M_2 : 0 \ge t > \infty \\ \downarrow \\ M_1 : 0 > t > \infty \\ \downarrow \\ U \to \begin{array}{l} t \le E_{Cell}; TC: M_3 \to U \\ t \sim E_{Human}; TC: U \to M_3 \end{array} \end{bmatrix}_{Time} \quad TC \to x_{Capacities\ of\ Self} \end{array} \right| \langle x_U | x_T \rangle$$

$\Rightarrow$

$\langle Space - Time - Energy - Gravity\ Material - Fabric\ Pre$
$\qquad - Genetic\ Code \rangle +$

$\langle Electromagnetic\ Spectrum\ \ Material - Fabric\ Pre - Genetic\ Code \rangle$
$\qquad +$

$\langle Quantum - Particle\ Material - Fabric\ Pre - Genetic\ Code \rangle +$

$\langle Atoms\ Material - Fabric\ Pre - Genetic\ Code \rangle +$

$\langle Molecules\ Material - Fabric\ Pre - Genetic\ Code \rangle + LSTE\ \langle \ldots \rangle$
$\qquad + PGGT < \cdots > +$

$Cells\ Genetic\ Code + LSTE < \cdots > +$

345

$$\left(\begin{array}{c} \sum Sig_{feelings} = Xa + \overline{Yb_{0-n}} \\ X \in [S_{System_N}] \\ where \left[Y \in [S_{System_{Pr}}, S_{System_P}, S_{System_K}, S_{System_N}] \right] \\ a, b \text{ are integers}; a > b \\ \sum Sig_{thoughts} = Xa + \overline{Yb_{0-n}} \\ X \in [S_{System_K}] \\ where \left[Y \in [S_{System_{Pr}}, S_{System_P}, S_{System_K}, S_{System_N}] \right] \\ a, b \text{ are integers}; a > b \\ \sum Sig_{urges} = Xa + \overline{Yb_{0-n}} \\ X \in [S_{System_P}] \\ where \left[Y \in [S_{System_{Pr}}, S_{System_P}, S_{System_K}, S_{System_N}] \right] \\ a, b \text{ are integers}; a > b \\ \sum Sig_{sensation} = Xa + \overline{Yb_{0-n}} \\ X \in [S_{System_{Pr}}] \\ where \left[Y \in [S_{System_{Pr}}, S_{System_P}, S_{System_K}, S_{System_N}] \right] \\ a, b \text{ are integers}; a > b \end{array} \right)$$

Eq. 7.3.7, Active Fundamental Capacities of Self Partial-Singularity Pre-Genetic and Genetic Code

Note that the final code segment following 'LSTE <...>' contains the code for all possible ($\sum x$) sensations, urges,

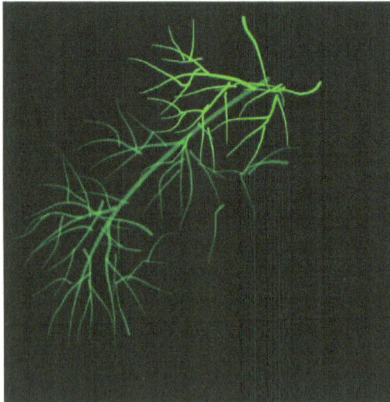

feelings, and thought, respectively, and depicts the growing biography by which the singular light-based edifice expresses its materialization, and at this stage, through the fundamental capacities of self partial-singularity.

Chapter 7.4: Generation of Truer Individuality Partial-Singularity

At the core individuals can be thought of as projections of seeds formed in a vast continent of Light. So that core must always be there but it is often covered by surface dynamics of the physical, the vital, and the mental. These too are formed from properties of Light, but since the light has been separated from its source these movements and dynamics are incomplete in their nature.

It is easy for these dynamics and movements to completely occupy all individual processing power. This can happen to such an extent that the individual acting on the surface can become subject to these surface movements. But so long as there is the remembrance that even in separation and smallness - whether selfishness, myopia, fundamentalism, or any other ism - these movements are still light, then that can become the means by which the apparent chains around individuals can be loosened.

For in its essence these small movements are trying to extend into something other than what they are. By connecting the light in them with the larger Light this extension can yield something different than what is possible through any other means. And in the process the hold that these smaller movements have is diminished and one can begin to see or experience larger vistas of light.

Such a passage where one is always moving to larger vistas of light, more easily allows entry into a field of original seeds and it is so that each can begin to enter into conscious communion with truer individuality.

Light's Emergence as Truer Individuality

As discussed previously the Light-Space-Time Emergence equation (3.1.3) being iterative, can be used to model emergence as it proceeds from simpler four-fold to more complex four-fold manifestations. Hence (3.1.3) has already been applied to suggest the emergence of the space-time-energy-gravity quadrumvirate, the electromagnetic spectrum, quantum particles in general, bosons as a further instance of a particular kind of quantum particle, atoms, living cells, and basic capacities of the self.

Here, a revised form of it building off the truer individuality equation (4.2.6) will be applied to suggest the emergence of truer individuality, Equation 7.4.1, Generation of Truer Individuality Partial-Singularity. The architecture and details of truer individuality can be seen to be the result of the application of the Light, Space, and Time matrices as will be elaborated. But further, as implied by (3.6.5), the Potential Effect of Levels of Light on Genetic-Type Information equation, any process of deeper organization, such as is responsible for the architecture and cohesiveness of truer individuality, has the possibility of altering FBLEE and post material-fabric housing structures so long as the bases involved are driven primarily by a meta-level.

$$Light - Space - Time\ Emergence_{Truer\ Individuality} =$$

$$\left| \begin{bmatrix} \begin{bmatrix} c_\infty : [Pr, Po, K, H] \\ \left(\downarrow R_{C_K} = f\left(R_{C_\infty}\right) \right) \\ c_K : [S_{Pr}, S_{Po}, S_K, S_H] \\ \left(\downarrow R_{C_N} = f\left(R_{C_K}\right) \right) \\ c_N : f(S_{Pr} \times S_{Po} \times S_K \times S_H) \\ \left(\downarrow R_{C_U} = f\left(R_{C_N}\right) \right) \\ c_U : [P, V, M, C] \\ \Uparrow \\ c_{0 : [D, W, I, C]} \end{bmatrix}_{Light} \begin{bmatrix} M_3 \to System_X \\ (\uparrow F \to I) \\ M_2 \to S_{System_X} \\ (\uparrow Sig \to F) \\ M_1 \to Sig_x \\ (\uparrow > P_x) \\ U \to x_{Humans} \end{bmatrix}_{Space} \\ \begin{bmatrix} M_3 : -\infty \le t \le \infty \\ \downarrow \\ M_2 : 0 \ge t > \infty \\ \downarrow \\ M_1 : 0 > t > \infty \\ \downarrow \\ U \to \begin{array}{l} t \le E_{Cell}; TC: M_3 \to U \\ t \sim E_{Human}; TC: U \to M_3 \end{array} \end{bmatrix}_{Time} \quad TC \to x_{Truer\ Individuality} \end{bmatrix} \right| \langle x_U | x_T \rangle$$

$\Rightarrow$

$\langle Space - Time - Energy - Gravity\ Material - Fabric\ Pre - Genetic\ Code \rangle +$

$\langle Electromagnetic\ Spectrum\ Material - Fabric\ Pre - Genetic\ Code \rangle +$

$\langle Quantum - Particle\ Material - Fabric\ Pre - Genetic\ Code \rangle +$

$\langle Atoms\ Pre - Genetic\ Code \rangle$
$\quad + \langle Molecules\ Material - Fabric\ Pre - Genetic\ Code \rangle +$

$LSTE \langle ... \rangle + PGGT < \cdots > + \langle Cells\ Genetic\ Code \rangle + LSTE \langle ... \rangle +$

$(Fundamental\ Capacities\ of\ Self\ Genetic\ Code) + FBLEEE \langle ... \rangle +$

$(Humans\ Genetic\ Code) + LSTE \langle ... \rangle + FBLEEE \langle ... \rangle +$

$Truer\ Individuality\ Genetic\ Code$

Eq 7.4.1: Generation of Truer Individuality Partial-Singularity

Starting with the Light-Matrix, the top left-hand matrix in

(7.4.1), the first line from the top, $C_\infty : [Pr, Po, K, H]$, specifies the architecture of truer individuality as will be further discussed in this chapter. Hence, Knowledge-type individuals are an emergence of Light's property of Knowledge, Power-type individuals are an emergence of Light's property of Power, Service-type individuals are an emergence of Light's property of Presence, and Harmony-type individuals are an emergence of Light's property of Harmony. The fundamental architecture of these aspects hence, is an emergence of the properties of Light at ∞.

Line 3 in the Light-Matrix, $C_K : [S_{Pr}, S_{Po}, S_K, S_H]$, elaborates the sets for Presence, Power, Knowledge, and Harmony, each containing multiple elements. For example, as will be explored in the subsequent subsection, various elements derived from the four sets would define the

architecture of Harmony-type individuals and could be functions such as 'driven to connect people together', amongst others, hence collectively describing Harmony-type individuals' way of being. Specifically, Line 5, $C_{N:} f(S_{Pr} \; x \; S_{Po} \; x \; S_K \; x \; S_H)$, suggests that unique seeds are created from a combination of such elements from all four sets, with a particular element leading or having more weight, that in effect creates the distinctness possible at the level of individuals.

Line 6, $(\downarrow R_{C_U} = f(R_{C_N}))$, specifies quantization between the layer where the seeds are formed, and the physical layer, and as explored in Chapter 3.5 and 3.6, will result in Line 7, C_U: $[P, V, M, C]$, hence changing the post material-fabric housing structure. Note that as in the process describing the generation of the code-segments for space-time-energy-gravity quantization at the time of the Big Bang, the FBLEE (four-base logic-encoding ecosystem) process is apparently transparent.

The possibilities represented by Lines 1 through 5 hence concretize through the quantization represented by Line 6 to become or further enhance the unique individuality with its subtle physical (related to Presence), vital (related to Power), mental (related to Knowledge), and connection (related to Harmony) aspects now existing in material reality typified by Light moving at c. Note that just as Line 6 represents a process of quantization relating the layer of reality created by Light traveling at c with the antecedent layers, so too Lines 2 and 4 as previously discussed, also represent quantization of a more subtle kind that ultimately plays a critical part in allowing FBLEE and post material-fabric housing structure to express infinite diversity.

Typically it is the process as captured by the Space-Matrix that will determine if Line 6 is activated. Specifically patterns at the untransformed layer, U, will need to be overcome, as specified by the second-line from the bottom of the Space-Matrix: $(\uparrow > P_x)$. But as specified by the bottom-line of the Time-Matrix, reproduced below, it is only with the advent of the human-system that the automaticity of the action of meta-levels is reversed:

$$U \rightarrow \begin{array}{l} t \leq E_{Cell}; \text{TC: } M_3 \rightarrow U \\ t \sim E_{Human}; \text{TC: } U \rightarrow M_3 \end{array}$$

Hence in the case of truer individuality, which in this emergence is a post-human system, the fact that patterns do need to be overcome means that quantization requires effort to happen. Given this, it is useful to review Equation 3.6.5, Potential Effect of Levels of Light on Genetic-Type Information:

Potential Effect of Levels of Light on Genetic
— Type Information =

$$\begin{bmatrix} STATIC \ \langle | [L][S][T]TC \ \rightarrow \ x_T | \langle x_U | x_T \rangle \rangle \\ \times \\ \left((Y > U : Z_Q) \ \vee \ (Y \le U : Z_F) \ \vee \ (Y = U : Z_R) \right) \\ \ni \\ \left(\begin{array}{c} Z \in \mathbb{U} \ (Space, Time, Energy, Gravity) \\ Q: Quantization; F: Fragmentation; R: Random \end{array} \right) \end{bmatrix} \rightarrow h \ni$$

$$h \in \left(\begin{array}{c} [Q]: Constructive \ zone, \\ [Q]: Constructive \ zone \ \wedge \ Constructive \ mutation, \\ [F]: Destructive \ mutation, \\ [R]: Random \ mutation \end{array} \right)$$

Line 1 from the top in the matrix is simply a static form of (3.6.1) the Simplified Light-Space-Time Emergence equation. The static form is designated by 'STATIC', and implies that fundamental operations true of (3.6.1) are being highlighted in (3.6.5). In other words (3.6.1) already has all the operations highlighted in (3.6.5) in it, but by 'freezing' it by making it static, the essential dynamics leading to possible mutations at the genetic level can more clearly be highlighted.

Line 1 is then subjected ($\times$) to a determination of the dominant levels of light that may be active, designated by Line 2, ' $\left((Y > U : Z_Q) \ \vee \ (Y \le U : Z_F) \ \vee \ (Y = U : Z_R) \right)$ '. Unpacking this, 'Y > U' implies meta-levels are active and as a result it is possible that Z_Q is going to take place (the subscript 'Q' implies quantum-level action). This also implies activation and potential change of FBLEE. The call from below, as it were, may invoke some function that already exists in the subtle-libraries 'above', so that some already existing function may influence FBLEE through ∞ -entanglement, K-entanglement, or N-entanglement. This may be thought of as a key-and-lock mechanism, where a deep enough visceral urge from below acting as the key, opens an entangled lock to alter FBLEE as per the visceral urge. 'Y $\le$ U' implies that only the untransformed levels are active, and therefore also the sub-level where the speed of light is 0 is active and as a result Z_F is going to take place (the subscript 'F' implies 'fragmentation'). 'Y = U' implies that all levels are active and as a result Z_R is going to take place (the subscript 'R' implies 'random').

Line 3 elaborates the significance of Z_Q, Z_F, and Z_R. Hence Z is the union of potential quantum-operations of space, time, energy, and gravity, designated by '$Z \in \mathbb{U}$ ($Space, Time, Energy, Gravity$)'. But the nature of the operations, as suggested in the previous paragraph, is designated   by 'Q: $Quantization$; F: $Fragmentation$; R: $Random$'. Z_Q, then, implies that the full quantization originating from updated four-base logic-encoding ecosystems (FBLEE) can take place, and will result in lasting material change at the Genetic-Type level. Z_F implies that the essential set will be fragmented and that only libraries at the level of local cellular-level DNA or precipitated material-fabric logic can potentially be altered. Z_R also precludes full quantization, and that some partial local-library constructive or destructive mutation may take place.

The '$\rightarrow h \ni h \in$' segment resolves the outcome of the operations implied by Lines 1 – 3, suggesting that the outcome will be 'h' such that ($\ni$) 'h' is an element ($\in$) of the set specified by the members '[Q]: $Constructive\ zone$', '[Q]: $Constructive\ zone\ AND\ Constructive\ mutation$', '[F]; $Destructive\ mutation$', and '{R}: $Random\ mutation$'. '[Q]: $Constructive\ zone$' implies that the in-built buffer has been crossed and that access to the deeper four-base logic-encoding ecosystem (FBLEE) has been granted. '[Q]' in this segment implies that there is the possibility that full-quantization as specified by Line 3 of the previous matrix can take place. Access to this zone is a prerequisite for constructive mutation to occur, as designated by the element   '   [Q]: $Constructive\ zone \cap Constructive\ mutation$ ', which implies that full-quantization is going to take place and will result in material change. The '[F]' specifies the relationship

354

between 'Fragmentation' in Line 3 of the previous matrix and destructive mutation. The '[R]' specifies the relationship between 'Random' in Line 3 of the previous matrix and random mutation.

But as just summarized in the Time-Matrix in (7.4.1) Y is by definition not greater than U and hence quantization is not automatic. In terms of truer individuality such quantization would imply that increasing wholeness can become fully active through specific space, time, energy, and gravity quantization to create an holistic "ecosystem" with its own "truer individuality logic" as it were. Possible precipitation into the post material-fabric housing structure is captured by the action of 'PGGT $<...>$' that can be assumed to be triggered by Z_Q. This "logic" or more specifically genetic code is elaborated in the following sub-section and suggests the increasing sphere of influence of the truer individuality partial-singularity that now also contains light-engendered laws for deeper individuality.

Generation of Truer Individuality Partial-Singularity Genetic-Code

The truer individuality itself is some mathematical function of the four properties of Light - Harmony, Power, Knowledge, and Service, which in turn, are practically infinite sets of qualities related to one of the main properties of Light.

So it may be that one individual is primarily driven to connect people together, itself a variation of the property of Harmony. The individual may further want to do this by deeply understanding what makes these people tick, itself a variation of the property of Knowledge. So the seed or the truer individuality of this person can be thought of as a mathematical function consisting of some element or property from the set of Harmony and some element or quality from the set of Knowledge, combined together with possibly different elements from the same or different sets, all with possibly different weights, but with the first weight of the need to connect people, being the strongest.

Such an individual may have an essential form as expressed in Equation 7.4.2, Harmony-Type Individual:

$$Sig_{Harmony-type} = Xa +$$

$$\overline{Yb_{0-n}} \quad where \begin{bmatrix} X \in [S_{System_N}] \\ Y \in [S_{System_{Pr}}, S_{System_P}, S_{System_K}, S_{System_N}] \\ a, b \ are \ integers; a > b \end{bmatrix}$$

Eq 7.4.2: Harmony-Type Individual

But equally, and consistent with the emergent properties of Light there could be essential Service-Type, Power-Type, and Knowledge-Type individuals as well, with infinite variation in the precise nature of the core seed, as captured in Equations 7.4.3-5:

$$Sig_{Service-type} = Xa + \overline{Yb_{0-n}}$$

$$where \begin{bmatrix} X \in [S_{System_{Pr}}] \\ Y \in [S_{System_{Pr}}, S_{System_P}, S_{System_K}, S_{System_N}] \\ a, b \ are \ integers; a > b \end{bmatrix}$$

Eq 7.4.3: Service-Type Individual

$$Sig_{Power-type} = Xa + \overline{Yb_{0-n}}$$

$$where \begin{bmatrix} X \in [S_{System_P}] \\ Y \in [S_{System_{Pr}}, S_{System_P}, S_{System_K}, S_{System_N}] \\ a, b \ are \ integers; a > b \end{bmatrix}$$

Eq 7.4.4: Power-Type Individual

$$Sig_{Knowledge-type} = Xa + \overline{Yb_{0-n}}$$

$$where \begin{bmatrix} X \in [S_{System_K}] \\ Y \in [S_{System_{Pr}}, S_{System_P}, S_{System_K}, S_{System_N}] \\ a, b \text{ are integers}; a > b \end{bmatrix}$$

Eq 7.4.5: Knowledge-Type Individual

Note that (7.4.2 - 5) already implies that Lines 1 – 5 in the Light Matrix (7.4.1) have been activated, and that the logic of the truer-individuality- ecosystem will precipitate into the post material-fabric cell-based genetic structure through the action of Line 6-7 of (7.4.1). This action of precipitating into the cell-based structure as opposed to the material-fabric is designated by 'PGGT <...>' as specified in (7.4.1).

Summary of Truer Individuality Partial-Singularity Genetic Code

Summarizing, after the truer individuality iteration of the Light-Space-Time Emergence equation is complete, the following code-segments will have been generated as specified by Equation 7.4.6, Active Truer Individuality Partial-Singularity Pre-Genetic and Genetic:

357

$$Light - Space - Time\ Emergence_{Truer\ Individuality} =$$

$$
\left\|
\begin{array}{l}
\begin{bmatrix}
\begin{array}{c}
c_\infty: [Pr, Po, K, H] \\
\left(\downarrow R_{C_K} = f\left(R_{C_\infty}\right) \right) \\
c_K: [S_{Pr}, S_{Po}, S_K, S_H] \\
\left(\downarrow R_{C_N} = f\left(R_{C_K}\right) \right) \\
c_N: f(S_{Pr}\ x\ S_{Po}\ x\ S_K\ x\ S_H) \\
\left(\downarrow R_{C_U} = f\left(R_{C_N}\right) \right) \\
c_U: [P, V, M, C] \\
\Uparrow \\
c_{0:[D,W,I,C]}
\end{array}
\end{bmatrix}_{Light}
\begin{bmatrix}
M_3 \to System_X \\
(\uparrow F \to I) \\
M_2 \to S_{System_X} \\
(\uparrow Sig \to F) \\
M_1 \to Sig_x \\
(\uparrow > P_x) \\
U \to x_{Human}
\end{bmatrix}_{Space} \\[2em]
\begin{bmatrix}
M_3 : -\infty \le t \le \infty \\
\downarrow \\
M_2 : 0 \ge t > \infty \\
\downarrow \\
M_1 : 0 > t > \infty \\
\downarrow \\
U \to \begin{array}{l} t \le E_{Cell}; TC: M_3 \to U \\ t \sim E_{Human}; TC: U \to M_3 \end{array}
\end{bmatrix}_{Time}
\quad TC \to x_{Truer\ Individuality}
\end{array}
\right\| \langle x_U | x_T \rangle
$$

$\Rightarrow$

$\langle Space - Time - Energy - Gravity\ Material - Fabric\ Pre$
$\qquad - Genetic\ Code \rangle +$

$\langle Electromagnetic\ Spectrum\ Material - Fabric\ Pre - Genetic\ Code \rangle +$

$\langle Quantum - Particle\ Material - Fabric\ Pre - Genetic\ Code \rangle +$

$\langle Atoms\ Pre - Genetic\ Code \rangle +$

$\langle Molecules\ Material - Fabric\ Pre - Genetic\ Code \rangle + LSTE\ \langle ... \rangle +$

$PGGT < \cdots > + \langle Cells\ Genetic\ Code \rangle + LSTE\ \langle ... \rangle +$

$(Fundamental\ Capacities\ of\ Self\ Genetic\ Code) +$

$FBLEEE\ \langle ... \rangle + (Humans\ Genetic\ Code) + LSTE\ \langle ... \rangle + FBLEEE\ \langle ... \rangle +$

358

$$\left(\begin{array}{l} \sum Sig_{Harmony-type} = Xa + \overline{Yb_{0-n}} \\ \begin{bmatrix} X \in [S_{System_N}] \\ where \quad Y \in [S_{System_{Pr}}, S_{System_P}, S_{System_K}, S_{System_N}] \\ a, b \ are \ integers; a > b \end{bmatrix} \\ \sum Sig_{Knowledge-type} = Xa + \overline{Yb_{0-n}} \\ \begin{bmatrix} X \in [S_{System_K}] \\ where \quad Y \in [S_{System_{Pr}}, S_{System_P}, S_{System_K}, S_{System_N}] \\ a, b \ are \ integers; a > b \end{bmatrix} \\ \sum Sig_{Power-type} = Xa + \overline{Yb_{0-n}} \\ \begin{bmatrix} X \in [S_{System_P}] \\ where \quad Y \in [S_{System_{Pr}}, S_{System_P}, S_{System_K}, S_{System_N}] \\ a, b \ are \ integers; a > b \end{bmatrix} \\ \sum Sig_{Service-type} = Xa + \overline{Yb_{0-n}} \\ \begin{bmatrix} X \in [S_{System_{Pr}}] \\ where \quad Y \in [S_{System_{Pr}}, S_{System_P}, S_{System_K}, S_{System_N}] \\ a, b \ are \ integers; a > b \end{bmatrix} \end{array}\right)$$

Eq. 7.4.6, Active Truer Individuality Partial-Singularity Pre-Genetic and Genetic Code

Note that the final code segment following 'FBLEEE <...>' (recall that FBLEEE refers to FBLEE-entanglement) contains the code for all possible $(\sum x)$ human-types, and depicts the growing biography by which the singular light-based edifice expresses its materialization, and at this stage, through the truer individuality partial-singularity.

SECTION 8: GENERATION OF PARTIAL-SINGULARITIES IN THE SURFACING OF COMPLEX ORGANIZATION

Having traced the computations resulting in the partial-singularity emergences of space-time-energy-gravity, the electromagnetic spectrum, matter, life, and even human becoming as manifest in sensation, urges, feelings, and thoughts, and truer individuality, we now turn our attention to study the computation resulting in the emergence of partial-singularities related to complex organizations.

In the previous sections we have seen that there is a gradual addition to partial-singularity biographies as pre-genetic and then genetic code capturing more complex fourfold-functionality is generated. In these biographies the condition for constructive mutation is that some meta-level is primarily active. When this happens there is a possibility that the existing four-base logic-encoding ecosystems (FBLEE) are enhanced or that new ones are created. Note that it is such FBLEE action that alters the quantum-level interface between the material and antecedent layers of light in effect changing the basis by which matter materializes. In other words FBLEE changes what matter can be.

But at its heart it is good to get clear that it is the complexification of four-foldness that is the driver of changes to FBLEE, and therefore of possible changes to the material-fabric, the genetic substance of living cells, and of any post-genetic substance that were to arise. This is so because such complexification essentially necessitates that existing habits, of any kind, are continually broken and in the process can admit of greater and more sweeping actions of light from the meta-levels. All of light is ever-present, but requires the receptacle to alter in order that it may precipitate more of its infinite possibility. It is such action that makes clear how the emergence of complex organization actually links to changes such as to matter and to the make-up of genetic content.

Chapter 8.1 examines the generation of the stable mega-organization partial-singularity. Similarly Chapter 8.2 examines the generation of the sustainable global civilization partial-singularity. In both these cases the nature of the partial-singularity is FBLEE-based, as opposed to being both FBLEE and MF-based. This is so because both the stable mega-organization and the sustainable global civilization are post-human structures and will require

some post-cell genetic-type structure to house their respective ecosystem logic materially.

Such post-cell genetic structure, as will be further explored in Section 9 on super-matter based partial-singularities, requires a much higher threshold of functional richness to come into material reality.

Chapter 8.1: Generation of Stable Mega-Organization FBLEE-Based Partial-Singularity

We have traced the emergence of the four properties of Light through the primary fourfold emergence of space-time-energy-gravity, the electromagnetic spectrum, through matter, through life, and even through human becoming as manifest in sensation, urges, feelings, and thoughts, and truer individuality. At each of these levels it seems it is a balanced combination of all four of the properties that creates stability and makes that structure or organism or being, sustainable. And this has to be since the four properties occur simultaneously in light and must be impelled to hold to that relationship even when projected from it. Hence it should be possible to deductively arrive at an equation for organizational sustainability given the four-fold emergence of properties of light through subsequent layers of manifestation.

To triangulate, though, we briefly trace the four-fold emergence, reinforce the significance of this and inductively arrive at an equation for sustainability that we would have arrived at deductively.

Starting with the primary emergence of space-time-energy-gravity we see that space is defined as that in which seeds of knowledge are planted – hence, existing as a means to express Light's property of knowledge.

363

Time is defined as the inevitability of the seeds of knowledge maturing into fullness, in spite of any opposition. Hence time is seen as an emergence of Light's property of power. Energy is seen as that which accumulates to create matter. Hence it is an emergence of Light's property of presence. Gravity is that by which relationship between objects comes into being – hence an emergence of Light's property of harmony.

With the electromagnetic spectrum we see that the spectrum as a whole is defined or architected by the speed of light c, as an emergence of nurturing, the wavelength λ, as an emergence of knowledge, the frequency f, as an emergence of power, and mass-potential as an emergence of presence. These occur simultaneously and constitute the wholeness of the electromagnetic spectrum.

Similarly, at the level of the quantum particle it can be seen that the integrity and functioning of the atom also depends on the integration of all four properties of light. Hence quarks, an emergence of knowledge, leptons, an emergence of power, bosons, an emergence of harmony or nurturing, and the Higgs-boson, an emergence of presence, act together to create the structure and functionality of every single atom.

What this reinforces is that every single thing must have been created through some integration and balance of the four properties of light.

Going up the organizational scale to the next level of complexity, to the level of atoms as a whole, the same pattern is found again: every single atom belongs to one of four groups, and in unison these four groups orchestrate the set of combinations of known compounds. Hence, Alkali Metals and Alkali Earth Metals configured by the s-Group, emanate from the family of system-power. Metals, Metalloid, Non-Metals, Halogens, and Noble Gases configured by the p-Group, emanate from the family of system-knowledge. Transition Metals configured by the d-Group, emanate from the family of system-presence. Lanthanides and Actinides configured by the f-Group, emanate from the family of system-nurturing. These groups act together to create the complex array of compounds that form the entire material from which the physical and subsequently even the organic world around us is constructed.

Going up the organizational scale to another level of complexity, to the level of the cell, the same pattern is found again: basically every single cell of every single living creature that has ever been studied by humankind also has a similar balance and integration of these four sets of forces acting together. Hence, proteins, from the family of presence, nucleic acids, from the family of knowledge, polysaccharides, from the family of power, and lipids, from the family of nurturing, work together to create the balanced functioning of every living cell.

The hypothesis hence, is that even at larger scale, in considering the most innovative organization at the level of teams, corporations, or markets, a similar integration and balance of the four sets of properties or forces may yield the best results. Recent developments in sustainability investment models ranging from Socially Responsible Investing (Logue, 2008), the Global Reporting Initiative (Willis, 2003), the Dow Jones Sustainability Index (Hope & Fowler, 2007), the Principles of Responsible Investing (Harvard-Edu, 2014), all reinforce the concept of investment criteria becoming broader, to be based not only on economic, but on environmental, social, and governance factors as well. Further evidence suggests that financial returns on such broad-based investment models continually beat financial returns on regular funds such as the S&P 500, for instance (Openshaw, 2015). In their study of major transitions in evolution Smith and Szathmary (Smith & Szathmary, 1995) chronicle the development of life toward increasing complexity as an application of successful collaboration and even co-evolution by which species evolve by changing together as system pressures increase.

This hypothesis hence can be summarized by the following graph:

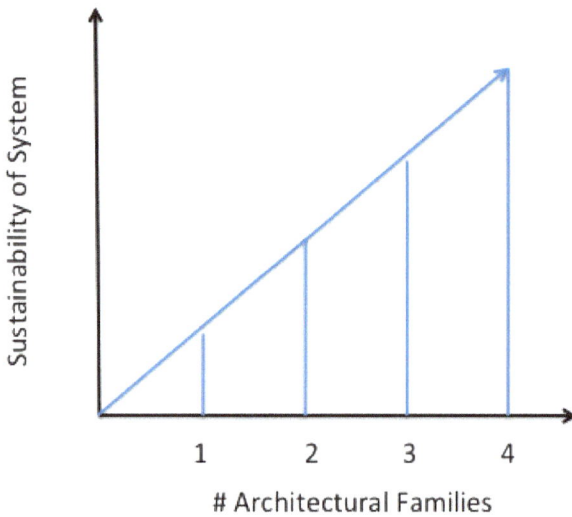

Figure 8.1.1 Sustainability of CAS

An equation for the sustainability of systems, $Sustainability_{Systems}$, where the interaction between the four families of forces is instrumental can be created. Hence, as in Equation 8.1.1:

$$Sustainability_{Systems}$$
$$\propto Interaction\ (S_{System_{Pr}}, S_{System_P}, S_{System_K}, S_{System_N})$$

366

Eq 8.1.1: Sustainability of Systems

This notion of the balance of the fourfold properties of Light as fundamental in creating stable mega-organization will be borne out in the next subsections of this chapter.

Light's Emergence as Stable Mega-Organization

As discussed previously the Light-Space-Time Emergence equation (3.1.3) being iterative, can be used to model emergence as it proceeds from simpler four-fold to more complex four-fold manifestations. Hence (3.1.3) has already been applied to suggest the emergence of the space-time-energy-gravity quadrumvirate, the electromagnetic spectrum, quantum particles in general, bosons as a further instance of a particular kind of quantum particle, atoms, living cells, basic capacities of the self, and truer individuality.

Here, a revised form of it leveraging the sustainable global civilization equation (4.2.7) introduced in Chapter 4.2 on post-Big Bang generation of light-based singularities, will be applied to suggest the emergence of stable mega-organization, Equation 8.1.2, Generation of Stable Mega-Organization FBLEE-Based Partial-Singularity. The architecture and details of a sustainable mega-organization can be seen to be the result of the application of the Light, Space, and Time matrices as will be elaborated. But further, as implied by (3.6.5), the Potential Effect of Levels of Light on Genetic-Type Information equation, any process of deeper organization, such as is responsible for the architecture and cohesiveness of sustainable mega-organizations, has the possibility of altering FBLEE, and in this case through FBLEEE influence, post-material fabric housing structures, so long as the bases involved are driven primarily by a meta-level.

$$Light - Space - Time\ Emergence_{Stable\ Mega-Orgnization}$$
$$=$$

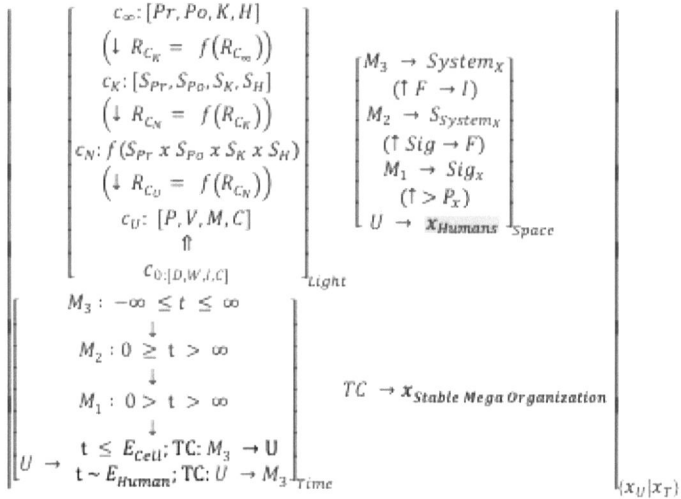

$$\Rightarrow$$

$$\langle Space - Time - Energy - Gravity\ Material - Fabric\ Pre$$
$$- Genetic\ Code \rangle +$$

$$\langle Electromagnetic\ Spectrum\ Material - Fabric\ Pre - Genetic\ Code \rangle +$$

$$\langle Quantum - Particle\ Material - Fabric\ Pre - Genetic\ Code \rangle +$$

$$\langle Atoms\ Pre - Genetic\ Code \rangle$$
$$+ \langle Molecules\ Material - Fabric\ Pre$$
$$- Genetic\ Code \rangle +$$

$$LSTE\ \langle \ldots \rangle + PGGT < \cdots$$
$$> + \langle Cells\ Genetic\ Code \rangle + LSTE\ \langle \ldots \rangle$$
$$+ FBLEEE\ \langle \ldots \rangle +$$

$$\langle Humans\ Genetic\ Code \rangle + LSTE\ \langle \ldots \rangle + FBLEEE\ \langle \ldots \rangle +$$

$$\langle Stable\ Mega - Organization\ FBLEE\ Genetic\ Code \rangle + PGCI\ \langle \ldots \rangle$$

Note that as indicated by the ' ⟨*Stable Mega – Organization FBLEE Genetic Code*⟩ ' code-segment, the code generated is at the level of FBLEE only. Further, as indicated by the 'PGCI ⟨...⟩' code-segment, a process referred to as Post Genetic Code Initiation is activated, by which some possibly hybrid genetic-type information housing-structure that integrates the subtle-libraries with matter more intimately, and allows an easy going back and forth between the two may materialize some time in the future.

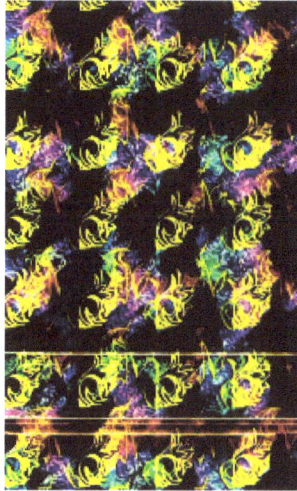

Starting with the Light-Matrix, the top left-hand matrix in (8.1.2), the first line from the top, $C_\infty : [Pr, Po, K, H]$, specifies the architecture of mega-organization as will be discussed shortly. Hence, Academic organizations are an emergence of Light's property of Knowledge, Political organizations are an emergence of Light's property of Power, Social organizations are an emergence of Light's property of Presence, and Commercial organizations can be thought of as an emergence of Light's property of

369

Harmony. The fundamental architecture of these aspects hence, is an emergence of the properties of Light at ∞.

Line 3 in the Light-Matrix, $C_K: [S_{Pr}, S_{Po}, S_K, S_H]$, elaborates the sets for Presence, Power, Knowledge, and Harmony, each containing multiple elements. For example, as will be explored in the following subsection, various elements derived from the four sets would define the architecture of a Silicon Valley type mega-commercial organization and could be functions such as 'commercial immunity', 'power of aesthetics', 'ease of informal meetings', amongst others, hence collectively describing a stable mega-commercial organization's way of being. Specifically, Line 5, $C_N: f(S_{Pr} \times S_{Po} \times S_K \times S_H)$, suggests that unique seeds are created from a combination of such elements from all four sets, with a particular element leading or having more weight, that in effect creates the distinctness possible at the level of a stable mega-organization.

Line 6, $(\downarrow R_{C_U} = f(R_{C_N}))$, normally specifies quantization between the layer where the seeds are formed, and the physical layer, and as explored in Chapter 3.5 and 3.6, will normally result in Line 7, $C_U: [P, V, M, C]$, hence potentially changing the post material-fabric housing structure. In a post-human organization such as a stable mega-organization as just suggested, however, the beginnings of a post-genetic hybrid structure will likely only begin to be initiated and would require far more functional-richness to materialize before it were to become a reality. Because such a post-genetic housing structure has yet to come into existence (8.1.2) can strictly only be a FBLEE-based partial-singularity.

The possibilities represented by Lines 1 through 5 hence concretize through the quantization represented by Line 6 to become or further enhance the mega-organization with its subtle physical (related to Presence), vital (related to Power), mental (related to Knowledge), and connection (related to Harmony) aspects now at least existing in FBLEE in the quantum-layer behind U   As a reminder FBLEE exists at that barrier where the speed of light is c, but at the quantum-level only. Note that just as Line 6 represents a process of quantization relating the layer of reality created by Light traveling at c with the antecedent layers, so too Lines 2 and 4 as previously discussed, also represent quantization of a more subtle kind that ultimately plays a critical part in allowing FBLEE to express infinite diversity.

Typically it is the process as captured by the Space-Matrix that will determine if Line 6 is activated. Specifically patterns at the untransformed layer, U, will need to be overcome, as specified by the second-line from the bottom of the Space-Matrix: $(\uparrow > P_x)$. But as specified by the bottom-line of the Time-Matrix, reproduced below, it is only with the advent of the human-system that the automaticity of the action of meta-levels is reversed:

$$U \rightarrow \begin{array}{l} t \leq E_{Cell}; \text{TC: } M_3 \rightarrow U \\ t \sim E_{Human}; \text{TC: } U \rightarrow M_3 \end{array}$$

Hence in the case of the stable mega-organization, which in this emergence is a post-human system, the fact that patterns do need to be overcome means that quantization requires effort to happen. Given this, it is useful to review Equation 3.6.5, Potential Effect of Levels of Light on Genetic-Type Information:

$$Potential\ Effect\ of\ Levels\ of\ Light\ on\ Genetic$$
$$- Type\ Information =$$
$$\begin{bmatrix} STATIC\ \langle |[L][S][T]TC \rightarrow x_T|_{\langle x_U|x_T\rangle}\rangle \\ \times \\ \left((Y > U: Z_Q) \vee (Y \leq U: Z_F) \vee (Y = U: Z_R) \right) \\ \ni \\ \begin{pmatrix} Z \in \mathbb{U}\ (Space, Time, Energy, Gravity) \\ Q: Quantization; F: Fragmentation; R: Random \end{pmatrix} \end{bmatrix} \rightarrow h \ni$$

$$h \in \begin{pmatrix} [Q]: Constructive\ zone, \\ [Q]: Constructive\ zone \wedge Constructive\ mutation, \\ [F]: Destructive\ mutation, \\ [R]: Random\ mutation \end{pmatrix}$$

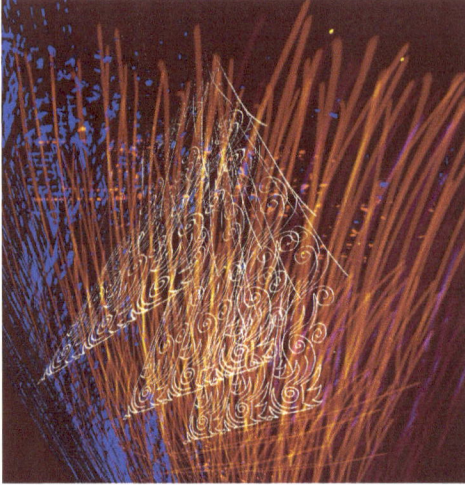

Line 1 from the top in the matrix is simply a static form of (3.6.1) the Simplified Light-Space-Time Emergence equation. The static form is designated by 'STATIC', and implies that fundamental operations true of (3.6.1) are being highlighted in (3.6.5). In other words (3.6.1) already has all the operations highlighted in (3.6.5) in it, but by 'freezing' it by making it static, the essential dynamics leading to possible mutations at the genetic level can more clearly be highlighted.

Line 1 is then subjected (×) to a determination of the dominant levels of light that may be active, designated by Line 2, ' $\left((Y > U: Z_Q) \vee (Y \leq U:\ Z_F) \vee (Y = U:\ Z_R) \right)$ '. Unpacking this, 'Y > U' implies meta-levels are active and as a result it is possible that Z_Q is going to take place (the subscript 'Q' implies quantum-level action). This also implies activation and potential change of FBLEE. The call from below, as it were, may invoke some function that already exists in the subtle-libraries 'above', so that some already existing function may influence FBLEE through ∞ -entanglement, K-entanglement, or N-entanglement. This may be thought of as a key-and-lock mechanism, where a deep enough visceral urge from below acting as the key, opens an entangled lock to alter FBLEE as per the visceral urge. '$Y \leq U$' implies that only the untransformed levels are active, and therefore also the sub-level where the speed of light is 0 is active and as a result Z_F is going to take place (the subscript 'F' implies 'fragmentation'). '$Y = U$' implies that all levels are active and as a result Z_R is going to take place (the subscript 'R' implies 'random').

Line 3 elaborates the significance of Z_Q, Z_F, and Z_R. Hence Z is the union of potential quantum-operations of space, time, energy, and gravity, designated by '$Z \in \mathbb{U}$ (*Space, Time, Energy, Gravity*)'. But the nature of the operations, as suggested in the previous paragraph, is designated by 'Q: *Quantization*; F: *Fragmentation*; R: *Random* '. Z_Q, then, implies that the full quantization originating from updated four-base logic-encoding ecosystems (FBLEE) can take place. Z_F implies that only a local library, as opposed to FBLEE, can potentially be altered. Z_R also precludes full quantization, and that some partial local-library constructive or destructive mutation may take place.

The '$\rightarrow h \ni h \in$' segment resolves the outcome of the operations implied by Lines 1 – 3, suggesting that the outcome will be 'h' such that ($\ni$) 'h' is an element ($\in$) of the set specified by the members '[Q]: *Constructive zone*', '[Q]: *Constructive zone AND Constructive mutation*', '[F]; *Destructive mutation*', and '{R}: *Random mutation*'. '[Q]: *Constructive zone*' implies that the in-built buffer has been crossed and that access to the deeper four-base logic-encoding ecosystem (FBLEE) has been granted. '[Q]' in this segment implies that there is the possibility that full-quantization as specified by Line 3 of the previous matrix can take place. Access to this zone is a prerequisite for constructive mutation to occur, as designated by the element $'$ $[Q\}$: *Constructive zone* ∩ *Constructive mutation'*, which generally implies that full-quantization is going to take place. In the case of an early post-human organization though, as already suggested in the discussion on (8.1.2), since there is not yet a post-genetic structure securely in place to house such full-quantization at the material level, this likely initiates a call to antecedent layers for such a structure to begin to materialize through the activation of PGCI. In the meanwhile quantization will affect FBLEE, and through FBLEEE may affect human-level genetic-code. The '[F]' specifies the relationship between 'Fragmentation' in Line 3 of the previous matrix and destructive mutation. The '[R]' specifies the relationship between 'Random' in Line 3 of the previous matrix and random mutation.

Generation of Stable Mega-Organization FBLEE-Based Partial-Singularity Genetic-Type Code

As suggested by inductively derived (8.1.1), a sustainable organization is the result of two dynamics: maturity along each of the architectural force dimensions, and a robust interaction between all for sets of forces. These dynamics are implicit in (8.1.2) as borne out in the previous subsection.

Consider the example of Silicon Valley.

374

While Silicon Valley may be primarily a commercial organization, yet we begin to glimpse something of what may become possible when the four emergent forces of light can begin to interact with one another in free fashion. That is, without the

motive of doing everything for the sake of generating money only. In the case of Silicon Valley we see that the climate and beauty of the Bay Area, thereby likely representing the impetus of nurturing and harmony, began to attract many talented people into the vicinity. Over time the talent pool became progressively diversified. Educational institutes, such as Stanford University and University of Berkeley cropped up and became centers of cutting-edge research, representing the impetus of knowledge. The Armed Forces were attracted to the area for the same reason, and came with their huge requirements for research for research's sake, and their funds in support of the area, representing the impetus of power. Graduates from the universities started companies that began to in turn support the universities with funds. Talent moved around from company to company like people from department to department. Thus we see that even when individual companies failed, Silicon Valley functioned as a larger organization and was able to retain the talent in the area, was able further, to buffer the shocks to some extent, hence preparing the ground for future waves of innovation. This reality of becoming a container for dynamic experimentation perhaps represents the impetus of service in that the container is serving the dynamic components existing within it.

Pragmatically speaking this dynamic of functioning as a larger mega-organization meant that some level of commercial-immunity began to develop, so that people losing their jobs was not necessarily considered as that stressful an event. There was a freeing, thus, from the purely commercial element. If the various powers of beauty, aesthetics, knowledge, power in the form of money, plus the myriad streams of talent – from engineers, scientists, managers, lawyers etc., did not exist in such close proximity, and further, if the various professionals could not thus support each other through their continued informal meetings, through the urge to create the next wave of innovation, through the urge to pursue progress for the sake of progress, then they would have left for other areas and the phenomenon of Silicon Valley would never have been.

Usually modern day organizations cannot provide these various buffers and opportunities for interaction that Silicon Valley provides, and hence when hit with adversity, more often than not, simply crumble. This is perhaps due to the fact that it is always only one-dimension that drives them, and when the health of that is threatened or falls, the organization resorts to tactics that will ensure that that dimension looks good at any cost, in the bargain sacrificing the development of the other dimensions, and therefore its longer term health.

The four-forces led organization will be something like a community. Having attained freedom from an exaggerated commercial impetus, people will 'live' their 'jobs' because it is what fulfills them. Such a freedom is what will allow the four primary powers to manifest to greater and greater degree within them and their environment. Seeing thus, their ability to become centers of knowledge and wisdom, or mutuality and harmony, or power and leadership and energy, or perfection and service, or some unique combination of these primary forces, so increase, a sense of satisfaction with life will more easily accompany all that they continue to do. Under the freer flow of these powers their uniqueness will be refined and flourish, and correspondingly, so too will the uniqueness of their respective organizations.

An organization may be political and hence be primarily driven by the power of leadership and courage, or it may be social and hence be primarily driven by the power of service and perfection, or it may be commercial and hence be primarily driven by the power of mutuality and harmony, or it may be research-oriented and academic and hence be primarily driven by the powers of knowledge and wisdom, but always each of the other powers will also be behind it, fulfilling and completing its primary urge.   Thus the dynamic of unique primary powers being supported by an increasing plethora of secondary powers is represented by the following four organizational equations, (8.1.3 – 6) and illustrates how the four powers of light emerge even in complex human-based collectivities:

$$Sig_{Political} = Xa + \overline{Yb_{0-n}}$$

$$where \begin{bmatrix} X \in [S_{System_P}] \\ Y \in [S_{System_{Pr}}, S_{System_P}, S_{System_K}, S_{System_N}] \\ a, b \ are \ integers; a > b \end{bmatrix}$$

Eq 8.1.3: Political Organization

$$Sig_{Social} = Xa + \overline{Yb_{0-n}}$$

$$where \begin{bmatrix} X \in [S_{System_{Pr}}] \\ Y \in [S_{System_{Pr}}, S_{System_P}, S_{System_K}, S_{System_N}] \\ a, b \ are \ integers; a > b \end{bmatrix}$$

Eq 8.1.4: Social Organization

$$Sig_{Commercial} = Xa + \overline{Yb_{0-n}}$$

$$where \begin{bmatrix} X \in [S_{System_N}] \\ Y \in [S_{System_{Pr}}, S_{System_P}, S_{System_K}, S_{System_N}] \\ a, b \ are \ integers; a > b \end{bmatrix}$$

Eq 8.1.5: Commercial Organization

$$Sig_{Academic} = Xa + \overline{Yb_{0-n}}$$

$$where \begin{bmatrix} X \in [S_{System_K}] \\ Y \in [S_{System_{Pr}}, S_{System_P}, S_{System_K}, S_{System_N}] \\ a, b \text{ are integers}; a > b \end{bmatrix}$$

Eq 8.1.6: Academic Organization

Note that (8.1.3 - 6) already implies that Lines 1 – 5 in the Light Matrix (3.1.3) have been activated, and that the logic of the stable mega-organization- ecosystem will precipitate into FBLEE and may also impact post material-fabric cell-based genetic structure through FBLEEE.

Summary of Stable Mega-Organization FBLEE-Based Partial-Singularity Genetic-Type Code

Summarizing, after the stable mega-organization iteration of the Light-Space-Time Emergence equation is complete, the following code-segments will have been generated as specified by Equation 8.1.7, Active Stable Mega-Organization Partial-Singularity Genetic-Type Code:

$$Light - Space - Time\ Emergence_{Stable\ Mega-Orgnization} =$$

379

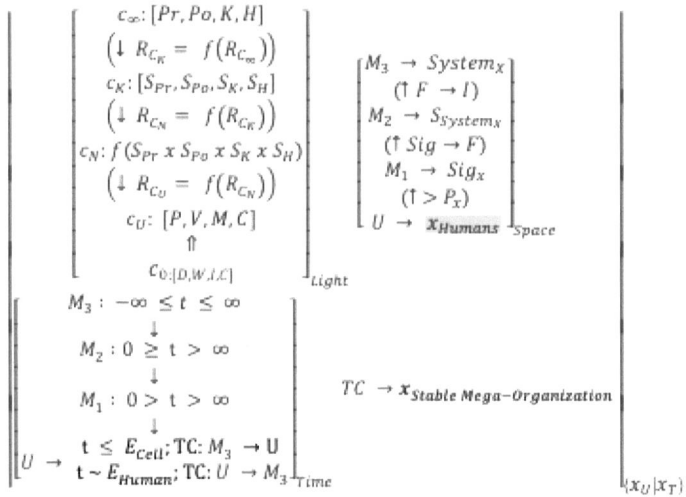

$$\begin{vmatrix} \begin{bmatrix} c_\infty: [Pr, Po, K, H] \\ \left(\downarrow R_{C_K} = f\left(R_{C_\infty}\right) \right) \\ c_K: [S_{Pr}, S_{Po}, S_K, S_H] \\ \left(\downarrow R_{C_N} = f\left(R_{C_K}\right) \right) \\ c_N: f\left(S_{Pr} \times S_{Po} \times S_K \times S_H\right) \\ \left(\downarrow R_{C_U} = f\left(R_{C_N}\right) \right) \\ c_U: [P, V, M, C] \\ \Uparrow \\ c_{0:[D,W,I,C]} \end{bmatrix}_{Light} \begin{bmatrix} M_3 \to System_X \\ (\uparrow F \to I) \\ M_2 \to S_{System_X} \\ (\uparrow Sig \to F) \\ M_1 \to Sig_X \\ (\uparrow > P_X) \\ U \to x_{Humans} \end{bmatrix}_{Space} \\ \begin{bmatrix} M_3: -\infty \leq t \leq \infty \\ \downarrow \\ M_2: 0 \geq t > \infty \\ \downarrow \\ M_1: 0 > t > \infty \\ \downarrow \\ U \to \begin{array}{l} t \leq E_{Cell}; TC: M_3 \to U \\ t \sim E_{Human}; TC: U \to M_3 \end{array} \end{bmatrix}_{Time} \qquad TC \to x_{Stable\ Mega-Organization} \end{vmatrix}_{\langle x_U | x_T \rangle}$$

$\Rightarrow$

$\langle Space - Time - Energy - Gravity\ Material - Fabric\ Pre$
$\qquad\qquad - Genetic\ Code \rangle +$

$\langle Electromagnetic\ Spectrum\ Material - Fabric\ Pre - Genetic\ Code \rangle +$

$\langle Quantum - Particle\ Material - Fabric\ Pre - Genetic\ Code \rangle +$

$\langle Atoms\ Pre - Genetic\ Code \rangle +$

$\langle Molecules\ Material - Fabric\ Pre - Genetic\ Code \rangle + LSTE \langle \dots \rangle +$

$PGGT < \dots > + \langle Cells\ Genetic\ Code \rangle + LSTE \langle \dots \rangle +$

$\langle Fundamental\ Capacities\ of\ Self\ Genetic\ Code \rangle + FBLEEE \langle \dots \rangle +$

$\langle Humans\ Genetic\ Code \rangle + LSTE \langle \dots \rangle + FBLEEE \langle \dots \rangle +$

.

$$\left(\begin{array}{c} \left[\begin{array}{c} \sum Sig_{Commercial} = Xa + \overline{Yb_{0-n}} \\ X \in [S_{System_N}] \\ where \quad Y \in [S_{System_{Pr}}, S_{System_P}, S_{System_K}, S_{System_N}] \\ a, b \ are \ integers; a > b \end{array} \right] \\ \left[\begin{array}{c} \sum Sig_{Academic} = Xa + \overline{Yb_{0-n}} \\ X \in [S_{System_K}] \\ where \quad Y \in [S_{System_{Pr}}, S_{System_P}, S_{System_K}, S_{System_N}] \\ a, b \ are \ integers; a > b \end{array} \right] \\ \left[\begin{array}{c} \sum Sig_{Political} = Xa + \overline{Yb_{0-n}} \\ X \in [S_{System_P}] \\ where \quad Y \in [S_{System_{Pr}}, S_{System_P}, S_{System_K}, S_{System_N}] \\ a, b \ are \ integers; a > b \end{array} \right] \\ \left[\begin{array}{c} \sum Sig_{Social} = Xa + \overline{Yb_{0-n}} \\ X \in [S_{System_{Pr}}] \\ where \quad Y \in [S_{System_{Pr}}, S_{System_P}, S_{System_K}, S_{System_N}] \\ a, b \ are \ integers; a > b \end{array} \right] \end{array} \right) + PGCI \langle ... \rangle$$

Eq. 8.1.7, Active Stable Mega-Organization FBLEE-Based Partial-Singularity Genetic-Type Code

Note that the final code segment preceding 'PGCI <…>' contains the code for all possible ($\sum x$) mega-organization types, and depicts the growing biography by which the singular light-based edifice expresses its materialization at FBLEE, and at this stage, through the stable mega-organization partial-singularity.

Chapter 8.2: Generation of Sustainable Global Civilization FBLEE-Based Partial-Singularity

This chapter will consider the quantum-level computation necessary in the creation of a sustainable global civilization partial-singularity and the genetic-type code segments that accompany that.

Light's Emergence as Sustainable Global Civilization

As discussed in previous chapters the Light-Space-Time Emergence equation (3.1.3) being iterative, can be used to model emergence as it proceeds from simpler four-fold to more complex four-fold manifestations. Hence (3.1.3) has already been applied to suggest the emergence of the space-time-energy-gravity quadrumvirate, the electromagnetic spectrum, quantum particles in general, bosons as a further instance of a particular kind of quantum particle, atoms, living cells, basic capacities of the self, truer individuality, and stable mega-organizations.

The sustainable global civilization equation (4.2.7) introduced in Chapter 4.2 on post-Big Bang generation of light-based singularities, will be leveraged as a starting point, and is reproduced below for convenience. The architecture and details of a sustainable global civilization

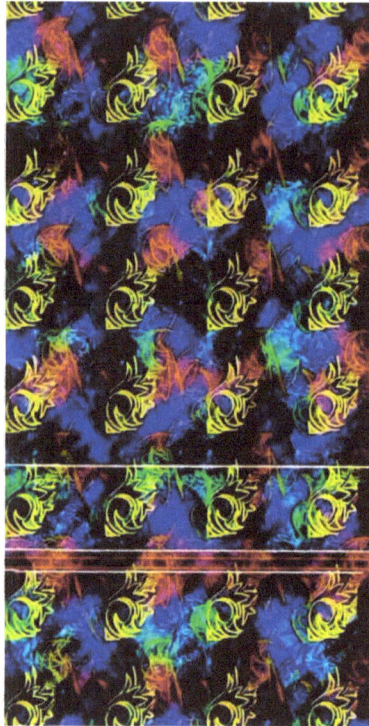

382

can be seen to be the result of the application of the Light, Space, and Time matrices as will be elaborated. But further, as implied by (3.6.5), the Potential Effect of Levels of Light on Genetic-Type Information equation, any process of deeper organization, such as is responsible for the architecture and cohesiveness of sustainable global civilization, has the possibility of altering FBLEE, and in this case through FBLEEE influence of altering post material fabric housing structures, so long as the bases involved are driven primarily by a meta-level.

$$Light - Space - Time\ Emergence_{Sustainable\ Global\ Civilization} =$$

$$
\left|
\begin{bmatrix}
\begin{bmatrix}
c_\infty \colon [Pr, Po, K, H] \\
\left(\downarrow R_{C_K} = f\left(R_{C_\infty}\right)\right) \\
c_K \colon [S_{Pr}, S_{Po}, S_K, S_H] \\
\left(\downarrow R_{C_N} = f\left(R_{C_K}\right)\right) \\
c_N \colon f(S_{Pr} \times S_{Po} \times S_K \times S_H) \\
\left(\downarrow R_{C_U} = f\left(R_{C_N}\right)\right) \\
c_U \colon [P, V, M, C] \\
\Uparrow \\
C_{0:[D,W,I,C]}
\end{bmatrix}_{Light}
\begin{bmatrix}
M_3 \to System_X \\
(\uparrow F \to I) \\
M_2 \to S_{System_X} \\
(\uparrow Sig \to F) \\
M_1 \to Sig_X \\
(\uparrow > P_X) \\
U \to x_{Humans}
\end{bmatrix}_{Space} \\
\begin{bmatrix}
U \to
\begin{bmatrix}
M_3 \colon -\infty \leq t \leq \infty \\
\downarrow \\
M_2 \colon 0 \geq t > \infty \\
\downarrow \\
M_1 \colon 0 > t > \infty \\
\downarrow \\
t \leq E_{Cell}; TC \colon M_3 \to U \\
t \sim E_{Human}; TC \colon U \to M_3
\end{bmatrix}_{Time}
\end{bmatrix}
\quad TC \to x_{Sustainbale\ Global\ Civilization}
\end{bmatrix}
\right|_{\langle x_U | x_T \rangle}
$$

$$\Rightarrow$$

$$\langle Space - Time - Energy - Gravity\ Material - Fabric\ Pre - Genetic\ Code \rangle +$$

$$\langle Electromagnetic\ Spectrum\ Material - Fabric\ Pre - Genetic\ Code \rangle +$$

$$\langle Quantum - Particle\ Material - Fabric\ Pre - Genetic\ Code \rangle +$$

$\langle Atoms\ Pre - Genetic\ Code \rangle$
$$+ \langle Molecules\ Material - Fabric\ Pre$$
$$- Genetic\ Code \rangle +$$

$LSTE\ \langle ... \rangle\ +\ PGGT\ <\ \cdots$
$$>\ +\ \langle Cells\ Genetic\ Code \rangle\ +\ LSTE\ \langle ... \rangle$$
$$+\ FBLEEE\ \langle ... \rangle\ +$$

$\langle Humans\ Genetic\ Code \rangle\ +\ LSTE\ \langle ... \rangle\ +\ FBLEEE\ \langle ... \rangle\ +$

$\langle Stable\ Mega - Organization\ FBLEE\ Genetic\ Code \rangle\ +\ \text{PGCI}\ \langle ... \rangle\ +$

$Sustainable\ Global\ Civilization\ Code$ FBLEE Genetic Code

Note that as indicated by the '$Sustainable\ Global\ Civilization\ Code$ FBLEE Genetic Code' code-segment, the code generated is at the level of FBLEE only.

Starting with the Light-Matrix, the top left-hand matrix in (4.2.7), the first line from the top, $C_\infty: [Pr, Po, K, H]$, specifies the architecture of sustainable global civilization as will be further discussed in this chapter. A key to such a civilization is also pointed to by the

previously derived equation (8.1.1), reproduced here for convenience:

$Sustainability_{Systems}$
$\propto Interaction\ (S_{System_{Pr}}, S_{System_{P}}, S_{System_{K}}, S_{System_{N}})$

This requires maturity by large organized parts of the world, whether nations or regional blocs. The maturity is such that the fourfold properties of Light are adequately expressed. Hence in the example of nations to be illustrated shortly, a country like India is in its deeper essence perhaps an emergence of Light's property of Knowledge, a country like Japan is in its deeper essence

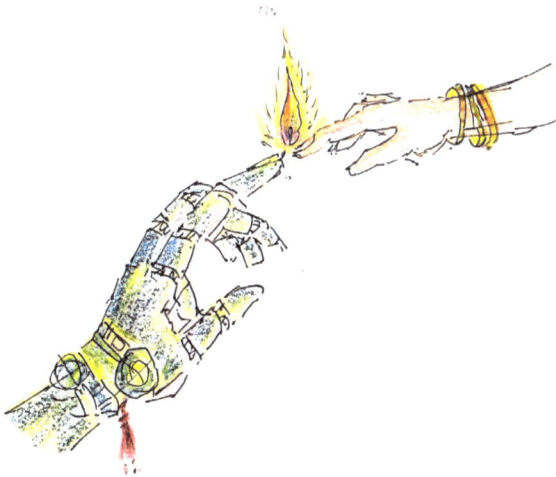

perhaps an emergence of Light's property of Power, a country like Thailand is in its deeper essence perhaps an emergence of Light's property of Presence, and a country like UK is in in its deeper essence perhaps an emergence of Light's property of Harmony. The fundamental architecture of the combination of such emergences, which as per (8.1.1) is required for sustainability, is hence an emergence of the properties of Light at ∞.

Line 3 in the Light-Matrix, C_K: $[S_{Pr}, S_{Po}, S_K, S_H]$, elaborates the sets for Presence, Power, Knowledge, and Harmony, each containing multiple elements. For example, as will be explored, various elements derived from the four sets define the architecture of a knowledge-essence country like India and could be functions such as 'exceptional capacity for penetrating behind the surface',

385

'meaningfully synthesizing many streams of development', amongst others, hence collectively describing different aspects of a knowledge-essence country's way of being. Specifically, Line 5, $C_N: f(S_{Pr} \times S_{Po} \times S_K \times S_H)$, suggests that unique seeds are created from a combination of such elements from all four sets, with a particular element leading or having more weight, that in effect creates the distinctness possible at the level of sustainable global civilization.

Line 6, $(\downarrow R_{C_U} = f(R_{C_N}))$, normally specifies quantization between the layer where the seeds are formed, and the physical layer, and as explored in Chapter 3.5 and 3.6, will normally result in Line 7, $C_U: [P, V, M, C]$, hence potentially changing the post material-fabric housing structure. In a post-human organization such as a sustainable global civilization, however, the beginnings of a post-genetic hybrid structure will likely be incrementally reinforced and would require far more functional-richness to materialize before it were to become a reality. Because such a post-genetic housing structure has yet to come into existence (4.2.7) can strictly only be a FBLEE-based partial-singularity.

The possibilities represented by Lines 1 through 5 hence concretize through the quantization represented by Line 6 to enhance sustainable global civilization with subtle physical (related to Presence), vital (related to Power), mental (related to Knowledge), and connection (related to Harmony) aspects now at least existing in FBLEE in the quantum-layer behind U. As a reminder FBLEE exists at that barrier where the speed of light is c, but at the quantum-level only. Note that just as Line 6 represents a process of quantization relating the layer of reality created by Light traveling at c with the antecedent layers, so too Lines 2 and 4 as previously discussed, also represent quantization of a more subtle kind that ultimately plays a critical part in allowing FBLEE to express infinite diversity.

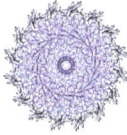

Typically it is the process as captured by the Space-Matrix that will determine if Line 6 is activated. Specifically patterns at the untransformed layer, U, will need to be overcome, as specified by the second-line from the bottom of the Space-Matrix: ($\uparrow > P_x$). But as specified by the bottom-line of the Time-Matrix, reproduced below, it is only with the advent of the human-system that the automaticity of the action of meta-levels is reversed:

$$U \rightarrow \begin{array}{l} t \leq E_{Cell}; TC: M_3 \rightarrow U \\ t \sim E_{Human}; TC: U \rightarrow M_3 \end{array}$$

Hence in the case of sustainable global civilization, which in this emergence is a post-human system, the fact that patterns do need to be overcome means that quantization requires effort to happen. Given this, it is useful to review Equation 3.6.5, Potential Effect of Levels of Light on Genetic-Type Information:

$$Potential\ Effect\ of\ Levels\ of\ Light\ on\ Genetic$$
$$-\ Type\ Information =$$
$$\begin{bmatrix} STATIC\ \langle|[L][S][T]TC \rightarrow x_T|_{\langle x_U | x_T \rangle} \rangle \\ \times \\ \Big((Y > U: Z_Q) \vee (Y \leq U: Z_F) \vee (Y = U: Z_R) \Big) \\ \ni \\ \big(\begin{array}{c} Z \in \mathbb{U}\ (Space, Time, Energy, Gravity) \\ Q: Quantization; F: Fragmentation; R: Random \end{array} \big) \end{bmatrix} \rightarrow h \ni$$

$$h \in \left(\begin{array}{c} [Q]: Constructive\ zone, \\ [Q]: Constructive\ zone \wedge Constructive\ mutation, \\ [F]: Destructive\ mutation, \\ [R]: Random\ mutation \end{array} \right)$$

Line 1 from the top in the matrix is simply a static form of (3.6.1) the Simplified Light-Space-Time Emergence equation. The static form is designated by 'STATIC', and implies that fundamental operations true of (3.6.1) are being highlighted in (3.6.5). In other words (3.6.1) already has all the operations highlighted in (3.6.5) in it, but by 'freezing' it by making it static, the essential dynamics leading to possible mutations at the genetic level can more clearly be highlighted.

Line 1 is then subjected ($\times$) to a determination of the dominant levels of light that may be active, designated by Line 2, ' $\left((Y > U\colon Z_Q) \vee (Y \leq U\colon Z_F) \vee (Y = U\colon Z_R) \right)$ '. Unpacking this, '$Y > U$' implies meta-levels are active and as a result it is possible that Z_Q is going to take place (the subscript 'Q' implies quantum-level action). This also implies activation and potential change of FBLEE. The call from below, as it were, may invoke some function that already exists in the subtle-libraries 'above', so that some already existing function may influence FBLEE through ∞ -entanglement, K-entanglement, or N-entanglement. This may be thought of as a key-and-lock mechanism, where a deep enough visceral urge from below acting as the key, opens an entangled lock to alter

FBLEE as per the visceral urge. '$Y \leq U$' implies that only the untransformed levels are active, and therefore also the sub-level where the speed of light is 0 is active and as a result Z_F is going to take place (the subscript 'F' implies 'fragmentation'). '$Y = U$' implies that all levels are active and as a result Z_R is going to take place (the subscript 'R' implies 'random').

Line 3 elaborates the significance of Z_Q, Z_F, and Z_R. Hence Z is the union of potential quantum-operations of space, time, energy, and gravity, designated by ' $Z \in \mathbb{U}$ ($Space, Time, Energy, Gravity$)'. But the nature of the operations, as suggested in the previous paragraph, is designated   by ' Q: $Quantization$; F: $Fragmentation$; R: $Random$ '. Z_Q, then, implies that the full quantization originating from updated four-base logic-encoding ecosystems (FBLEE) can take place. Z_F implies that only a local library, as opposed to FBLEE, can potentially be altered. Z_R also precludes full quantization, and that some partial local-library constructive or destructive mutation may take place.

The ' $\rightarrow h \ni h \in$ ' segment resolves the outcome of the operations implied by Lines 1 – 3, suggesting that the outcome will be 'h' such that ($\ni$) 'h' is an element ($\in$) of the set specified by the members '[Q]: *Constructive zone*', '[Q]: *Constructive zone AND Constructive mutation*', '[F]: *Destructive mutation*', and '{R}: *Random mutation*'. '[Q]: *Constructive zone*' implies that the in-built buffer has been crossed and that access to the deeper four-base logic-encoding ecosystem (FBLEE) has been granted. '[Q]' in this segment implies that there is the possibility that full-quantization as specified by Line 3 of the previous matrix can take place. Access to this zone is a prerequisite for constructive mutation to occur, as designated by the element   '   [Q]: *Constructive zone* ∩ *Constructive mutation* ', which generally implies that full-quantization is going to take place. In the case of an early post-human organization though, as previously suggested in since there is not yet a post-genetic structure securely in place to house such full-quantization, this likely reinforces the call to antecedent layers for such a structure to begin to materialize through the activation of PGCI. In the meanwhile quantization will affect FBLEE, and through FBLEEE may affect human genetic-code. The '[F]' specifies the relationship between 'Fragmentation' in Line 3 of the previous matrix and destructive mutation. The '[R]' specifies the relationship between 'Random' in Line 3 of the previous matrix and random mutation.

Generation of Global Sustainable Civilization FBLEE-Based Partial-Singularity Genetic-Type Code

A brief look at history will reinforce the idea that it is typically maturity along multiple and distinct dimensions, represented by the four properties of light, and their rich interaction that allows civilizations to endure.

Thus, those civilizations that have endured typically have a balance of all four families (Sri Aurobindo, 1971). Civilizations that have become extinct typically have had a focus on few drivers of innovation. Jared Diamond proposes five interconnected causes of collapse that may reinforce each other: non-sustainable exploitation of resources, climate changes, diminishing support from friendly societies, hostile neighbors, and inappropriate attitudes for change (Diamond, 2005). But these five sources may also be thought of as symptoms that arise due to the failure to adopt the catholicity of the sources of innovation emanating from each of the four sets or families. Further, the historian Toynbee suggested that societies decay because of their over-reliance on structures that helped them solve old problems (Toynbee, 1961). It can be interpreted that being thus biased they are unable to adopt the catholicity of the sources of innovation emanating from each of the four sets of families.

Approaching Civilization from a big-picture, global basis though, the sustainability of humankind will be ensured by a balance of development amongst the four sets of forces. This means that countries and global regions must be unique and in such a way that their primary emergence is distributed amongst all four sets of forces. Further, and based on this uniqueness, there must be an open and healthy interaction amongst these centers of uniqueness.

Hence, India, Japan, Thailand, and UK will be considered as representative examples of the four distinct and emergent properties that must be balanced in the whole.

391

Historically at least, India, for example, in its essence, may be thought of as having an exceptional capacity for penetrating behind the surface, and further of meaningfully synthesizing many streams of development. Its primary power may thus be from the family of knowledge, with a strong secondary driver being its ability to create living harmonies. The equation for India will then likely be represented by Equation 8.2.1:

$$Sig_{India} = Xa +$$
$$\overline{Yb_{0-n}} \quad where \begin{bmatrix} X \in [S_{System_K}] \\ Y \in [S_{System_{Pr}}, S_{System_P}, S_{System_K}, S_{System_N}] \\ a, b \ are \ integers; a > b \end{bmatrix}$$

Eq 8.2.1: India (Knowledge Family)

Japan, in its essence, may be thought of as having a strong and noble warrior nature, along with a refined sense of aesthetics, amongst other qualities. Its equation, Equation 8.2.2 would be of the form:

$$Sig_{Japon} = Xa +$$
$$\overline{Yb_{0-n}} \quad where \begin{bmatrix} X \in [S_{System_P}] \\ Y \in [S_{System_{Pr}}, S_{System_P}, S_{System_K}, S_{System_N}] \\ a, b \ are \ integers; a > b \end{bmatrix}$$

Eq 8.2.2: Japan (Power Family)

UK, in its essence, may be thought of as having a strong ability to create practical, materialistic harmonies, resulting in such things as working parliaments and advanced democracy, for example. Its equation, Equation 8.2.3, would be of the form:

$$Sig_{UK} = Xa +$$
$$\overline{Yb_{0-n}} \quad where \begin{bmatrix} X \in [S_{System_N}] \\ Y \in [S_{System_{Pr}}, S_{System_P}, S_{System_K}, S_{System_N}] \\ a, b \ are \ integers; a > b \end{bmatrix}$$

Eq 8.2.3: UK (Nurturing Family)

Thailand, in its essence, may be characterized by an exceptional sense of hospitality and sweet service, with an attention to detail in the practical arrangement of things. Its equation, Equation 8.2.4, would be of the form:

$$Sig_{Thailand} = Xa + \overline{Yb_{0-n}} \ \ where \begin{bmatrix} X \in [S_{System_N}] \\ Y \in [S_{System_{Pr}}, S_{System_P}, S_{System_K}, S_{System_N}] \\ a, b \ are \ integers; a > b \end{bmatrix}$$

Eq 8.2.4: Thailand (Service Family)

Note that (8.2.1 - 4) already implies that Lines $1 - 5$ in the Light Matrix (3.1.3) have been activated, and that the logic of the sustainable-global-civilization- ecosystem will precipitate into FBLEE and may also impact post material-fabric cell-based genetic structure through FBLEEE.

Similarly every nation on earth will have a uniqueness that can be represented by equations belonging to one of the four families.

As discussed in Chapter 2.5, A Possible Mutational-Sequence in Partial-Singularities, for uniqueness to emerge and mature is a process. The process is represented by Equation 2.5.1 reproduced here for convenience:

$$Sig_E = X \begin{vmatrix} C: Sig * mod \left(\int = 1 \right) \\ F: Sig \ mod \ (c) \\ I: Sig \ mod \left(\int \overline{G, e, \pi} \right) \\ M: Sig * mod \ (G) \\ V: Sig * mod \ (e) \\ P: Sig * mod \ (\pi) \end{vmatrix}$$

Equations 8.2.1 – 4 represent example of the essence of uniqueness. For these to become living practicalities requires work at many different levels within a nation. If a nation has not really done the work to animate itself with its uniqueness then it may exist in the physical or P-state, or vital of V-state, or mental or M-state, which by definition lack sufficient maturity, and interaction with other nations is then going to be compromised. Moving to the integral or I-state, will allow a nation to at least not be locked in to a point of view. A consistent I-state practiced by each nation then is the minimum requirement for a sustainable global civilization. But a consistent I-state also means that restricting patterns are always being broken and that the quantization effect of organization on the post material-fabric is actively in play. In other words, separations of light are continually uniting with the larger continent of Light and changing the very nature of possibility on larger and larger scale.

Summary of Sustainable Global Civilization FBLEE-Based Partial-Singularity Genetic-Type Code

Summarizing, after the sustainable global civilization stage iteration the Light-Space-Time Emergence equation is complete, the following code-segments will have been 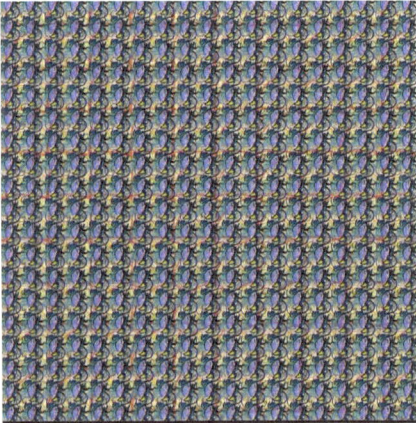 generated as specified by Equation 8.2.5, Active Sustainable Global Civilization FBLEE-Based Partial-Singularity Genetic-Type Code:

394

$$Light - Space - Time\ Emergence_{Sustainable\ Global\ Civilization} =$$

$$\begin{Vmatrix} \begin{bmatrix} \begin{bmatrix} c_\infty : [Pr, Po, K, H] \\ \left(\downarrow R_{C_K} = f\left(R_{C_\infty}\right) \right) \\ c_K : [S_{Pr}, S_{Po}, S_K, S_H] \\ \left(\downarrow R_{C_N} = f\left(R_{C_K}\right) \right) \\ c_N : f(S_{Pr}\ x\ S_{Po}\ x\ S_K\ x\ S_H) \\ \left(\downarrow R_{C_U} = f\left(R_{C_N}\right) \right) \\ c_U : [P, V, M, C] \\ \Uparrow \\ c_{0:[D,W,I,C]} \end{bmatrix}_{Light} \\ \begin{bmatrix} M_3 : -\infty \leq t \leq \infty \\ \downarrow \\ M_2 : 0 \geq t > \infty \\ \downarrow \\ M_1 : 0 > t > \infty \\ \downarrow \\ U \to \begin{array}{l} t \leq E_{Cell};\ TC:\ M_3 \to U \\ t \sim E_{Human};\ TC:\ U \to M_3 \end{array} \end{bmatrix}_{Time} & \begin{matrix} \begin{bmatrix} M_3 \to System_X \\ (\uparrow F \to I) \\ M_2 \to S_{System_X} \\ (\uparrow Sig \to F) \\ M_1 \to Sig_X \\ (\uparrow > P_X) \\ U \to x_{Stable\ Mega-Organization} \end{bmatrix}_{Space} \\ \\ TC \to x_{Sust.Global\ Civilization} \end{matrix} \end{bmatrix} \end{Vmatrix} \langle x_U | x_T \rangle$$

$$\Rightarrow$$

$\langle Space - Time - Energy - Gravity\ Material - Fabric\ Pre - Genetic\ Code \rangle +$

$\langle Electromagnetic\ Spectrum\ Material - Fabric\ Pre - Genetic\ Code \rangle +$

$\langle Quantum - Particle\ Material - Fabric\ Pre - Genetic\ Code \rangle +$

$\langle Atoms\ Pre - Genetic\ Code \rangle +$

$\langle Molecules\ Material - Fabric\ Pre - Genetic\ Code \rangle + LSTE\ \langle \ldots \rangle +$

$PGGT < \cdots > + \langle Cells\ Genetic\ Code \rangle + LSTE\ \langle \ldots \rangle +$

$(Fundamental\ Capacities\ of\ Self\ Genetic\ Code) + FBLEEE\ \langle \ldots \rangle +$

$(Humans\ Genetic\ Code) + LSTE\ \langle \ldots \rangle + FBLEEE\ \langle \ldots \rangle +$

$$(Stable\ Mega - Organization\ Genetic\ Code) + PGCI\ \langle...\rangle + LSTE < \cdots$$
$$> +$$

$$\left(\begin{array}{c} \sum Sig_{Nurturing-Nation} = Xa + \overline{Yb_{0-n}} \\ where \begin{bmatrix} X \in [S_{System_N}] \\ Y \in [S_{System_{Pr}}, S_{System_P}, S_{System_K}, S_{System_N}] \\ a, b\ are\ integers; a > b \end{bmatrix} \\ \sum Sig_{Knowlegd??-Nation} = Xa + \overline{Yb_{0-n}} \\ where \begin{bmatrix} X \in [S_{System_K}] \\ Y \in [S_{System_{Pr}}, S_{System_P}, S_{System_K}, S_{System_N}] \\ a, b\ are\ integers; a > b \end{bmatrix} \\ \sum Sig_{Power-Nation} = Xa + \overline{Yb_{0-n}} \\ where \begin{bmatrix} X \in [S_{System_P}] \\ Y \in [S_{System_{Pr}}, S_{System_P}, S_{System_K}, S_{System_N}] \\ a, b\ are\ integers; a > b \end{bmatrix} \\ \sum Sig_{Service-Nation} = Xa + \overline{Yb_{0-n}} \\ where \begin{bmatrix} X \in [S_{S??stem_{Pr}}] \\ Y \in [S_{System_{Pr}}, S_{System_P}, S_{System_K}, S_{System_N}] \\ a, b\ are\ integers; a > b \end{bmatrix} \end{array} \right)$$

Eq. 8.2.5, Active Sustainable Global Civilization **FBLEE-Based Partial-Singularity** *Genetic-Type Code*

Note that the final code segment contains the code for all possible ($\sum x$) nation types, and depicts the growing biography by which the singular light-based edifice expresses its materialization, and at this stage, through the sustainable global civilization FBLEE-Based partial-singularity.

SECTION 9: EXTRAPOLATION TO SUPER-MATTER BASED PARTIAL-SINGULARITIES

Sections 2 and 3 proposed a mathematical framework for light-based singularities. Sections 4 through 8 leveraged the mathematical framework to explore the biographies of light-based singularities, including both the seed-singularity and numerous partial-singularities. Such biographies are entirely generated by aspects of the light-space-time emergence equation, which with the emergence of partial-singularities is also known as the partial-singularity generation equation. Components of such biographies include the essential architecture of the singularity generated from the four principal characteristics of light, the four sets of properties related to the principal characteristics of light, unique seeds created from combinations of elements derived from the four sets, and the essential dynamics of maturation culminating in possible quantization.

Previous sections elaborated the broad lines of emergent biographies of partial-singularities. This section will go a step further and examine micro-level dynamics (Chapter 9.1 will look at insect adaptation leveraging a discussion on cells in Chapter 7.2, and Chapter 9.2 will look at sustainable global civilization adaptation leveraging a discussion in Chapter 8.2) leading to meta-level induced change in specific biographies, and macro-implications that will result in possible creation of super-matter (Chapter 9.3).

Super-matter is suggested to be that type of matter created through the intervention of conscious will or cohesive want. While complexification of matter as summarized in previous sections is often an "automatic" process of Nature, super-matter is a foundation based on will or cohesive want and sets the stage for a potentially unending willed development in which functional-richness existing in Light in its native state traveling at infinite speed can manifest in this material universe. From a macro-level, meta-level bases are what will ensure the development of super-matter.

Creation of such super-matter is tied to the meaning of space, and to cosmic observations of the expansion of the universe (Chapter 9.4). Chapter 9.4 hence focuses on the nature of matter by drawing on observations from the field of astrophysics.

The gestalt of such an analyses may engender a visceral sense of the intimacy and oneness of the light-based adventure that results in the vastness of the cosmos tied to human-precipitated determinable dynamics at the quantum-level. As such the persistent quantum-level computation resulting in a constant stream of genetic-type information alters the code and the very light-based computational machinery emergent as more and more sophisticated pre-matter, matter, and post or super-matter structures.

Such an analyses prepares us for a comparison between any AI-based singularity and the Second Singularity (to be discussed in Section 10). The Second Singularity can be thought of as a further extrapolation of super-matter based partial-singularities.

Chapter 9.1: Micro-Lever Based Adaptation Dynamics in an Insect Partial-Singularity

Leveraging the discussion on the broad lines of the emerging living cell partial-singularity from Chapter 7.2,

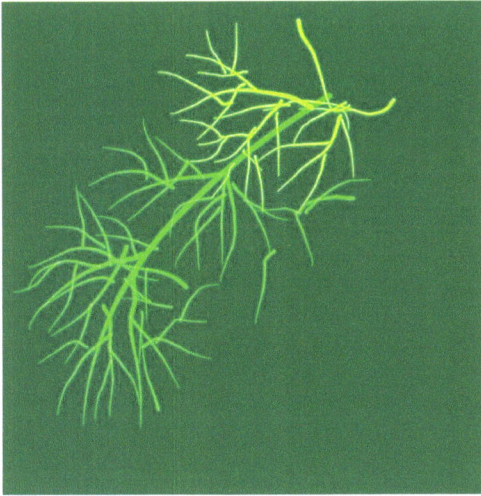

Generation of Living Cell Partial-Singularity, this chapter will specifically consider how aspects of this biography are determined in the case of the adaptation of an insect. Such adaptation signals the successful change in biography of the insect partial-singularity, and in the case elaborated in this chapter, due to the interaction of sufficient urge that precipitates a process of quantization. Such an urge-to-quantization process sheds insight into the dynamics of partial-singularities.

Recall that the 'quantum' is a window into layers of reality behind the surface layer U. This insight was captured by Schrodinger's Equation (discussed in Chapter 3.3) and by Heisenberg's Uncertainty Principle (discussed in Chapter 3.4), the former suggesting that matter, or what is going to appear materially, is a function of some incredible number of superposed states containing infinite possibility, and the latter suggesting that quantum fluctuation due to an always buzzing pregnant-infinity, lies behind everything even when seemingly still.

Recall that the basis of cell adaptability was discussed in Chapter 7.2, and (7.2.1) modeling generation of cells partial-singularity is reproduced here:

$Light - Space - Time\ Emergence_{Cells} =$

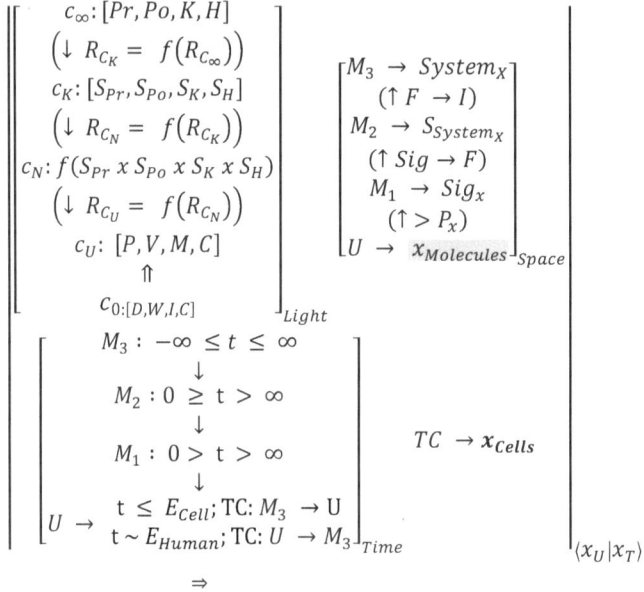

$$\left\| \begin{bmatrix} \begin{array}{c} c_\infty: [Pr, Po, K, H] \\ \left(\downarrow R_{C_K} = f\left(R_{C_\infty}\right) \right) \\ c_K: [S_{Pr}, S_{Po}, S_K, S_H] \\ \left(\downarrow R_{C_N} = f\left(R_{C_K}\right) \right) \\ c_N: f(S_{Pr}\ x\ S_{Po}\ x\ S_K\ x\ S_H) \\ \left(\downarrow R_{C_U} = f\left(R_{C_N}\right) \right) \\ c_U: [P, V, M, C] \\ \Uparrow \\ c_{0:[D,W,I,C]} \end{array} \end{bmatrix}_{Light} \begin{bmatrix} M_3\ \rightarrow\ System_X \\ (\uparrow F\ \rightarrow I) \\ M_2\ \rightarrow\ S_{System_X} \\ (\uparrow Sig \rightarrow F) \\ M_1\ \rightarrow\ Sig_x \\ (\uparrow > P_x) \\ U\ \rightarrow\ x_{Molecules} \end{bmatrix}_{Space} \right.$$

$$\begin{bmatrix} M_3 :\ -\infty\ \le t\ \le\ \infty \\ \downarrow \\ M_2 :\ 0\ \ge t\ >\ \infty \\ \downarrow \\ M_1 :\ 0\ >\ t\ >\ \infty \\ \downarrow \\ U\ \rightarrow \begin{array}{l} t\ \le\ E_{Cell}; TC: M_3\ \rightarrow U \\ t\ \sim E_{Human}; TC: U\ \rightarrow M_3 \end{array} \end{bmatrix}_{Time} \left. \begin{array}{c} TC\ \rightarrow\ x_{Cells} \end{array} \right\| \langle x_U | x_T \rangle$$

$$\Rightarrow$$

$\langle Space - Time - Energy - Gravity\ Material - Fabric\ Pre$
$\qquad\qquad - Genetic\ Code \rangle +$

$\langle Electromagnetic\ Spectrum\ \ Material - Fabric\ Pre - Genetic\ Code \rangle$
$\qquad\qquad +$

$\langle Quantum - Particle\ \ Material - Fabric\ Pre - Genetic\ Code \rangle +$

$\langle Atoms\ \ Material - Fabric\ Pre - Genetic\ Code \rangle +$

$\langle Molecules\ Material - Fabric\ Pre - Genetic\ Code \rangle + LSTE\ \langle \dots \rangle$
$\qquad\qquad + PGGT\ < \dots > +$

$Cells\ Genetic\ Code$

The Light-Matrix, the top-left matrix in (7.2.1), suggests the possibilities in the pregnant-infinity. These possibilities, recall, have been set up by realities so created by light traveling at different speeds. While prevalent dynamics at the surface layer, U, are always influenced by dynamics from the deeper layers, there is a state that can be created that will allow a more focused and intentional activation of the meta-functions resident in the deeper layers. Such activation will potentially trigger a series of processes culminating in a state of quantum-certainty (Malik, 2017e), introduced in Chapter 3.5 when taking a deeper look at S-T-E-G quantization, and as will be explored through the rest of this chapter.

The Space-Matrix, the top-right matrix in (7.2.1), suggests the dynamics involved in creating such an activation-state. Essentially this involves the overcoming of patterns at U.

Insect Adaptation Case

Consider the case of a hypothetical insect, for example, that is always prey to other predators. At some point a visceral urge to overcome some of these predators may arise in this species of insect. This visceral urge is a breaking of habitual patterns common to that species of insect. If it is deep enough and pervasive enough, it can be thought of as an activation-state and will allow the opening of a quantum-window so that there is now a connection with the quantum worlds, Q (as introduced in Chapter 3.2), behind.

It is proposed that for successful adaptability to come about it is only such a pervasive state that will open a quantum-window to effectively stimulate the interaction between layers that will ultimately allow adaptability to occur. This activation-state can be further specified by leveraging (3.7.4) – The Generalized Equation for Direction of Mutation, that elaborates some Space-Matrix dynamics and was derived in Chapter 3.7, on qualified determinism.

Reproducing Equation 3.7.4:

$$Org\_Dir = DI \left(\begin{bmatrix} M_3 \rightarrow System_X \\ (\uparrow F \rightarrow I) \\ M_2 \rightarrow S_{System_X} \\ (\uparrow Sig \rightarrow F) \\ M_1 \rightarrow Sig_x \\ (\uparrow > P_P) \\ U \rightarrow x_U \end{bmatrix}_{x=p,v,m,i} \right) \rightarrow$$

$$x\_matrix_{strongest} @ level_{strongest}$$

The activation-state can be thought of as being invoked when $level_{strongest}$ is at least M_1. Hence, as in Equation 9.1.1:

$$Activation - State_{condition} \geq M_1$$

Eq 9.1.1: Condition for Activation-State

If the activation-state is invoked, that implies access to the four-base logic-encoding ecosystem (FBLEE).

Note that the Time-Matrix, in the lower-left side of (7.2.1), specifies that approximately, adaptability may tend to be more automatic up to the emergence of cellular organisms. This is specified by the direction $M_3 \rightarrow U$, which indicates that it is the meta-level that organizes activity at U. At approximately the human level the direction is flipped as specified by $U \rightarrow M_3$, and suggests that will, intention, feeling and the like are more important in stimulating the process of adaptability.

The visceral urge, then, may stimulate interaction with a collective-intelligence or 'specific-species-intelligence meta-function' at FBLEE, which allows the generation of a new and specific 'predator-overcoming meta-function' so that the hypothetical insect in question can now go through an adaptation to survive at least some kinds of predator attacks. This may take the form of this species of insect creating a hard-shell around it. Mathematically this new meta-function may take the following form as suggested by Equation 9.1.2:

$$Sig_{hard-protective-shell} = Xa + \overline{Yb_{0-n}}$$

$$where \begin{bmatrix} X \in [S_{System_P}] \\ Y \in [S_{System_{Pr}}, S_{System_P}, S_{System_K}, S_{System_N}] \\ a, b \ are \ integers; a > b \end{bmatrix}$$

Eq 9.1.2: Hard Protective Shell Meta-Function

In this example, S_{System_P} relates to a set of Polysaccharides, as introduced in Chapter 7.2, and suggests that the chains of sugar molecules will adapt to become a shell to protect the insect. The primary element X, is therefore an element of the set or Polysaccharides.

$S_{System_{Pr}}$ refers to the set of Proteins. S_{System_K} refers to the set of Nucleic Acids. S_{System_N} refers to the set of Lipids. Y as the secondary element will invoke the action of some proteins, some existing polysaccharides, some nucleic acids, and some lipids in bringing about the adaptation as specified by $Sig_{hard-protective-shell}$. Perhaps the nucleic acids will code or coordinate how the proteins will work with the existing polysaccharides and lipids to create a new arrangement of polysaccharides that becomes the protective layer for the insect. This new arrangement of polysaccharides then becomes the purpose of a particular new type of specialized cell that exists to protect the insect against certain types of predators.

So while the breaking of patterns as specified by the Space-Matrix was a first step in the process, the subsequent creation of a new meta-function can be thought of as a second step. This new meta-function can be thought of as happening due to some combined functioning of the Light and Space Matrices. Specifically in the Space-Matrix, the breaking of patterns, $(\uparrow > P_x)$, allows the dynamics of M_1 to become active: $M_1 \rightarrow Sig_x$. This allows the ever-present layers of light, and specifically $C_{N:} f(S_{Pr} \times S_{Po} \times S_K \times S_H)$ in the Light-Matrix to become more consciously active so that a new meta-function that may add to the insect-specific FBLEE is generated.

The third step is suggested by ($\downarrow R_{CU} = f(R_{CN})$) that allows specific quantization to occur. It is proposed that the hard-protective-shell adaptability becomes real through a series of tightly coordinated space, time, energy, and gravity quantization as specified by (3.5.2-7) derived in Chapter 3.5. A relevant subset of equations is reproduced here for convenience and suggests the mechanics for the adaptability along this specific line of development.

Equation 3.5.2 models space-quantization:

$$Space_{quantization} = h_{UK}\left(Xa + \overline{Yb_{0-n}}\right)$$

$$where: \begin{bmatrix} X \in [S_{System_K}] \\ Y \in [S_{System_{Pr}}, S_{System_P}, S_{System_K}, S_{System_N}] \\ a, b\ are\ integers; a > b \end{bmatrix}$$

Recall that space-quantization further structures space so that seeds of knowledge or knowledge-potential to do with the new meta-function are now resident in the material ecosystem associated with the meta-function. In this case, the material ecosystem is the cell-based genetic structure, as indicated by the action of PGGT in (7.2.1). In such a manner the species that set this meta-function into action will more and more tap into the possibilities to bring about the new protective mechanism it seeks.

Equation 3.5.4 models time-quantization:

$$Time_{quantization} = h_{UP}\left(Xa + \overline{Yb_{0-n}}\right)$$

$$where: \begin{bmatrix} X \in [S_{System_P}] \\ Y \in [S_{System_{Pr}}, S_{System_P}, S_{System_K}, S_{System_N}] \\ a, b \ are \ integers; a > b \end{bmatrix}$$

Recall that time-quantization alters the structure of time so that phases of maturity associated with seeds in space are now encoded in time. Slow and faster periods of change are in this way connected to the species seeking to protect itself against predators.

Equation 3.5.6 models energy-quantization:

$$Energy_{quantization} \ = \ h_{UPr}\left(Xa + \overline{Yb_{0-n}}\right)$$

$$where: \begin{bmatrix} X \in [S_{System_{Pr}}] \\ Y \in [S_{System_{Pr}}, S_{System_P}, S_{System_K}, S_{System_N}] \\ a, b \ are \ integers; a > b \end{bmatrix}$$

Recall that energy-quantization allows matter to be formed, and in this case, will have to do with the incremental material changes that the species will go through in forming a hard protective shell.

Equation 3.5.7 models gravity-quantization:

$$Gravity_{quantization} \ = \ h_{UN}\left(Xa + \overline{Yb_{0-n}}\right)$$

$$where: \begin{bmatrix} X \in [S_{System_N}] \\ Y \in [S_{System_{Pr}}, S_{System_P}, S_{System_K}, S_{System_N}] \\ a, b \ are \ integers; a > b \end{bmatrix}$$

Recall that gravity-quantization has to do with inter-relation between the species and surrounding objects and will change the very nature of gravity to allow a subtle new balance in the species interaction with its surrounding so that the deep urge of the species can more easily be fulfilled.

As a result of such adaptation FBLEE is changed and this changes the way in which matter can materialize. It is such changes initiated by deep urge or will made at the margin that accumulate to enhance the materialization of fourfold richness and therefore of the functionality of matter. An accumulated or threshold volume of such changes will gradually change the very way matter materializes, approaching the reality of materialization of functionally rich super-matter.

Chapter 9.2: Micro-Lever Based Development Dynamics in a Sustainable Global Civilization Partial-Singularity

Leveraging the discussion on the broad lines of the emerging sustainable global civilization partial-singularity from Chapter 8.2, Generation of a Sustainable Global Civilization FBLEE-Based Partial-Singularity, this

chapter will specifically consider how aspects of this biography are determined in the case of the development of a sustainable global civilization. Such development signals the successful change in biography of the sustainable global civilization partial-singularity, and in the case elaborated in this chapter, due to the interaction of sufficient urge that precipitates a process of quantization. Such an urge-to-quantization process sheds insight into an increasingly important aspect of the dynamics of partial-singularities, especially when humans begin to play a part.

The Light-Space-Time Emergence equation (4.2.7) leveraged in Chapter 8.2 is reproduced here for convenience:

$$Light - Space - Time \; Emergence_{Sustainable \; Global \; Civilization} =$$

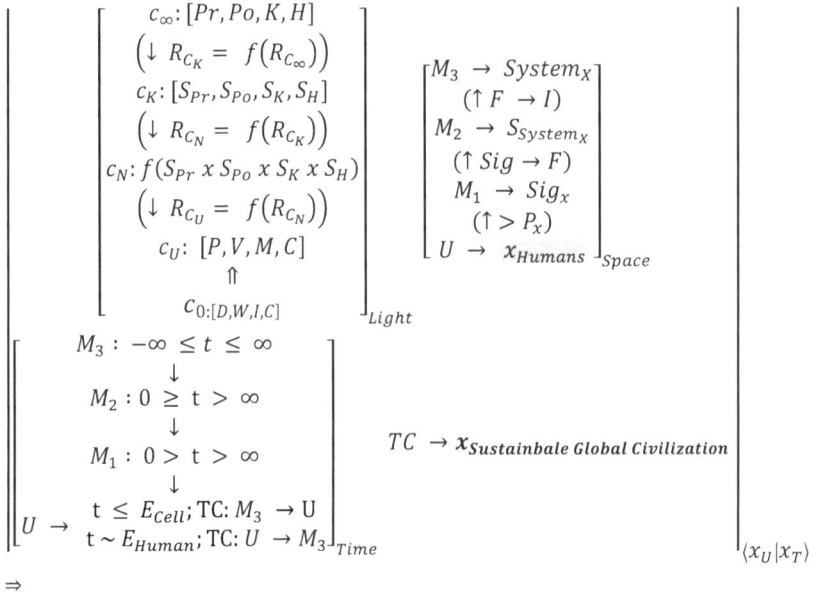

$$
\left|
\begin{array}{l}
\left[
\begin{array}{c}
c_\infty : [Pr, Po, K, H] \\
\left(\downarrow R_{C_K} = f(R_{C_\infty}) \right) \\
c_K : [S_{Pr}, S_{PO}, S_K, S_H] \\
\left(\downarrow R_{C_N} = f(R_{C_K}) \right) \\
c_N : f(S_{Pr} \; x \; S_{PO} \; x \; S_K \; x \; S_H) \\
\left(\downarrow R_{C_U} = f(R_{C_N}) \right) \\
c_U : [P, V, M, C] \\
\Uparrow \\
c_{0:[D,W,I,C]}
\end{array}
\right]_{Light}
\quad
\left[
\begin{array}{c}
M_3 \rightarrow System_X \\
(\uparrow F \rightarrow I) \\
M_2 \rightarrow S_{System_X} \\
(\uparrow Sig \rightarrow F) \\
M_1 \rightarrow Sig_x \\
(\uparrow > P_x) \\
U \rightarrow x_{Humans}
\end{array}
\right]_{Space}
\\[4em]
\left[
\begin{array}{c}
M_3 : -\infty \leq t \leq \infty \\
\downarrow \\
M_2 : 0 \geq t > \infty \\
\downarrow \\
M_1 : 0 > t > \infty \\
\downarrow \\
U \rightarrow \begin{array}{l} t \leq E_{Cell}; TC: M_3 \rightarrow U \\ t \sim E_{Human}; TC: U \rightarrow M_3 \end{array}
\end{array}
\right]_{Time}
\quad
TC \rightarrow x_{Sustainbale \; Global \; Civilization}
\end{array}
\right| \langle x_U | x_T \rangle
$$

$\Rightarrow$

$\langle Space - Time - Energy - Gravity \; Material - Fabric \; Pre - Genetic \; Code \rangle +$

$\langle Electromagnetic \; Spectrum \; Material - Fabric \; Pre - Genetic \; Code \rangle +$

$\langle Quantum - Particle \; Material - Fabric \; Pre - Genetic \; Code \rangle +$

$\langle Atoms \; Pre - Genetic \; Code \rangle$
$\quad + \langle Molecules \; Material - Fabric \; Pre - Genetic \; Code \rangle +$

$LSTE \; \langle ... \rangle \; + \; PGGT <$
$\quad\quad > + \langle Cells \; Genetic \; Code \rangle \; + \; LSTE \; \langle ... \rangle$
$\quad\quad + FBLEEE \; \langle ... \rangle \; +$

$\langle Humans \; Genetic \; Code \rangle \; + \; LSTE \; \langle ... \rangle + FBLEEE \; \langle ... \rangle +$

$\langle Stable \; Mega - Organization \; FBLEE \; Genetic \; Code \rangle + PGCI \; \langle ... \rangle +$

409

The Light-Matrix, the top-left matrix in (4.2.7), recall, has been set up by realities so created by light traveling at different speeds. While prevalent dynamics at the surface layer, U, are always influenced by dynamics from the deeper layers, there is an activation-state that can be created that will allow a more focused and intentional activation or creation of the meta-functions in the deeper layers.

The Space-Matrix, the top-right matrix in (4.2.7), suggests the dynamics involved in creating such an activation-state. Essentially this involves the overcoming of patterns at U. Note that by definition such an activation-state necessitates moving in the direction where the speed of light increases. Note that speed of light increasing implies that the sense of separation is diminishing, and entities can have a more complete access to the fullness that they are. This can only happen when patterns get elevated. If patterns are driven primarily by the U realm, being therefore untransformed, that will by definition not allow a quantum-window to open.

This activation-state can be further specified by leveraging (3.7.4) – The Generalized Equation for Direction of Mutation, derived in Chapter 3.7, on Qualified Determinism.

Reproducing Equation 3.7.4:

$$Org\_Dir = DI \left(\begin{bmatrix} M_3 \rightarrow System_X \\ (\uparrow F \rightarrow I) \\ M_2 \rightarrow S_{System_X} \\ (\uparrow Sig \rightarrow F) \\ M_1 \rightarrow Sig_x \\ (\uparrow > P_{P)} \\ U \rightarrow x_U \end{bmatrix}_{x=p,v,m,i} \right) \rightarrow$$

$$x\_matrix_{strongest} @ level_{strongest}$$

Activation-state can be thought of as being invoked when $level_{strongest}$ is at least M_1. Hence, as suggested by the previously derived (9.1.1), reproduced here for convenience:

$$Activation - state_{condition} \geq M_1$$

If the activation-state is invoked, that implies access to the four-base logic-encoding ecosystem (FBLEE).

Deeper Nation Essence Foundation Case

Similar to the discussion in the previous chapter, assume that there is a deep wanting felt by a threshold number of people. Perhaps this wanting, a step towards a global sustainable civilization, is for each nation to operate more from their true essence, or I-state, as opposed to fundamentalist form. In other words a meta-function, $Sig_{Nation-essence-foundation}$, must be created that will then organize its own reality, and as indicated by Equation 9.2.1.

$$Sig_{Nation-essence-foundatiuon} = Xa + \overline{Yb_{0-n}}$$

$$where \begin{bmatrix} X \in [S_{System_{Pr}}] \\ Y \in [S_{System_{Pr}}, S_{System_P}, S_{System_K}, S_{System_N}] \\ a, b \text{ are integers}; a > b \end{bmatrix}$$

Eq 9.2.1: Nation-Essence-Foundation Meta-Function

While the key aspect, or primary X-element in the nation-essence-foundation meta-function may be an element from the set of Presence, such as 'nation-essence appreciation', there will be many parts or sub-functions, that could be represented by Y-elements, such as the structure, processes, institutions, culture, cohesion that will also need to be created to ensure such nation essence foundation. Cohesion, as a sub-function, represents an important denominator in breaking patterns at the individual and collective levels and as in Equation 9.2.2, Nation-Essence-Cohesion, is represented with the primary element belonging to the set of Harmony or Nurturing:

$$Sig_{Nation-essence-cohesion} = Xa + \overline{Yb_{0-n}}$$

$$where \begin{bmatrix} X \in [S_{System_N}] \\ Y \in [S_{System_{Pr}}, S_{System_P}, S_{System_K}, S_{System_N}] \\ a, b \ are \ integers; a > b \end{bmatrix}$$

Eq 9.2.2: Nation-Essence-Cohesion Meta-Function

412

Hence, S_{System_K} relates to the set of Thoughts required to contextualize and frame nation-essence. $S_{System_{Pr}}$ refers to the set of Sensations required to assess and interpret the constantly arising signs of the new development – the sensory cues as it were that will allow any individual or collectivity

to sense that they were on the right path. S_{System_P} refers to the set of Urges and Wills constantly required to ensure that the goal of cohesion was attained. S_{System_N} refers to the set of Feelings and Emotions that will need to be generated to attain cohesion.

As such cohesion becomes a reality, older patterns will break down and there will be easier and longer activation-state periods, potentially allowing the additional related sub-functions to also become increasingly active.

As suggested by the previous chapter, while the breaking of patterns as specified by the Space-Matrix was a first step in the process to making 'cohesion' a reality, the subsequent creation of a new meta-function of cohesion can be thought of as a second step. This new meta-function can be thought of as happening due to some combined functioning of the Light and Space Matrices. Specifically in the Space-Matrix, the breaking of patterns, $(\uparrow > P_x)$, allows the dynamics of M_1 to become active: $M_1 \rightarrow Sig_x$. This allows the ever-present layers of light,

413

and specifically $C_{N:}\ f(S_{Pr}\ x\ S_{Po}\ x\ S_K\ x\ S_H)$ in the Light-Matrix to become more consciously active so that new meta-function that may add to the sustainable global civilization FBLEE becomes available or is generated.

The third step is suggested by ($\downarrow R_{C_U} = f(R_{C_N})$) that allows specific quantization to occur. Quantization is of fundamental importance because in this model it is what allows the fabric of experienced reality to change. The new thought to do with cohesion is quantized as seeds of knowledge in space, fundamentally changing the structure of space. Hence, reproducing Equation 3.5.2, space-quantization is modeled as:

$$Space_{quantization} = h_{UK}(Xa + \overline{Yb_{0-n}})$$

$$where: \begin{bmatrix} X \in [S_{System_K}] \\ Y \in [S_{System_{Pr}}, S_{System_P}, S_{System_K}, S_{System_N}] \\ a, b\ are\ integers; a > b \end{bmatrix}$$

The phases of maturity that the seeds of knowledge will go through give time a new reality or structure. In other words time too is re-oriented to promote the outcome of the new cohesion meta-function. This quantization of time, essentially giving time in this new fabric a different meaning is modeled by Equation 3.5.4, reproduced here for convenience:

$$Time_{quantization} = h_{UP}(Xa + \overline{Yb_{0-n}})$$

$$where: \begin{bmatrix} X \in [S_{System_P}] \\ Y \in [S_{System_{Pr}}, S_{System_P}, S_{System_K}, S_{System_N}] \\ a, b\ are\ integers; a > b \end{bmatrix}$$

The new meta-function changes the very nature of matter, which is quantized to allow the material fabric of existence to promote the new meta-function of cohesion in unimaginable ways. This quantization of energy,

414

resulting in a subtly different matter is modeled by Equation 3.5.6, reproduced here for convenience:

$$Energy_{quantization} = h_{UPr}\left(Xa + \overline{Yb_{0-n}}\right)$$

$$where: \begin{bmatrix} X \in [S_{System_{Pr}}] \\ Y \in [S_{System_{Pr}}, S_{System_P}, S_{System_K}, S_{System_N}] \\ a, b \ are \ integers; a > b \end{bmatrix}$$

As a result the very force of gravity is altered locally as well, so that its quantization promotes subtle interaction in the related ecosystem that will be predisposed to making the meta-function a reality. This quantization of gravity is modeled by Equation 3.5.7, reproduced here for convenience:

$$Gravity_{quantization} = h_{UN}\left(Xa + \overline{Yb_{0-n}}\right)$$

$$where: \begin{bmatrix} X \in [S_{System_N}] \\ Y \in [S_{System_{Pr}}, S_{System_P}, S_{System_K}, S_{System_N}] \\ a, b \ are \ integers; a > b \end{bmatrix}$$

As a result of such adaptation FBLEE is changed and this changes the way in which matter can materialize. It was

415

proposed in Section 8 that in the case of post-human organizations, such as a sustainable global civilization, the genetic structure of cells might change primarily through FBLEEE. This will have to be the case until such time as some future hybridized structure allows more sophisticated multi-layer genetic-type information to materialize.

It is such changes initiated by deep urge or will made at the margin that accumulate to enhance the materialization of fourfold richness and therefore of the functionality of matter. An accumulated or threshold volume of such changes will gradually change the very way matter materializes, approaching the reality of materialization of functionally rich super-matter.

Chapter 9.3: Macro-Conditions for Transitioning to Super-Matter Based Partial-Singularities

Super-Matter is positioned as being fundamentally different from matter. While matter can be thought of as the result of Nature's automatic working, super-matter can be thought of as the result of a conscious will and cohesive wanting that causes a deliberate process of quantization by which the very fabric of matter is changed. Quantization occurs in any case even when Nature's automatic working drives change. But with conscious will this process can be accelerated, broadened, and heightened, so that a wider and higher possibility of function may consciously materialize.

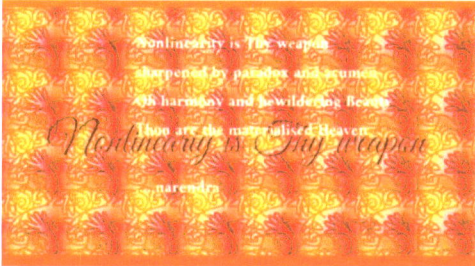

As suggested in the previous sections such quantization is essentially related to enhancing the functional-richness of matter. Imagine through a cohesive-wanting matter being able to become more flexible, more durable, more stretchable, light at will, heavy at will, change color at will, expand in size, contract in size, amongst infinite other functions and possibilities contained in Light in its native state at infinite speed.

The previous sections explored numerous examples of the process by which matter becomes functionally rich. Such increase in functional richness of matter is synonymous with an increase in a partial-singularities' sphere of influence as recorded by its genetic-type information based biography. In cases getting closer to, and even beyond the human-system, such increase in functional richness involves a visceral urge that can open the entity to the influence of meta-levels of Light. Such

417

opening also gives access to FBLEE, and as further elaborated in the previous two chapters, an enhanced function can then materialize through a process of quantization. This increasing possibility of FBLEE means that matter can transition to richer versions of itself.

Further, in considering the vast information embedded in the ubiquitous-point-instant pre-genetic ∞-entanglement library, the architectural forces pre-genetic K-entanglement library, and the organizational-uniqueness pre-genetic N-entanglement library, discussed in Chapters 2.2, 2.3, and 2.4 respectively, it is clear that many possible functions already exist in layers of light. The call from below, whether a pattern-breaking will, urge, or desire, likely causes some meta-function or precipitation into FBLEE, from where materialization becomes possible through the process of quantization.

Another way to look at this is summarized in the previously derived (4.2.8) which models super- matter:

$$Light - Space - Time\ Emergence_{Suoer-Matter} =$$

$$\left|\left|\begin{bmatrix} c_\infty:[Pr,Po,K,H] \\ \left(\downarrow R_{C_K} = f\!\left(R_{C_\infty}\right)\right) \\ c_K:[S_{Pr},S_{Po},S_K,S_H] \\ \left(\downarrow R_{C_N} = f\!\left(R_{C_K}\right)\right) \\ c_N: f(S_{Pr}\ x\ S_{Po}\ x\ S_K\ x\ S_H) \\ \left(\downarrow R_{C_U} = f\!\left(R_{C_N}\right)\right) \\ c_U:[P,V,M,C] \\ \Uparrow \\ c_{0:[D,W,I,C]} \end{bmatrix}_{Light} \begin{bmatrix} M_3\ \rightarrow\ System_X \\ (\uparrow F\ \rightarrow I) \\ M_2\ \rightarrow\ S_{System_X} \\ (\uparrow Sig\ \rightarrow F) \\ M_1\ \rightarrow\ Sig_X \\ (\uparrow > P_X) \\ U\ \rightarrow\ x_{Molecules} \end{bmatrix}_{Space}\right.\right.$$

$$\left|\left|\begin{bmatrix} M_3:\ -\infty\ \leq t\ \leq\ \infty \\ \downarrow \\ M_2:0\ \geq\ \mathrm{t}\ >\ \infty \\ \downarrow \\ M_1:\ 0 > \mathrm{t}\ >\ \infty \\ \downarrow \\ U\ \rightarrow\ \begin{matrix} \mathrm{t}\ \leq\ E_{Cell};\mathrm{TC:}\ M_3\ \rightarrow \mathrm{U} \\ \mathrm{t}\sim E_{Human};\mathrm{TC:}\ U\ \rightarrow M_3 \end{matrix} \end{bmatrix}_{Time} \quad TC\ \rightarrow x_{Super-Matter} \right.\right|_{\langle x_U|x_T\rangle}$$

$$\Rightarrow$$

$$\langle Space - Time - Energy - Gravity\ Material - Fabric\ Pre$$
$$- Genetic\ Code \rangle +$$

$$\langle Electromagnetic\ Spectrum\ \ Material - Fabric\ Pre - Genetic\ Code \rangle$$
$$+$$

$$\langle Quantum - Particle\ Material - Fabric\ Pre - Genetic\ Code \rangle +$$

$$\langle Atoms\ Material - Fabric\ Pre - Genetic\ Code \rangle +$$

$$\langle Molecules\ Material - Fabric\ Pre - Genetic\ Code \rangle + LSTE\ \langle ... \rangle$$
$$+\ FBLEEE\ \langle ... \rangle +$$

$$LSTE\ \langle ... \rangle +\ Super - Matter\ Post - Genetic\ Code$$

Equation 4.2.8 suggests computational realities yet to emerge. The essential action of space-time-energy-gravity

quantization can cause the materialization of potentially infinite four-base logic-encoding ecosystems. It is conceivable that such a variation of space-time-energy-gravity quantization coupled with an increasing ability to easily move back and forth between the material and antecedent realms, perhaps the outcome of an enhanced function driven by an initiating will itself, makes the need to house genetic information in a form such as DNA alone, burdensome. It is conceivable that four-base logic-encoding ecosystems, or even the ∞ -entanglement, K-entanglement, and N-entanglement libraries may be able to more directly act at the material level in some composite subtle-material post-genetic form that also includes DNA. This possibility is in general referred to as post-genetic code.

This chapter takes a more macro-level mathematical view based on a cosmology of light, to summarize the aggregate conditions required for the creation of functionally rich super-matter. In the process the distinction between three distinct phases of matter emerge: that of established-matter, matter-in-transition, and super-matter. The development of super-matter will naturally also lead to a post-genetic housing structure for genetic-type information.

It makes sense for the structure of genetic-type information-storage to change in each of these cases since it is something different in each case. In such a view the material-fabric is the genetic-type information storage medium for "established matter". Hence, code for the electromagnetic spectrum, quantum particles, and atoms, is stored in the material-fabric as reviewed in Sections 5 and 6, on the generation of the electromagnetic spectrum partial-singularity and generation of partial-singularities in the surfacing of matter, respectively.

Matter-in-transition would be associated with the emergence of life and as has been suggested in some detail in my previous book on genetics (Malik, 2019a), matter becomes more dynamic and adaptable as a result of this. Cellular-based genetic structure is the vehicle for storage for code associated with the cell. But further, as discussed, with the emergence of basic human capabilities, and truer individuality, the genetic information is further enhanced. These were reviewed in detail in Section 7 on the generation of partial-singularities in the surfacing of life. It has also been argued in Section 8, that the emergence of complex organization, which incudes mega-organization and a sustainable global civilization, will likely make further changes to the cellular-based genetic code. Such changes are only possible because in the cosmology of light view that animates this book, matter is essentially quantized light, and any precipitated functional richness as manifest in mega-organizations, for example, means that the genetic structure of matter has had to have changed due to the activation of such functional richness. All is one, and matter and light are two poles of that oneness.

When we arrive at super-matter, which is inevitable with the increase in functional-richness, then it stands to reason that there will likely come into existence some post-genetic structure that will house essential genetic-type information in a very different way.

The Light-Space-Time Emergence equation (3.1.3) derived in Chapter 3.1, and reproduced here for convenience, will be leveraged to derive summary equations to distinguish between the different phases of matter. The following derivations run parallel to equations created in Chapter 3.6 that focused on levels of light on genetic-mutation. Reproducing (3.1.3):

$$Emergence_{light-space-time} =$$

$$
\left| \begin{bmatrix} \begin{bmatrix} c_\infty : [Pr, Po, K, H] \\ (\downarrow R_{C_K} = f(R_{C_\infty})) \\ c_K : [S_{Pr}, S_{Po}, S_K, S_H] \\ (\downarrow R_{C_N} = f(R_{C_K})) \\ c_N : f(S_{Pr} \times S_{Po} \times S_K \times S_H) \\ (\downarrow R_{C_U} = f(R_{C_N})) \\ c_U : [P, V, M, C] \\ \Uparrow \\ c_{0:[D,W,I,C]} \end{bmatrix}_{Light} \begin{bmatrix} M_3 \to System_X \\ (\uparrow F \to I) \\ M_2 \to S_{System_X} \\ (\uparrow Sig \to F) \\ M_1 \to Sig_X \\ (\uparrow > P_{x)} \\ U \to x_U \end{bmatrix}_{Space} \\ \left[U \to \begin{bmatrix} M_3 : -\infty \leq t \leq \infty \\ \downarrow \\ M_2 : 0 \geq t > \infty \\ \downarrow \\ M_1 : 0 > t > \infty \\ \downarrow \\ t \sim E_{Cell}; TC: M_3 \to U \\ t \sim E_{Human}; TC: U \to M_3 \end{bmatrix}_{Time} TC \to x_T \right] \end{bmatrix} \right|_{\langle x_U | x_T \rangle}
$$

Simplifying each of the main matrices yielded (3.6.1), the simplified version of the Light-Space-Time Emergence equation, reproduced here for convenience:

$$Emergence_{light-space-time} = |[L][S][T]TC \to x_T|_{\langle x_U | x_T \rangle}$$

422

As explained by Neil Turok in his book The Universe Within (Turok, 2012), Euler's formula reproduced below, can be used to model many naturally occurring

phenomena because of its sinusoidal oscillation between narrow bounds as x increases. Reproducing Euler's formula:

$$e^{ix} = \cos x + i \sin x$$

The sum of the squares of the ordinary and complex parts, on the right side of the equation, is one. In quantum theory this ensures that the probabilities for all possible outcomes add up to one. Hence this formula is useful when summarizing the macro-level effects of (3.6.1) which necessarily has diverse drivers of phenomenon.

Further, in modeling matter a modified notation of the Schrodinger wavefunction as interpreted by Feynman is leveraged. This version features the integral sign, $\int$, meaning that all terms to the right of it have to be summed up for all space and time till the moment when the wavefunction is required to be known.

Summary Equations for Established-Matter

Hence, combining (3.6.1) with Feynman's interpretation of the Schrodinger wavefunction, with the Euler formula yields the following equations, 9.3.1-6 for matter:

$$\psi_{established-matter} = \left| \iint e^{i \int |[L][S][T]|} \right|_U$$

Eq. 9.3.1: Wavefunction for Established-Matter

(9.3.1) suggests that so long as the basis of matter is untransformed, specified by the U following the vertical-brackets, the outcome is not going to be any different from established matter, specified by $\psi_{established-matter}$. U implies that there is not any action of the conscious will yet. This makes sense since conscious will supposedly arises with close-to human-systems. Note also that the presumption here is that matter as we perceive it – comprising of quantum particle, atoms, molecules – is perhaps only at the start of its journey. This observation is reinforced through the possibility of infinite information existing in meta-layers of Light that has yet to precipitate materially.

Further, as specified by Equation 9.3.2, the Probability-View of Established-Matter, the very basis of matter is going to be either the untransformed physical (P_U), the untransformed vital (V_U), the untransformed mental (M_U), or the untransformed integral (I_U):

$$|\psi_{established-matter}|^2 = P_U^2 + V_U^2 + M_U^2 + I_U^2 = 1$$

Eq. 9.3.2: Probability-View for Established-Matter

As specified by (9.3.2) the probability that any of these bases will be leveraged in the creation of matter adds up to one.

Another way of viewing established matter is by leveraging the non-wavefunction form as in (3.6.2) and adapting it to create Equation 9.3.3, Simplified Light-Space-Time Matrix Form for Established-Matter:

$$Established - Matter = \left| \left| [L][S][T]TC \rightarrow x_T \right|_{\langle x_U | x_T \rangle} \right|_U$$

Eq. 9.3.3: Simplified Light-Space-Time Matrix Form for Established-Matter

The notion of U as the bases in the formation of established-matter simply implies that this is a baseline and is subject to all the adaptability yet to come about through the play of dynamics such as urge and will which increases with the generation of subsequent partial-singularities.

Summary Equations for Matter-in-Transition

Equation 9.3.4, Wavefunction for Matter-in-Transition, suggests the mixed bases for matter, as specified by dynamics of both the untransformed (U) and the meta-levels (M_x), which therefore results in matter-in-transition.

$$\psi_{matter-in-transition} = \left| \int e^{i \int |[L][S][T]|} \right|_{U \& M_x}$$

Eq. 9.3.4: Wavefunction for Matter-in-Transition

The influence of meta-level, M_x, implies that action of visceral urge or desire or will. This is mixed with Nature's existing results, specified by U. As already discussed, such matter would be the result of the dynamics of life and its emergences.

As specified by Equation 9.3.5, Probability-View for Matter-in-Transition, the probability that any of the

425

untransformed (U) and transformed (T) bases will be leveraged in the creation of matter adds up to one.

$$|\psi_{matter-in-transition}|^2$$
$$= P_U^2 + P_T^2 + V_U^2 + V_T^2 + M_U^2 + M_T^2 + I_U^2$$
$$+ I_T^2 = 1$$

Eq. 9.3.5: Probability View for Matter-in-Transition

Note also that another way of viewing matter in transition is by leveraging the non-wavefunction form as in (3.6.3) and adapting it to create Equation 9.3.6, Simplified Light-Space-Time Matrix Form for Matter-in-Transition:

$$Matter - in - transition$$
$$= \left| |[L][S][T]TC \rightarrow x_T|_{\langle x_U|x_T\rangle} \right|_{U\&M_x}$$

Eq. 9.3.6: Simplified Light-Space-Time Matrix Form for Matter-in-Transition

Summary Equations for Super-Matter

Equation 9.3.7, Wavefunction for Super-Matter, suggests some transformed bases for matter, as specified by dynamics of the meta-levels (M_x), which therefore results in super-matter. It is essentially more consciousness, whether of visceral urge, desire, or will, that is driving change.

$$\psi_{super-matter} = \left| \int e^{i \int |[L][S][T]|} \right|_{M_x}$$

Eq. 9.3.7: Wavefunction for Super-Matter

As specified by Equation 9.3.8, Probability-View for Super-Matter, the probability that the transformed (T) bases will be leveraged in the creation of matter adds up to one.

426

$$\left|\psi_{super-matter}\right|^2 = P_T^2 + V_T^2 + M_T^2 + I_T^2 = 1$$

Eq. 9.3.8: Probability View for Super-Matter

Note also that another way of viewing super-matter is by leveraging the non-wavefunction form for as in (3.6.4) and adapting it to create Equation 9.3.9, Simplified Light-Space-Time Matrix Form for Super-Matter:

$$Super - Matter = \left|\|[L][S][T]TC \rightarrow x_T|_{\langle x_U|x_T\rangle}\right|_{M_x}$$

Eq. 9.3.9: Simplified Light-Space-Time Matrix Form for Super-Matter

Chapter 9.4: The Emerging Link Between Matter and Cosmos in Partial-Singularities

The previous chapters in this section further discussed the effect of functional richness in the creation of super-matter. The question then, is what is matter? Further, the previous chapters also suggested a tight relationship or correlation between the type of matter and the type of information-storage required for genetic-type information. This chapter will take a deeper look at the nature of matter by drawing on some observations from the field of astrophysics.

From discussion in the previous chapter it is clear that if the cumulative effect of overcoming habitual patterns increases the likelihood of super-matter being formed, this implies the increase too of four-fold quantization. Space, as the seeding ground of knowledge, may therefore also require expansion to continue to allow seeding to take place. But if space needs to expand in such a scenario, then the fact that scientists have observed that the universe is expanding may be related to this. This chapter hence also explores a possible link between the expansion of the universe and an increase in functional-richness that implies the expansion of space. Such a link may provide further insight into the nature of matter.

The entertainment of such an idea perhaps can be strengthened through consideration of the cosmological principle that suggests that the spatial distribution of matter in the universe is homogenous and isotropic when viewed on a threshold scale of 250 million light years. The nine-year cosmic microwave background image created with the Wilkinson Microwave Anisotropy Probe (WMAP) depicts 13.77 Billion year old temperature differences that correspond to the seeds that became the galaxies (NASA-WMAP, 2014). This homogeneity in temperature difference in the early universe may suggest too a process of homogenous four-fold quantization that set into motion the development of galaxies.

Any instance of quantization though can never be equal to another instance of quantization as suggested by the light-based-singularity dynamics in this book. On the surface and in their first emergence phenomena may appear to be homogenous, but that must only hide the vast diversity that is seeded in infinite uniqueness in Light. Materialized diversity with all its infinite uniqueness will become more apparent as matter tends toward super-matter, due to the increasing materialization of possibility driven by the unfolding of innate fourfold functional richness.

Complexity of Matter, Dark Matter, & Dark Energy

According to NASA only 5% of the matter in the universe is visible, while as much as 27% is 'dark matter'. It is the 5% of 'visible' matter though that forms the familiar stars, galaxies, or the atoms (NASA-darkmatter, 2016). But since, as explored in Chapter 3.7 on Qualified Determinism, the architectural sets at M_2 are infinite the question is why would all functions need to manifest in the same way?

As stated in an article 'Dark Energy: The Biggest Mystery in the Universe' (Panek, 2010), "Sight itself has blinded us to the Universe". It should be possible for other matter-based constructions to come into being based on the functionality they exist for. For instance there could be other fields that create other kinds of elementary particles, that create other kinds of atoms and that results in other kinds of structures not visible with our current instrumentation.

In reference to the four sets at M_2 (as discussed in Chapter 2.3 on The Seed-Singularity Architectural Forces K-Entanglement Pre-Genetic Library & Dynamics) it may also be that as the basis of matter complexifies in emergent partial-singularities as it journeys through the electromagnetic spectrum, quantum particles, the atom, cellular life, complex individual and further organizational development, the sets which have been positioned to each contain infinite elements, concretely manifest more of their function-elements.

This suggestion has been the basis of the equations derived in previous chapters. The complexity of matter may therefore be related to the number of manifested function-elements of the set of four sets. This idea is consistent with the notion of the "adjacent possible" suggested by Kaufmann (Kaufmann, 2003) in which innovation is positioned as a recombination of existing parts to create new value – or of existing sets to combine parts of themselves to create new elements based on new circumstance. If MS signifies manifested-set, so that the cardinality or number of elements in the combined set is the union of the four manifested-sets, this may be summarized by the following equation, Equation 9.4.1, Complexity of Matter in Terms of Manifested Set:

$$Matter_{Complexity} \propto$$
$$\left| MS_{System_{Pr}} \cup MS_{System_{P}} \cup MS_{System_{K}} \cup MS_{System_{N}} \right|$$

Eq 9.4.1: Complexity of Matter in Terms of Manifested Set

It is also possible to hypothesize an Equation, 9.4.2, relating complexity of matter in terms of visible and dark matter:

$$Matter_{Complexity} \propto [Visible\ Matter + Dark\ Matter]$$

Eq. 9.4.2: Complexity of Matter in Terms of Visible and Dark Matter

Further, as suggested by the previous discussions on the quantization of space, space is not empty but is a seeding ground for a variety of emergences. The notion of space having 'amazing' properties was, according to a report on 'dark energy' by the Harvard-Smithsonian Center for Astrophysics, first suggested by Einstein (Harvard-Smithsonian Center for Astrophysics, 2004). Quoting: "Einstein was the first person to realize that empty space is not nothingness. Space has amazing properties, many of which are just beginning to be understood." Einstein had suggested the existence of a 'dark energy' about 100 years ago (NASA-Supernova, 2001) as a property of space that caused the expansion of the universe. Dark Energy is estimated to comprise as much as 68% of the universe.

Reinterpretation of Cosmological Expansion-Contraction Dynamics

Keeping in mind the complexification of matter as it journeys from the field-level through the quantum-, atomic-, and cellular-levels, and beyond, it may be possible to re-interpret the supposed expansion-

431

contraction dynamics of cosmology in relation to the mathematical model presented in this book.

First, flipping the left and right sides of the equation (9.4.1) on $Matter_{Complexity}$ to yield Equation 9.4.3:

$$\left| MS_{System_{Pr}} \; \cup \; MS_{System_P} \; \cup \; MS_{System_K} \; \cup MS_{System_N} \right|$$
$$\propto \; Matter_{Complexity}$$

Eq 9.4.3: Flipped Matter-Complexity Equation

This implies that the manifested-set, MS, is growing at a certain threshold level. Assuming this threshold level, $MS\_Growth_{Threshold}$, is a property of space related to dark energy it may be possible to restate the condition of cosmological expansion and contraction. Hence, so long as the MS is increasing at a certain rate that exceeds

$MS\_Growth_{Threshold}$ the level of dark energy is such that the acceleration of galaxies, $Acceleration_{Galaxies}$ exceeds the contracting force of gravity at the universal level, $Gravity_{Universe}$.

Therefore, as in Equations 9.4.4 and 9.4.5:

$$MS\_Growth_{Threshold}: Acceleration_{Galaxies}$$
$$> Gravity_{Universe}$$

Eq. 9.4.4: Expansion of Universe Related to Manifest Set Growth Threshold

Conversely:

$$!MS\_Growth_{Threshold}: Acceleration_{Galaxies}$$
$$< Gravity_{Universe}$$

Eq. 9.4.5: Contraction of Universe Related to Manifest Set Growth Threshold

The relation between $MS\_Growth_{Threshold}$ and Dark Energy is suggested by Equation 9.4.6:

$$MS\_Growth_{Threshold} \propto Dark\ Energy$$

Eq. 9.4.6: Relation Between Manifested Set Growth Threshold and Dark Energy

In this interpretation so long as more of the infinite set at M_2 results in the manifested-set, MS_{System}, the universe keeps growing. In other words, so long as matter tends towards super-matter, the universe continues with expansion. It seems therefore that matter is a dynamic state linked to the very life of the universe.

Further if the relative distances between some of the emergences discussed in this book are considered a pattern emerges. The quantum level is dealing in distances of the order of 10^{-35}m. The cellular level is dealing with distances of the order of 10^{-6}m. The astronomical level is dealing with distances as large as 10^{26}m – the estimated size of the universe. It can be observed therefore that the cell is approximately in the center with roughly 30 orders of magnitude on either side.

Such symmetry that we are at now with the human being at the mid-point between two known material extremes, seems to suggest a cosmic balance of scales with human-will and other faculties such as awareness as a potential arbiter. But this we already know is the condition for the activation of meta-levels and also therefore for the creation of super-matter. Lack of will for continued functional-richness will in this view lead to a reversal in the Manifested Set due to a reversal in the quantization trend and consequently to a return of this universe to a state of collapse or unfulfilled absorption into the initiating Light.

Matter hence collapses into naught. In that case it will be another possible Big Bang initiated through the interplay of multiple layers of Light, and a birth following such death as depicted by the ancient Ouroboros symbol of a dragon or snake eating its own tail. The eating of its own tail can be mathematically depicted by a condition of x_T reverting to x_U hence resulting in collapse as captured by Equation 9.4.7, Light-Space-Time Collapse:

$Light - Space - Time\ Collapse =$

$$\left\| \begin{bmatrix} \begin{array}{c} c_\infty: [Pr, Po, K, H] \\ (\downarrow\ R_{C_K} =\ f(R_{C_\infty})) \\ c_K: [S_{Pr}, S_{Po}, S_K, S_H] \\ (\downarrow\ R_{C_N} =\ f(R_{C_K})) \\ c_N: f(S_{Pr}\ x\ S_{Po}\ x\ S_K\ x\ S_H) \\ (\downarrow\ R_{C_U} =\ f(R_{C_N})) \\ c_U: [P, V, M, C] \\ \Uparrow \\ c_{0:[D,W,I,C]} \end{array} \end{bmatrix}_{Light} \begin{bmatrix} M_3\ \rightarrow\ System_X \\ (\uparrow F\ \rightarrow I) \\ M_2\ \rightarrow\ S_{System_X} \\ (\uparrow Sig \rightarrow F) \\ M_1\ \rightarrow\ Sig_x \\ (\uparrow > P_x) \\ U\ \rightarrow\ x_U \end{bmatrix}_{Space} \right.$$

$$\left. \begin{bmatrix} U\ \rightarrow\ \begin{bmatrix} M_3 :\ -\infty \le t\ \le\ \infty \\ \downarrow \\ M_2 : 0 \ge t\ >\ \infty \\ \downarrow \\ M_1 :\ 0 > t\ >\ \infty \\ \downarrow \\ t \sim E_{Cell};\ TC: M_3\ \rightarrow U \\ t \sim E_{uman};\ TC: U\ \rightarrow M_3 \end{bmatrix}_{Time} \end{bmatrix} \quad TC\ \rightarrow x_U \quad \right\| \langle x_U | x_U \rangle$$

Eq. 9.4.7: Light-Space-Time Collapse

On the other hand the will for continued functional-richness may lead to continuous quantization, space expansion, and to an increasing reality of super-matter as was suggested in (4.2.8) reproduced here for convenience:

$Light - Space - Time\ Emergence_{Suoer-Matter} =$

$$\left[\begin{array}{c}\left[\begin{array}{c}c_\infty:[Pr,Po,K,H]\\\left(\downarrow R_{C_K}=f(R_{C_\infty})\right)\\c_K:[S_{Pr},S_{Po},S_K,S_H]\\\left(\downarrow R_{C_N}=f(R_{C_K})\right)\\c_N:f(S_{Pr}\ x\ S_{Po}\ x\ S_K\ x\ S_H)\\\left(\downarrow R_{C_U}=f(R_{C_N})\right)\\c_U:[P,V,M,C]\\\Uparrow\\c_{0:[D,W,I,C]}\end{array}\right]_{Light}\quad\left[\begin{array}{c}M_3\ \to\ System_X\\(\uparrow F\ \to\ I)\\M_2\ \to\ S_{System_X}\\(\uparrow Sig\ \to\ F)\\M_1\ \to\ Sig_x\\(\uparrow >P_x)\\U\ \to\ x_{Molecules}\end{array}\right]_{Space}\right.$$

$$\left.\left[\begin{array}{c}M_3:\ -\infty\ \le t\ \le\ \infty\\\downarrow\\M_2:0\ \ge\ t\ >\ \infty\\\downarrow\\M_1:\ 0>\ t\ >\ \infty\\\downarrow\\U\ \to\ \begin{array}{c}t\ \le\ E_{Cell};TC:M_3\ \to U\\t\ \sim\ E_{Human};TC:U\ \to\ M_3\end{array}\end{array}\right]_{Time}\qquad TC\ \to\ x_{Super-Matter}\right|\langle x_U|x_T\rangle$$

$\Rightarrow$

$\langle Space-Time-Energy-Gravity\ Material-Fabric\ Pre$
$\qquad -Genetic\ Code\rangle\ +$

$\langle Electromagnetic\ Spectrum\ \ Material-Fabric\ Pre-Genetic\ Code\rangle$
$\qquad +$

$\langle Quantum-Particle\ Material-Fabric\ Pre-Genetic\ Code\rangle\ +$

$\langle Atoms\ Material-Fabric\ Pre-Genetic\ Code\rangle\ +$

$\langle Molecules\ Material-Fabric\ Pre-Genetic\ Code\rangle\ +LSTE\ \langle\ldots\rangle$
$\qquad +\ FBLEEE\ \langle\ldots\rangle\ +$

$LSTE\ \langle\ldots\rangle\ +\ PGCI\ \langle\ldots\rangle\ +\ Super-Matter\ Post-Genetic\ Code$

436

Further iterations of (4.2.8) will lead to a possibility of all patterns in the Space-Matrix being broken, thus leading to the Second Singularity. This possibility will be taken up in Section 10, The Second Singularity.

This Section summarizes some key aspects of the Second Singularity by highlighting the difference with an AI-based singularity, suggesting a set of representative operators that will expedite its emergence, and tracing its emergence.

Hence:

Chapter 10.1, Summarizing Fourfold Adherence and Implicit Wholeness in Light-Based Singularities, summarizes key mechanisms in light-based singularities and highlights the essential connection between one light-based singularity and another. There is "singleness" in the emergence of any light-based singularity that binds it with all previous light-based singularity so that a subsequent partial-singularity can be thought of as having a greater sphere of influence than preceding ones by virtue of all previous emergent laws being active also in it.

Chapter 10.2, AI-Like Fragmentation of Light-Based Singularities, by contrast elaborates hypothetical fragmentation of a light-based singularity to provide further insight into the limit of AI-based singularities. This is done by considering the space-time-energy-gravity fourfold schema generated by the Big Bang partial-singularity and fragmenting this to explore what reality might have looked like had an AI-like singularity been adopted in a fourfold schema design. In this way the difference between AI-based singularities and light-based singularities will become even clearer.

Chapter 10.3, Leveraging Mathematical Operators to Expedite the Arrival of the Second Singularity, explores the use of a set of mathematical operators in expediting the arrival of the Second Singularity. These operators are non-exhaustive, but rather are indicative of the nature of dynamics that will increasingly arise in the extrapolation of partial-singularities to a super-matter and beyond basis, due to the increasing presence of meta-level seed-singularity entanglement and precipitated functional-richness. The use of such operators assist in perceiving reality as derivate from Light, and hence their active adoption will accelerate the overcoming of limiting patterns and subsequently the integration of the multiple layers of light in a light-based singularity.

Chapter 10.4, Altering the Microcosmic-Macrocosmic Balance, examines on-going conditions for the generation of the Second Singularity. A key condition is the audacity of the human-species made possible because the human is the result of 30 orders of microcosmic magnitude implicit to it. Fields, quanta, atoms, cells, the incredible four-fold emergences of properties codified in Light have all contributed to the emergence of the human. Now in a possible reversal of causality the human appears to be positioned to similarly influence 30 orders of magnitude explicit to it and even structure the further development of the macrocosm by mastery of the microcosm.

Chapter 10.5, The Second Singularity Biography, reviews key insights that have surfaced through tracing the emergence of the series of light-based singularities.

Essentially the materialization of potentiality in Light may lead to The Second Singularity, the singularity whose power will far surpass any AI-based singularity precisely because of the level and sophistication of integration, and the continual real-time re-computation of reality and generation of genetic-type information driven by enlightened heart, will, and mind dynamics.

Chapter 10.1: Summarizing Fourfold Adherence and Implicit Wholeness in Light-Based Singularities

The preceding sections have elaborated the architecture of light-based singularities that are all built from the four essential properties of Light. In this elaboration the connection between one partial-singularity emergence and the next is clear in that there is an essential complexification of the underlying four-foldness. The same four organizational properties resident in Light create more complex forms of themselves with the emergence of subsequent partial-singularities. Further, the four-foldness that had emerged in a previous partial-singularity is implicit in a succeeding partial-singularity. There is hence, "singleness" in the emergence of any singularity that binds it with all previous light-based singularity so that a subsequent partial-singularity can be thought of as having a greater sphere of influence than preceding ones by virtue of all previous emergent laws being active also in it.

Figure 10.1.1, Fourfold Adherence and Levels of Implicit Wholeness in Representative Partial-Singularities, summarizes the two architectural devices of light-based singularities that distinguish it from any fragmented singularity such as an AI-based singularity. Subsequent sub-sections in this chapter will review the notion of fourfold adherence for each row in the following figure:

441

Level	Fourfold Adherence	Minimum Levels of Implicit Wholeness	Minimum Number of Levels
Space-Time-Energy-Gravity (S-T-E-G)	Seeds	• S-T-E-G	1
EM Spectrum	C, v or λ	• S-T-E-G • EM	2
Particle Level	Quarks, Leptons, Bosons, Higgs-Boson	• S-T-E-G • EM • Particle	3
Atomic Level	s-Group, p-Group, d-Group, f-Group	• S-T-E-G • EM • Particle • Atom	4
Cellular Level	Proteins, Nucleic Acids, Lipids, Polysaccharides	• S-T-E-G • EM • Particle • Atom • Cell	5
Level of Capacities of Self	Sensations, Wills, Emotions, Thoughts	• S-T-E-G • EM • Particle • Atom • Cell • Capacities of Self	6
Truer Individuality Level	Service-type-individual, Knowledge-type-individual, Harmony-type-individual, Power-type-individual	• S-T-E-G • EM • Particle • Atom • Cell • Capacities of Self • Truer Individuality	7

Sustainable Global Civilization Level	Service-type-region, Knowledge-type-region, Harmony-type-region, Power-type-region	• S-T-E-G • EM • Particle • Atom • Cell • Capacities of Self • Truer Individuality • Sustainable Global Civilization	8

Figure 10.1.1 Fourfold Adherence and Levels of Implicit Wholeness in Representative Partial-Singularities

Space-Time-Energy-Gravity Four-Foldness

Hence in thinking through the first apparent space-time-energy-gravity emergence at the time of the Big Bang partial-singularity, these may be summarized by the equations, Equations 10.1.1 through 10.1.4. In each case there is a dependence on the single fundamental attribute of 'seeds'.

Equation 10.1.1, System-Knowledge (Space) at Space-Time-Energy-Gravity Level, summarizes how knowledge is related to 'infinity of seeds' with each seed representing, as it were, a different arrangement of knowledge:

$$System_{Knowledge} \propto [f(infinity\ of\ seeds)]$$

Eq 10.1.1: System-Knowledge (Space) at Space-Time-Energy-Gravity Level

Note that the infinity of seeds requires 'space' to be expressed.

443

Equation 10.1.2, System-Power (Time) at Space-Time-Energy-Gravity Level, summarizes how power is related to 'maturation of seeds', since anything that gets in the way of the maturation will generally be overcome:

$$System_{Power} \propto [f(maturation\ of\ seeds)]$$

Eq 10.1.2: System-Power (Time) at Space-Time-Energy-Gravity Level

Note that maturation of seeds is felt in 'time'.

Equation 10.1.3, System-Presence (Energy/Matter) at Space-Time-Energy-Gravity Level, summarizes how presence is related to 'materialization of seeds', since the accumulation of energy will express itself in material form:

$$System_{Presence} \propto [f(materialization\ of\ seeds)]$$

444

Eq 10.1.3: System-Presence (Energy/Matter) at Space-Time-Energy-Gravity Level

Equation 10.1.4, System-Harmony (Gravity) at Space-Time-Energy-Gravity Level, summarizes how harmony is related to 'relationship of seeds', since seeds will generally have a relationship to one another and likely exist in collectivities:

$$System_{Harmony} \propto [f(relationship\ of\ seeds)]$$

Eq 10.1.4: System-Harmony (Gravity) at Space-Time-Energy-Gravity Level

Note that such relationship may be expressed as 'gravity'.

The creation of space-time-energy-gravity is a fundamental device that facilitates the process of quantization thereby allowing the accumulated material-fabric to change.

Electromagnetic (EM) Spectrum Four-Foldness

In the case of the EM spectrum the number of basic attributes to express four-foldness is two as will be observed.

Equation 10.1.5, System-Harmony at Electromagnetic Level, is related to the speed of light, c at U:

$$System_{Harmony} \propto c_U$$

Eq 10.1.5: System-Harmony at Electromagnetic Level

Equation 10.1.6, System-Power at Electromagnetic Level, is related to the frequency, v, of the EM spectrum:

$$System_{Power} \propto hv$$

Eq 10.1.6: System-Power at Electromagnetic Level

Equation 10.1.7, System-Knowledge at Electromagnetic Level, is related to the wavelength, λ, of the EM spectrum:

$$System_{Knowledge} \propto [f(\lambda)]$$

Eq 10.1.7: System-Knowledge at Electromagnetic Level

And since, $E = mc^2$ or $m = \frac{E}{c^2}$, and substituting hv for E, yields Equation 10.1.8, System-Presence at Electromagnetic Level:

$$System_{Presence} \propto {hv}/{c^2}$$

Eq 10.1.8: System-Presence at Electromagnetic Level

But, $C \propto {1}/{h}$

Hence, the general relationships of the four-fold wholeness may be understood by knowing just two variables, C, and either, λ or v.

The emergence of the EM Spectrum adds another fold as it were to the partial-singularity biographies that can progressively be seen as a complexification of four-foldness.

Quantum Particle Four-Foldness

As per the model of early cosmic development presented by the Particle Data Group (Particle Data Group, 2015) at time, $t = 10^{-10}$ s, the four-fold fullness begins to express itself in a series of particles. Quarks, leptons, bosons, which also may imply the presence of the Higgs-boson that gives quarks their mass are observed. But as discussed quarks are a precipitation of Light's property of Knowledge, leptons of Power, bosons of Harmony, and the Higgs-boson of Presence. Hence it is again observed that the four-fold fullness has expressed itself at a different order of complexity.

The Standard Model suggests that quarks, leptons, bosons, and the Higgs-boson are different fundamental particles, which implies that the fourfold wholeness has now a more complex basis of its operation, since as compared with the fourfold wholeness of the EM spectrum the number of independent bases appears to have increased.

447

Summarizing, as in Equations 10.1.9 through 10.1.12:

$$System_{Harmony} \propto f(bosons)$$

Eq 10.1.9: System-Harmony at Particle Level

$$System_{Power} \propto f(leptons)$$

Eq 10.1.10: System-Power at Particle Level

$$System_{Knowledge} \propto f(quarks)$$

Eq 10.1.11: System-Knowledge at Particle Level

$$System_{Presence} \propto f(Higgs\_boson)$$

Eq 10.1.12: System-Presence at Particle Level

Hence the quantum-particle ecosystem logic is also now added to the quantum particle partial-singularity, which progressively continues to complexify along the lines set up by the underlying fourfold organizational principles as dictated by Light's properties.

Periodic Table/Atoms Four-Foldness

As discussed in Chapter 6.4 the Periodic Table itself is also an expression of the four-fold wholeness. Hence, d-Group elements are an instance of Light's property of Presence, p-Group elements are an instance of Light's property of Knowledge, s-Group elements are an instance of Light's property of Power, and f-Group elements are an instance of Light's property of Harmony or Nurturing. As per the Particle Data Group model (Particle Data Group, 2015) this emergence likely began at $t = 3 \times 10^5$ years when the atom is suggested to have emerged, and continued at least to $t = 10^9$ years with the emergence of stars which it is already known are the furnaces in which heavier atoms were created.

Leveraging the fundamentally different groups yields Equations 10.1.13 through 10.1.16:

$$System_{Harmony} \propto f(f\_group)$$

Eq 10.1.13: System-Harmony at Atomic Level

$$System_{Power} \propto f(s\_group)$$

Eq 10.1.14: System-Power at Atomic Level

$$System_{Knowledge} \propto f(p\_group)$$

Eq 10.1.15: System-Knowledge at Atomic Level

$$System_{Presence} \propto f(d\_group)$$

Eq 10.1.16: System Presence at Atomic Level

Atoms therefore have at least four four-fold wholenesses implicit in them.

Living Cell Four-Foldness

At time, $t = 13.8 \times 10^9$ years, there is a fifth clear

expression of the same four-fold order as the bases of an even more complex organization, that of cellular life and all that is founded on it. This can be summarized as in Equations 10.1.17 through 10.1.20:

449

$$System_{Harmony} \propto f(lipids)$$

Eq 10.1.17: System-Harmony at Cellular Level

$$System_{Power} \propto f(polysaccharides)$$

Eq 10.1.18: System-Power at Cellular Level

$$System_{Knowledge} \propto f(nucleic\ acids)$$

Eq 10.1.19: System-Knowledge at Cellular Level

$$System_{Presence} \propto f(proteins)$$

Eq 10.1.20: System-Presence at Cellular Level

Cellular life therefore seems to have at least five four-fold-wholenesses implicit in it to therefore further complexify the architecture of matter and matter-based partial-singularities.

Fundamental Capacities of Self Four-Foldness

Equations 10.1.21 through 10.1.24 summarizes a sixth clear expression of four-foldness that will change the structure life-based partial-singularities.

$$System_{Harmony} \propto f(Emotions, Feelings)$$

Eq 10.1.21: System-Harmony at Level of Capacities of Self

$$System_{Power} \propto f(Urges, Wills, Desires)$$

Eq 10.1.22: System-Power at Level of Capacities of Self

$$System_{Knowledge} \propto f(Thoughts)$$

Eq 10.1.23: System-Knowledge at Level of Capacities of Self

$$System_{Presence} \propto f(Sensations)$$

Eq 10.1.24: System-Presence at Level of Capacities of Self

Truer Individuality Four-Foldness

Similarly as truer individuality expresses itself S-T-E-G quantization initiated by quantum-certainty will further complexify the bases of matter.

Equations 10.1.25 through 10.1.28 therefore summarize a seventh clear expression of four-foldness that will further complexify the structure of the material-fabric and change the structure of matter.

$$System_{Harmony} \propto f(Harmony - type - individuals)$$

Eq 10.1.25: System-Harmony at Level of Truer Individuality

$$System_{Power} \propto f(Power - type - individuals)$$

Eq 10.1.26: System-Power at Level of Truer Individuality

$$System_{Knowledge} \propto f(Knowledge - type - individuals)$$

Eq 10.1.27: System-Knowledge at Level of Truer Individuality

$$System_{Presence} \propto f(Service - type - individuals)$$

Eq 10.1.28: System-Presence at Level of Truer Individuality

Sustainable Global Civilization Four-Foldness

Similarly as more complex organizational levels, such as a sustainable global civilization expresses itself, S-T-E-G quantization initiated by quantum-certainty will further complexify such complex organization partial-singularities.

Equations 10.1.29 through 10.1.32 therefore summarize an eighth clear expression of four-foldness. Note that these expressions are indicative only and there are other missed levels that would need to be expressed for completeness.

451

$$System_{Harmony} \propto f(Harmony - type - region)$$

Eq 10.1.29: System-Harmony at Level of Nations or Blocs

$$System_{Power} \propto f(Power - type - region)$$

Eq 10.1.30: System-Power at Level of Nations or Blocs

$$System_{Knowledge} \propto f(Knowledge - type - region)$$

Eq 10.1.31: System-Knowledge at Level of Nations or Blocs

$$System_{Presence} \propto f(Service - type - region)$$

Eq 10.1.32: System-Presence at Level of Nations or Blocs

Chapter 10.2: AI-Like Fragmentation of Light-Based Singularities

The previous chapter summarized fourfold adherence and implicit levels of wholeness in partial-singularities.

By contrast AI-based singularities are fragmented in that there is no connection with any technology that is not mind-based. The essential elements of AI-based technologies are all derivatives from capacities created with the emergence of a "mind" partial-singularity and would include sensing, memory, and the use of logic and empiricism.

This chapter will consider the space-time-energy-gravity fourfold schemas generated by the Big Bang partial-singularity and fragment this to explore what reality might have looked like had an AI-like singularity been adopted in their designs. In this way the difference between AI-based singularities and partial-singularities and the Second Singularity will become clearer.

Fragmentation of Big Bang Partial-Singularity Space-Time-Gravity-Energy Four-Foldness

In the previous chapter equations (10.1.1 – 4) summarized an implicit reality of 'seeds' that gave space, time, energy, and gravity meaning. The essence of these equations are summarized here for convenience as Equations 10.2.1 – 4:

Hence, Equation 10.2.1, Space:

$$Space: System_{Knowledge} \propto [f(infinity\ of\ seeds)]$$

Equation 10.2.2, Time:

$$Time: System_{Power} \propto [f(maturation\ of\ seeds)]$$

Equation 10.2.3, Energy / Matter:

$$Energy/Matter: System_{Presence} \propto [f(materialization\ of\ seeds)]$$

Equation 10.2.4, Gravity:

$$Gravity: System_{Harmony} \propto [f(relationship\ of\ seeds)]$$

In a reality of fragmentation however, there would be a de-linking with 'seeds', which recall from (3.1.3) are a result of unique combinations of implicit function resident in several antecedent layers of light. This would mean that space, time, energy, and gravity are no longer emergent from light, become devoid of deeper significance, and would exist as purely independent physical phenomenon.

In this deeper significance space, time, energy, and gravity are processes by which the infinite possibility in Light begins to materialize. In a fragmented mode these phenomenon by contrast now become independent sets of elements that at best may combine together to entirely determine all possibility within a universe. A universe that arose from a space-set, time-set, energy-set, and gravity-set would be a function of only these sets and would become devoid of all the possibility that exists in a universe generated from a light-based singularity.

Exploring such a fragmented universe mathematically, the following sets would be starting points.

454

Hence, space could have elements such as 'point', 'line', 'plane', 'cube', 'sphere', amongst possibly infinite other space-elements. The Set of Space, Set_{Space}, is summarized by Equation 10.2.5, Set_{Space}:

$$Set_{Space} \ni [Point, Line, Plane, Cube, Sphere, ...]$$

Equation 10.2.5: Set_{Space}

Time could have elements such as 'slow', 'fast', 'stagnant', 'accelerating', amongst possibly infinite other time-elements. The Set of Time, Set_{Time}, is summarized by Equation 10.2.6, Set_{Time}:

$$Set_{Time} \ni [Slow, Fast, Stagnant, Accelerating, ...]$$

Equation 10.2.6: Set_{Time}

Energy could have elements such as 'concentrated', 'diffuse', 'intense', 'subtle', amongst possibly infinite other energy-elements. The Set of Energy, Set_{Energy}, is summarized by Equation 10.2.7, Set_{Energy}:

$$Set_{Energy} \ni [Concentrated, Diffuse, Intense, Subtle, ...]$$

Equation 10.2.7: Set_{Energy}

455

Gravity could have elements such as 'close', 'far', 'focused', 'encompassing', amongst possibly infinite other energy-elements. The Set of Gravity, $Set_{Gravity}$, is summarized by Equation 10.2.8, $Set_{Gravity}$:

$$Set_{Gravity} \ni [Close, Far, Focused, Encompassing, ...]$$

Equation 10.2.8: $Set_{Gravity}$

A Fragmented Space-Time-Gravity-Energy Universe

A universe created from any combination of the elements from these four sets can be depicted by Equation 10.2.9, Space-Time-Energy-Gravity Universe, where the elements of such a universe is at least the Cartesian product of the four infinite sets:

$$Space - Time - Energy - Gravity\ Universe$$
$$\geq Set_{Space} \times Set_{Time} \times Set_{Energy} \times Set_{Gravity}$$

Equation 10.2.9: Space-Time-Energy-Gravity Universe

Such a universe would have infinite "elements" but there would not be any order to it. Order is a function of implicit design. But such design cannot exist in a fragmented universe where there are no attendant meta-layers. Anything that is created or that arises has to arise from the combination of elements in the four foundational sets. So such a universe may give rise to waves and particles through various combinations of elements from the space, time, energy, and gravity sets. But these waves and particles would be random and chaotic with no reason to begin to generate one kind of element over another or to maintain one kind of pattern over another. The notion of particles settling into agreeable combinations to create atoms, and then molecules will therefore not be possible in such a universe. There would be a lot of activity, but there would be neither rhyme nor reason to it. Rhyme and

456

reason imply the existence of function around which elements anchor or settle. But this can only be if such functions have a meta-significance, implying meta-layers, which by definition cannot exist in a fragmented universe.

Further, even if the elements in the sets in a space-time-energy-gravity universe were to create waves or something akin to the electromagnetic spectrum, there would be no continuity of fourfold-ness from the space-time-energy-gravity layer to the electromagnetic spectrum layer. This means that even though multiple "layers" may exist there would still be no implicit design or meaning to anything that emerges. It also means that nothing that is not of the nature of the initial set of elements can conceivably emerge.

This is in stark contrast to a light-based singularity that to begin with contains everything in potentiality at Light's native state traveling at infinite speed. Therefore starting from the four properties implicit in Light of Knowledge, Presence, Power, and Harmony, radically different capacity and possibility of diverse instrumentation are already seeded as information and emerge in cohesive ways as this treatise has already explored. There is further a continuity in emergence so that all previous "laws", beginning with the emergence of the space-time-energy-gravity quadrumvirate, continuing through to the functioning of the four molecular plans in every cell, to the four macro-organizational principles that determine nation uniqueness and set the foundation for a sustainable global civilization, are available in ever-present information-form to any subsequent partial-singularity emergence.

In the absence of such light-based singularity design and implicit order, a space-time-energy-gravity universe would become increasingly chaotic, as more and more permutations and combinations involving the elements in the space-set, time-set, energy-set, and gravity-set were to emerge.

Contrasting an AI-Based Singularity with the Second Singularity

An AI-based singularity would permute and combine a vast number of mind-generated elements, such as sensory inputs, conceptual constructs, logical constructs, learning algorithms, possibly from a vast global field, and all in relative real-time. As a result of sifting through myriad possibilities in microseconds, seeing patterns, and possibly selecting the unexpected based on this, it would possibly overleap current human cognitive ability. But still, all it created would at best be based on permuting and combining what already exists. There would not be any fundamentally new type of creation.

In the Second Singularity emergences are characterized by integrated light-based functioning and would have deeper access to the founts of innovation resident in deeper libraries at the antecedent layers of light. Emergences would be the tip of a sophisticated light-based edifice where no characteristics of previous partial-singularities would be absent. The ability to effectively concretize knowledge in real-time, based on a foundation of harmony and detailed consideration on the effect of things would be the norm of operation. Life and possibilities would quickly become radically different as a result.

The next chapter explores some functions that would further accelerate the formation of the Second Singularity.

Chapter 10.3: Leveraging Mathematical Operators to Expedite the Arrival of the Second Singularity

Based on the mathematics developed in Section 2, on modeling the mathematical structure of light-based-singularities, twelve general 'mathematical operators' are arrived at deductively. These operators are the result of considering some of the dynamics of each of the levels in the seed-singularity, and integration of the multiple levels of the overall light-based-singularity mathematical model. These operators are non-exhaustive, but rather are indicative of the nature of dynamics that will increasingly arise in the extrapolation of partial-singularities to a super-matter and beyond basis, due to the increasing presence of meta-level seed-singularity entanglement and precipitated functional-richness.

The use of such operators assist in perceiving reality as derivate from Light, and hence their active adoption will accelerate the overcoming of limiting patterns and subsequently the integration of the multiple layers of light in a light-based singularity. As a result the arrival of the Second Singularity will be expedited.

Note that in a previous book on quantum computing in a cosmology of light (Malik, 2018c) these operators were explored as a possible basis for alternative quantum-related computational models. Such quantum-level computation is a dynamic embedded in any light-based-singularity and such mathematical operators have to arise as the practical union of matter and light continues.

Presence-Based Mathematical Operators

Considering Light's dimension of Presence, Equation 10.3.1, Presence-Based Mathematical Operators, summarizes a set of representative presence-based mathematical operators:

$$Presence\_based_{Mathematical\_operators} \ni$$
$$[Fullness, Equality, Uniqueness, ...]$$

Eq. 10.3.1: Presence-Based Mathematical Operators

Fullness refers to the possibility that every single point in any time-space continuum is informed by Light's four-fold fullness. That is, the reality of ∞-entanglement is fully active. Note that ∞-entanglement is going to become more effective as the very fabric of existence is changed by partial-singularities with increasing spheres of influence. Hence, in equation form it may be suggested that Fullness is the union (U) of $System_{Pr}$, $System_{P}$, $System_{K}$, and $System_{N}$ as in Equation 10.3.2, Fullness:

$$Fullness = \bigcup \begin{bmatrix} System_{Pr} \\ System_{P} \\ System_{K} \\ System_{N} \end{bmatrix}$$

Eq. 10.3.2: Fullness

To see such 'fullness' in everything is to begin to see as perhaps Light sees, and this will assist in closing the gap between the material layer and layers of Light antecedent to it.

461

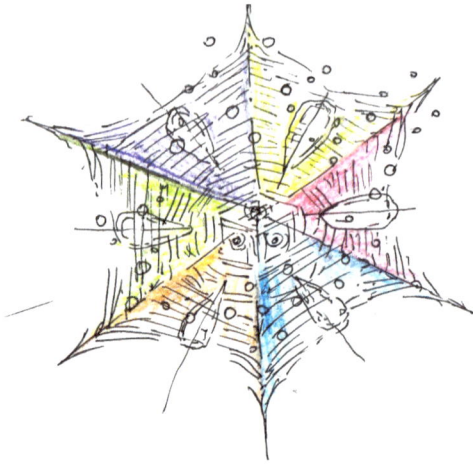

In general for any two points 'A' and 'B' it can be further suggested that the Fullness behind A is the same as the Fullness behind B, as in Equation 10.3.3, Equivalence of Fullness. This suggestion may also be arrived at by considering Einstein's General Theory of Relativity (Einstein, 1995) that states: "All bodies of reference K, K1 etc. are equivalent for the description of natural phenomena (formulation of the general laws of nature), whatever may be their state of motion." If we shrink these bodies of reference or coordinate systems to infinitesimal dimensions, thus approaching a 'point', this suggests that there is equivalence in that the general laws of nature are equally valid at any two points. Hence, Equation 10.3.3:

$$Fullness_A \equiv Fullness_B$$

Eq. 10.3.3: Equivalence of Fullness

Equality refers to the possibility that since every point in a time-space continuum is always some expression of Light's four-fold fullness, hence every point is equal to any other point. As in the likely practically felt effectiveness of the previous operator, such equality will also be felt more tangibly as the very fiber of existence tends towards super-matter based partial-singularities and beyond. This implies that all expressions and developments share a fundamental equality with all other points.

Depicting a point 'a' by $Point_A$ and a point 'b' by $Point_B$, then an equation, Equation 10.3.4, Equality, would be:

Equality: $Point_A \equiv Point_B$

Eq. 10.3.4: Equality

Uniqueness refers to the possibility that any point is fundamentally unique. This is of course the very meaning of infinite information in Light, as elaborated in some detail in Section 2, and specifically in the further elaboration of seed-singularity dynamics involving ∞-entanglement, K-entanglement, N-entanglement, and subsequently FBLEE-entanglement, that necessitate the reality of uniqueness. As matter tends toward super-matter and beyond, the very fiber of existence reflects antecedent functional-richness more fully, and the underlying reality of uniqueness becomes materially more visible.

Practically though, perhaps an easy way to envision this is to see that any point in a time-space continuum is a result of a unique time-space intersection. This becomes more meaningful knowing that time and space are in reality the summary result of a persistent computation resulting in light-based quantization. Hence, two points within a time-space continuum, A and B, can always be envisioned as having unique time and space coordinates and hence characteristics, as in Figure 10.3.1:

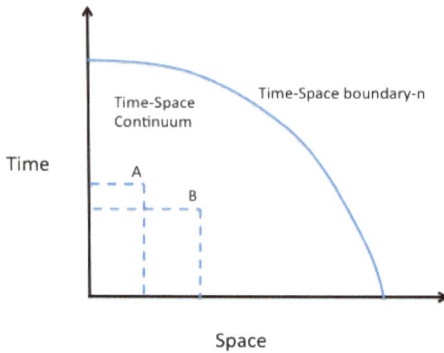

Figure 10.3.1: Uniqueness in Time-Space Continuum

But even for any organization, which can itself be considered a further development of a point or points through the action of time and space, the already proposed equation, (2.4.5) that governs organizational signatures may be restated as an equation for uniqueness, as in Equation 10.3.5:

$$Uniqueness: Sig = Xa +$$

$$\overline{Yb_{0-n}} \quad where \quad \begin{bmatrix} X \in [S_{System_{Pr}}, S_{System_P}, S_{System_K}, S_{System_N}] \\ Y \in [S_{System_{Pr}}, S_{System_P}, S_{System_K}, S_{System_N}] \\ a, b \ are \ integers; a > b \end{bmatrix}$$

Eq. 10.3.5: Uniqueness

Power-Based Mathematical Operators

464

Power-based mathematical operators pertain to the fount of dynamism that will tend to determine an organization's practical action in super-matter type partial-singularities. Mathematical operators related to 'power' give insight into how an organization will tend to meet circumstance and other organizations.

Representative operators include Direction, Fractal, and Intersection and in contrast to Knowledge-based operators discussed in the next sub-section, tend to create the more visceral and immediate reactions to circumstance. Knowledge-based operators on the other hand tend to create action more in line with long-term or strategic plans. This power-based set can be summarized by Equation 10.3.6:

$$Power\_based_{Mathematical\_operators} \ni$$
$$[Direction, Fractal, Intersection \dots]$$

Eq. 10.3.6: Power-Based Mathematical Operators

Direction refers to the possibility that direction at any possible system bifurcation point is not random and not determined either. Rather it can be thought of as a qualified determinism, as introduced in Chapter 3.7, and is a function of applying DI_V and DI_H to the set of possibilities existing at a bifurcation point. As per the functioning of DI_V and DI_H, it is the strongest or most 'powerful' possibility that will tend to determine what will emerge as an organization meets circumstance and other organizations. Hence leveraging (3.7.4) creates Equation 10.3.7:

$$\text{Direction: } Org\_Dir = DI \left(\begin{bmatrix} M_3 \rightarrow System_x \\ (\uparrow F \rightarrow I) \\ M_2 \rightarrow S_{System_x} \\ (\uparrow Sig \rightarrow F) \\ M_1 \rightarrow Sig_x \\ (\uparrow > P_P) \\ U \rightarrow x_U \end{bmatrix}_{x=p,v,m,i} \right) \rightarrow$$

$$x\_matrix_{strongest} @ level_{strongest}$$

Eq. 10.3.7: Direction

Fractal refers to the possibility that as organizational complexity increases, the base orientation, orientation-x, where x could be physical, vital, mental, or integral, of an average organization at some level of complexity 'n', will tend to determine the orientation of an organization at a level of complexity 'n+1'. Likewise the orientation of an organization at level of complexity 'n+1' will tend to determine the orientation of an organization at level of complexity 'n'. This is summarized by Equation 10.3.8:

$$Fractal: \; Orientation_x @ Complexity_n$$
$$\leftrightarrow Orientation_x @ Complexity_{n+1}$$

Eq. 10.3.8: Fractal

Organizational complexity refers to an order of magnitude change as in from a person to a team, or from a team to a business unit, and so on, for example. In Nature such fractal arrangements abound in the way the human body is constructed to the very structure of galaxies (Briggs, 1992). This notion has been suggested to exist in complex Fig 156:

behavioral systems as well as described in some detail in books such as The Fractal Organization (Hoverstadt, 2008), The Fractal Organization (Malik, 2015), and The Misbehavior of Markets (Mandelbrot, 2006). This notion relates to 'power' in that it is the patterns at one level that will tend to determine the patterns at another level, often preempting what may be a more logical choice based on reason. This kind of behavior has been suggested as causing cyclic fluctuations in stock and other markets where greed and fear often trump more intelligent and rational choices (Frost, 2005). Greed rises until fear sets in. Fear rises until greed sets in.

Intersection occurs when two organizations shift orientations to the next successive level due to the shock of interaction. Hence if an organization is at a physical orientation, it may be shifted to a vital orientation when intersection occurs, as in Equation 10.3.9, Intersection, where the function 'Next Element' extracts the next element from the Set S comprising the elements (physical,

467

vital, mental, integral). Examples of such phenomena abound where failure to make a shift results in shock of conflict repeating itself endlessly. Such a process with applicability at multiple levels of complexity has been captured by the series of books on crucial conversations (Patterson, 2011).

Intersection: *Organization Orientation → Next Element (S)*

Eq. 10.3.9: Intersection

Knowledge-Based Operators

Knowledge-based mathematical operators have to do with how organizations tend to develop by existing or toward increasing knowledge over time. Representative knowledge properties discussed in this section include Alternative, Flowering, and Higher, which as suggested in the previous section are of a different nature than 'power' or dynamism-based properties that tend to determine an organization's visceral or immediate reaction to its environment. This knowledge-based set can be summarized by Equation 10.3.10:

$$Knowledge\_based_{Mathematical\_operators} \ni [Alternative, Flowering, Higher ...]$$

Eq. 10.3.10: Knowledge-Based Mathematical Operators

Alternative refers to an alternative narrative that an organization will tend to embed itself in. These alternative narratives relate to the physical, the vital, the mental and the integral orientations. These narratives can easily become fixed and can strongly influence the entire internal and external orientation of an organization. The book 'The Fractal Organization: The Future of Enterprise' (Malik, 2015) suggests a theory of such narratives with their consequent effect on practical action. The alternative narratives are best described using the

468

generalized equation (2.6.6) modified as Equation 10.3.11. Hence:

$$Alternative:\ Innovation_{orientation-x}$$

$$= \begin{bmatrix} M_3 \rightarrow System_X \\ (\uparrow F \rightarrow I) \\ M_2 \rightarrow S_{System_X} \\ (\uparrow Sig \rightarrow F) \\ M_1 \rightarrow Sig_x \\ (\uparrow > P_{x)} \\ U \rightarrow x_U \end{bmatrix} TC \rightarrow x_T\ ,where\ \begin{bmatrix} x_U \ni [\ldots] \\ x_T \ni [\ldots] \end{bmatrix}$$

Eq. 10.3.11: Alternative

Where, 'x' can be thought of as an element from the Set of Orientations:

$$x \in (physical, vital, mental, integral)$$

'X' can be thought of as an element from the Set of System-level architectural forces. Hence:

$$X \in (Presence, Power, Knowledge, Nurturing)$$

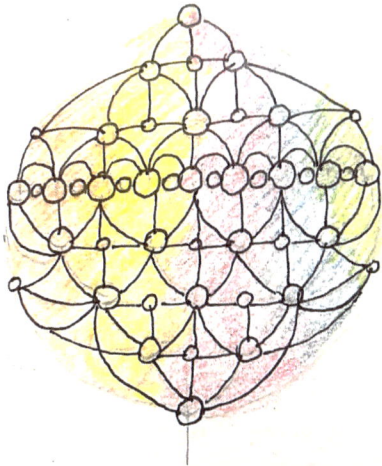

Flowering refers to the possibility that any time-space boundary-n, depicted as TS_n (as in Fig. 10.3.1) will have more potential or possibility associated with it than a time-space boundary-(n-1). Putting this into equation format yields Equation 10.3.12:

Flowering: Possibility$_{TS_n}$ > Possibility$_{TS_{n-1}}$

Eq. 10.3.12: Alternative

Higher refers to the possibility that over time the direction will always tend to move to a higher meta-level. This can be summarized by using the notion of the core-matrix (2.6.7) yielding Equation 10.3.13:

$$\text{Higher: Upward} \begin{bmatrix} M_3 \rightarrow System_X \\ (\uparrow F \rightarrow I) \\ M_2 \rightarrow S_{System_X} \\ (\uparrow Sig \rightarrow F) \\ M_1 \rightarrow Sig_x \\ (\uparrow > P_x) \\ U \rightarrow x_U \end{bmatrix}$$

Eq. 10.3.13: Higher

Note also that while many organizations are practically at the untransformed level for a long time there is still movement within that level that can generally be depicted as a change from a predominantly physical-orientation to a more mental-orientation. In general this shift in orientation is implied, as more and more fixed patterns are overcome: $\uparrow > P_x$, where P_x refers to patterns along an orientation 'x' where x is an element from the set: (physical, vital, mental, integral).

Nurturing or Harmony-Based Mathematical Operators

Nurturing-based mathematical operators have to do with the nature of relationship within systems. These relationships are posited as being of a nurturing nature and emanate from the notion of 'nurturing' as an organizing class. Representative mathematical-operators include Remember, Linking, and Relate. This nurturing-based set can be summarized by Equation 10.3.14:

$$Nurturing\_based_{Mathematical\_operators} \ni [Remember, Linking, Relate ...]$$

Eq. 10.3.14: Nurturing-Based Mathematical Operators

Remember has to do with remembering that there is something in each organization that existed before the existence of any organization, and further, that there is something in each organization that exists in every other organization. This captures a key dynamic that accompanies changing biography in any partial-singularity. This dynamic can be thought of as going back to a time-space moment of zero, and subsequently of expanding into the Time-Space Continuum keeping that connection in mind.

There is something in each organization that exists in every organization and highlights a special way to relate to the underlying system as in Figure 10.3.2:

471

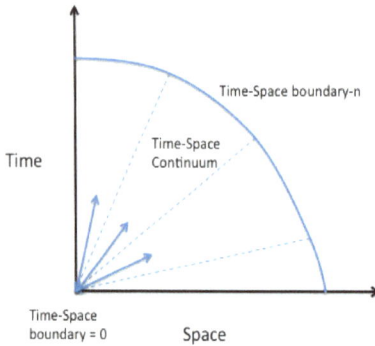

Figure 10.3.2: Remembrance in Time-Space Continuum

This may be depicted by Equation 10.3.15:

$$Remember:\ Ubiquity\ \begin{bmatrix} \overleftarrow{TS=0} \\ \overrightarrow{TS>0} \end{bmatrix}$$

Eq. 10.3.15: Remember

Where the condition of going back ($\leftarrow$) before any organization existed, TS = 0, is invoked as all developments proceed ($\rightarrow$), TS > 0, to create a sense of ubiquity. The sense of ubiquity is a remembrance. This notion may be akin to the concept that everything that is in the universe emanated from the Big Bang and that we are all recycled stardust (Swimme, 2001).

Link refers to the condition whereby in any time-space coordinate, irrespective of the level of untransformed reality (U), any present state can be consciously linked to the underlying ubiquitous system. This can be depicted by Equation 10.3.16, Link, where the conditions that usually need to be in place for a meta-layer to actively influence layer U disappear. There may be an attitude or receptiveness on the part of the element at U that allows such linking to take place. Hence:

472

$$Link: \begin{bmatrix} M_3 & \rightarrow & System_X \\ & \uparrow\downarrow & \\ M_2 & \rightarrow & S_{System_X} \\ & \uparrow\downarrow & \\ & M_1 & \rightarrow & Sig_X \\ & & \uparrow\downarrow & \\ & U & \rightarrow & x_U \end{bmatrix}$$

Eq. 10.3.16: Link

Relate is a way to relate to the System so as to offer or surrender activities any kind of organization is involved in, to the System, and hence the Fullness or intelligence embedded in every point. This can be depicted by an Offer function, such that the first or relatively untransformed element in the function, depicted by x_U, is being offered to the second one, the union of the four-fold intelligence embedded in each point and depicted by the union function U[ ], as in Equation 10.3.17:

$$Relate: \ Offer \left(x_U, \bigcup \begin{bmatrix} System_{Pr} \\ System_P \\ System_K \\ System_N \end{bmatrix} \right)$$

Eq. 10.3.17: Relate

In the previous section a possible link was made between the expansion of the universe and the changing nature of matter due to continued precipitation of fourfold functional richness. Perhaps it is that if the expansion continues the apparent gaps between galaxies filled with extraordinarily high dark-energy, possibly correlated with an increasing "manifested set", becomes ripe for a series of inter-galactic-bangs. And perhaps such a series of inter-galactic-bangs in such a functionally-rich universe precipitates cycles of development based on other speeds of light, that in turn sets up a more conscious fusion of the multiple layers of existence so created, into one incredible Universe where an entirely new type of super-matter can form.

The fruition of such a possibility signifies a new Ouroboros where richness builds on richness and new types of super-matter on previous super-matter. In Martin Nowak's book, Super Cooperators, he proposes a third principle of evolution: that of cooperation. A first principle is that of mutation, responsible for generating genetic diversity. A second principle is that of selection, that focuses on individuals best suited a certain environments. He calls cooperation the master architect of evolution since from it emerges the constructive side of evolution 'from genes to organisms to language to complex social behavior' (Nowak, 2011). But as suggested here, evolution will occur driven by Light itself until emergent constructs see as Light sees, and become as Light is, even materially. This is the condition of the Second Singularity. Meta-layers have to become overtly active at the material layer.

The human has in a certain sense been the result of 30 orders of microcosmic magnitude implicit to it. Fields, quanta, atoms, cells, the incredible four-fold emergences of properties codified in Light have all contributed to the emergence of the human. Now in a possible reversal of

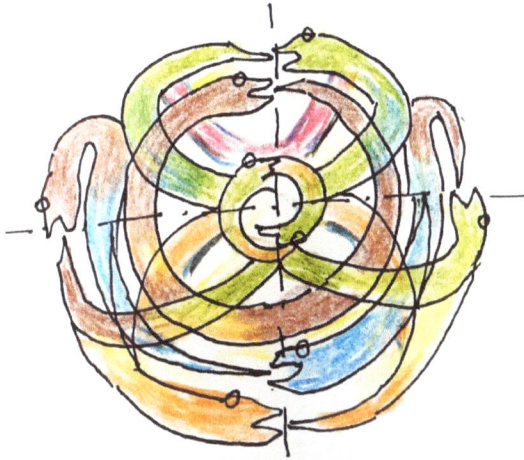

causality the human appears to be positioned to similarly influence and even structure the further development of the macrocosm by mastery of the microcosm. Such mastery will involve leveraging an increasing set of functional operators of the type introduced in the previous chapter.

Through will, human faculties, and the use of functional operators continued states of quantum-certainty resulting in four-fold quantization will occur. Such quantization changes the very nature of space, time, energy, and gravity and can alter the development of the 30 orders of macrocosmic magnitude.

Such audacity is only possible because the human is a master emergence, a promising partial-singularity in which by definition the micro and the macro can increasingly become extraordinary reflections of each other, in a universe of Light where all is Light and all is of Light. Light in material form turning on itself can alter the very nature of what it is by what it is. Such is the future tale of Ouroboros changing from swallowing its tail to entwining itself through a Cosmos in a never-ending spiral toward unforeseen flowerings.

The activation of meta-layers at "U" in (3.1.3) is the condition of the Second Singularity and is depicted by Equation 10.4.1, The Second Singularity:

The Second Singularity

$$
=
\begin{Vmatrix}
\begin{bmatrix}
c_\infty: [Pr, Po, K, H] \\
(\downarrow R_{C_K} = f(R_{C_\infty})) \\
c_K: [S_{Pr}, S_{Po}, S_K, S_H] \\
(\downarrow R_{C_N} = f(R_{C_K})) \\
c_N: f(S_{Pr} \times S_{Po} \times S_K \times S_H) \\
(\downarrow R_{C_U} = f(R_{C_N})) \\
c_U: [P, V, M, C] \\
\Uparrow \\
c_{0:[D,W,I,C]}
\end{bmatrix}_{Light}
\begin{bmatrix}
M_3 \rightarrow System_X \\
\updownarrow \\
M_2 \rightarrow S_{System_X} \\
\updownarrow \\
M_1 \rightarrow Sig_x \\
\updownarrow \\
U \rightarrow x_U
\end{bmatrix}_{Space} \\
\begin{bmatrix}
M_3 : -\infty \le t \le \infty \\
\downarrow \\
M_2 : 0 \ge t > \infty \\
\downarrow \\
M_1 : 0 > t > \infty \\
\downarrow \\
U \rightarrow \begin{matrix} t \sim E_{Cell}; TC: M_3 \rightarrow U \\ t \sim E_{Human}; TC: U \rightarrow M_3 \end{matrix}
\end{bmatrix}_{Time}
\quad TC \rightarrow x_T
\end{Vmatrix}_{\langle x_U | x_T \rangle}
$$

Eq. 10.4.1: Second Singularity

476

Note that prerequisite conditions for activation of meta-layers have been entirely transcended in the Space-Matrix as depicted by (↕) in (10.4.1).

Equation 10.4.2, Future Code, suggests genetic-type code generated in the Second Singularity. Note that the starting point in this iteration is assumed to be some approximate form of super-matter, depicted as '~Super-Matter' in x_u:

Future Code =

$$
\left|
\begin{bmatrix}
\begin{bmatrix}
c_\infty : [Pr, Po, K, H] \\
\left(\downarrow R_{C_K} = f(R_{C_\infty}) \right) \\
c_K : [S_{Pr}, S_{Po}, S_K, S_H] \\
\left(\downarrow R_{C_N} = f(R_{C_K}) \right) \\
c_N : f(S_{Pr} \times S_{Po} \times S_K \times S_H) \\
\left(\downarrow R_{C_U} = f(R_{C_N}) \right) \\
c_U : [P, V, M, C] \\
\Uparrow \\
c_{0:[D,W,I,C]}
\end{bmatrix}_{Light}
\begin{bmatrix}
M_3 \rightarrow System_X \\
\updownarrow \\
M_2 \rightarrow S_{System_X} \\
\updownarrow \\
M_1 \rightarrow Sig_X \\
\updownarrow \\
U \rightarrow x_{\sim Super-Matter}
\end{bmatrix}_{Space} \\
\left[U \rightarrow
\begin{matrix}
M_3 : -\infty \le t \le \infty \\
\downarrow \\
M_2 : 0 \ge t > \infty \\
\downarrow \\
M_1 : 0 > t > \infty \\
\downarrow \\
t \le E_{Cell}; TC: M_3 \rightarrow U \\
t \sim E_{Human}; TC: U \rightarrow M_3
\end{matrix}
\right]_{Time}
\quad TC \rightarrow x_{Future}
\end{bmatrix}
\right|_{\langle x_U | x_T \rangle}
$$

$\Rightarrow$

$\langle Space - Time - Energy - Gravity\ Material - Fabric\ Pre$
$- Genetic\ Code \rangle +$

$\langle Electromagnetic\ Spectrum\ Material - Fabric\ Pre - Genetic\ Code \rangle +$

$\langle Quantum - Particle\ Material - Fabric\ Pre - Genetic\ Code \rangle +$

$\langle Atoms\ Pre - Genetic\ Code \rangle$
$$+ \langle Molecules\ Material - Fabric\ Pre$$
$$- Genetic\ Code \rangle +$$

$LSTE \langle ... \rangle + PGGT < \cdots > + \langle Cells\ Genetic\ Code \rangle + LSTE \langle ... \rangle +$

$(Fundamental\ Capacities\ of\ Self\ Genetic\ Code) + FBLEEE \langle ... \rangle +$

$(Humans\ Genetic\ Code) + LSTE \langle ... \rangle + FBLEEE \langle ... \rangle +$

$(Stable\ Mega - Organization\ Post - Genetic\ Code) + LSTE < \cdots > +$

$(Sustainable\ Global\ Civilization\ Post - Genetic\ Code) + LSTE < \cdots$
$$> +$$

$(\sim Super - Matter\ Post - Genetic\ Code) + LSTE < \cdots > +$

$$\sum_{n}^{\infty} \left(\begin{array}{c} \sum Sig_I = Xa + \overline{Yb_{0-n}} \\ where \left[\begin{array}{c} X \in [S_{System_N}] \\ Y \in [S_{System_{Pr}}, S_{System_P}, S_{System_K}, S_{System_N}] \\ a, b\ are\ integers; a > b \end{array} \right] \\ \sum Sig_M = Xa + \overline{Yb_{0-n}} \\ where \left[\begin{array}{c} X \in [S_{System_K}] \\ Y \in [S_{System_{Pr}}, S_{System_P}, S_{System_K}, S_{System_N}] \\ a, b\ are\ integers; a > b \end{array} \right] \\ \sum Sig_V = Xa + \overline{Yb_{0-n}} \\ where \left[\begin{array}{c} X \in [S_{System_P}] \\ Y \in [S_{System_{Pr}}, S_{System_P}, S_{System_K}, S_{System_N}] \\ a, b\ are\ integers; a > b \end{array} \right] \\ \sum Sig_P = Xa + \overline{Yb_{0-n}} \\ where \left[\begin{array}{c} X \in [S_{System_{Pr}}] \\ Y \in [S_{System_{Pr}}, S_{System_P}, S_{System_K}, S_{System_N}] \\ a, b\ are\ integers; a > b \end{array} \right] \end{array} \right)$$

Eq. 10.4.2 Future Code

The final code-segment in Equation 10.4.2, Future Code, is a generalization (Σ) of all the development that can take place along known dimensions of physicality, vitality, mentality, and integrality. This equation implies a constant stream of materialized functional-richness, an evolving super-matter, and an evolving post-genetic structure for encapsulating such future code.

In a society driven by rapid technological advancement it is easy to come to believe that the future of life will be synonymous with such advancement. There has been much talk of an AI-based singularity – an event or point in time in which human capability will be marginalized by a singular, global, vastly intelligent AI. But this ignores the fundamental light-based nature of Life.

This book has proceeded to summarize a perception of Life based on the nature and play of Light. In this play everything that is, emerges from Light. Light's essential nature, housing infinite potentiality within it has already created this extra-ordinary, massively intelligent universe, run by a persistent quantum-level computation that generates a continuous stream of genetic-type data that functions as relevant "law" as it were, to allow more of Light's potentiality to continually materialize.

Essentially the materialization of potentiality in Light may lead to the Second Singularity, the singularity whose power will far surpass any AI-based singularity precisely because of the level and sophistication of integration, and the continual real-time re-computation of reality driven by enlightened heart, will, and mind dynamics.

This chapter summarizes the Second Singularity Biography by reviewing key insights that have surfaced through considering the structure and generation of light-based singularities:

1. When Light travels at speed c, 186,000 miles per second in vacuum it creates past, present, future, and an accumulation of possibility due to the finite speed that becomes matter.
2. There appears to be no reason for light to travel at c, than for reality to be experienced the way it is when light travels at c.
3. The necessity for light to travel at c to create a material universe is also the foundation for Einstein's Theory of Relativity in that for speed to remain constant, time and space have to be allowed to fluctuate.
4. In Light's native state it travels infinitely fast. This means it is present everywhere instantaneously, or is omnipresent. It means that nothing that is not of the nature of Light can persist against the reality of Light, and hence it is omnipotent. Light being everywhere is aware of all that is happening in it, and it is therefore omniscient. All is connected in one embrace in the substance of Light – therefore it is omniloving.

5. These four implicit properties of Light – presence, power, knowledge, harmony – hint at the potentiality resident in Light.

6. The pre-existent complexity of Life derives from the fact that Light has infinite potentiality embedded in it.

7. The essence of light traveling at c can be seen to be mathematically symmetrical to the four implicit properties of Light in its native state at infinity.

8. A past implies a status quo – the result of a working out of things into established reality. Hence, symmetrical to the notion of 'presence'.

9. A present implies a vitality or dynamism with the stronger influence expressing itself in the present. Hence, symmetrical to the notion of 'power'.

10. A future implies the working out of deeper ideas or seed concepts. Hence, symmetrical to the notion of 'knowledge'.

11. Interaction between material islands created in a reality of light traveling at c implies working out connection. Hence, symmetrical to the notion of 'harmony'.

12. For the four implicit properties of Light at infinity to become the reality of vast diversity in a material universe generated when light travels at c, means that some essential mathematical transformations have to occur to diversify the implicit four, between these two layers of Light.

13. The first transformation occurs when each of the four properties become variations based on their essence, to create four sets of infinite elements each.

14. The second transformation occurs when elements from the four sets combine together in unique combinations to create an infinite number of unique seeds.

15. The first transformation can be envisioned as a field. A field means that the essential binding

factor that creates matter is relaxed so that there is a vaster spreading out of information. This will occur when light is traveling quite a bit faster than c.

16. The second transformation is envisioned as a wave. A wave means that the form of the information being represented is closer to being a particle than a field. This will occur when light is traveling faster than c but slower than the speed required to create a field.

17. Light traveling at c creates an upper bound in this known material universe. The inverse of c creates a lower bound and is proportional to Planck's constant h.

18. Planck's constant h is significant in that it provides insight into the least amount of energy required for matter to express itself.

19. When light travels faster than c, then the inverse of this speed is a constant smaller than h. This smaller constant implies that the energy will not be able to express itself in matter, but as a more spread-out wave.

20. When light travels faster than c, but closer to being infinitely fast, than the inverse of that speed is a constant, significantly smaller than h. This miniscule constant implies that energy will not even be able to express itself in a more contained wave, but will require a field to express itself.

21. The down-shifting of information in Light, from its native state of traveling infinitely fast and housing infinite potential, to each subsequent layer where light travels slower, implies the accumulation of some of that information so that it can express itself in more material form in the layer of slower moving light. This downshifting requires the device of quanta.

22. Quanta are the device by which information in a faster-moving layer of light materializes in a slower-moving layer of light.

23. The speed at which light travels to create a reality of fields of information is referred to as c_k. The corresponding layer of reality is referred to as K.

24. K can also be thought of as a partial-seed light-based singularity in that there is wholeness and a complete backward integration with light in its native state traveling infinitely fast.

25. A light-based singularity occurs when the most current emergent form has all previously generated code embedded or available to it thereby inherently abiding with all "laws" that have thus far emerged in previous light-based singularities.

26. The seed-singularity itself will have had a progressive biography resulting in partial-seed-singularities along the way, until the seed-singularity has itself emerged.

27. The speed at which light travels to create a reality of waves of information is referred to as c_n. The corresponding layer of reality is referred to as N.

28. N can also be thought of as the seed-singularity in that there is wholeness and an integration of with K and the layer representing Light's native state at infinity. This seed-singularity is the seed of all other light-based singularities.

29. There is a light-based wholeness in the seed-singularity that results in a first meaningful sphere of cohesion with the creation of myriad light-based seeds of uniqueness.

30. The layer of reality created when light is traveling at c is referred to as U.

31. The pre-existent complexity in Life is further reinforced by different information fields existing in each subsequent layer of slower light, relative to Light in its native state at infinity.

32. The origin of genetics is the vast amount of information embedded in Light in the seed-singularity.

33. If there is a history of singularities, or a biography that relates the seed-singularity with partial-singularities, with the Second Singularity, then genetics can be thought of as the language in which it is written.

34. Singularity-biographies are the process by which a singularity materializes possibility contained in the seed-singularity through a trajectory of partial-singularities culminating in the Second Singularity.

35. In its native state with light traveling infinitely fast the libraries of information generated are subject to a class of entanglement referred to as ∞-entanglement. This implies that this information is subtly available in all of existence.

36. At layer K, libraries of information generated from the four sets are subject to a class of entanglement referred to as K-entanglement.

37. At layer N, libraries of information generated from the infinite number of unique seeds, is subject to a class of entanglement referred to as N-entanglement.

38. Entanglement is what ensures uniqueness as information materializes.

39. Quantum downshifts occur between subsequent layers of light until that point when information is ready to materialize at U, where light travels at c.

40. When light exists at speed zero, the opposite of its native state of being infinitely fast, that reality also becomes the opposite of omnipresence, omnipotence, omniscience, and omniloving. Such a reality would therefore be characterized by properties such as weakness, ignorance, darkness, and hate.

41. Materialization of information in the form of matter involves superposition of different possibilities resident at different layers of Light.

42. The influence of the layer of light specified by zero speed will impregnate practical existence with an essential inability and cause obstinate habit to form.

43. Quantum-level computation is the process that arbitrates the material expression of different states existing simultaneously in different layers of light.

44. The continual shifting of information in layers of light necessitates the reality of a persistent quantum-level computation.

45. Schrodinger's equation summarizes matter in its wave-aspect and models how infinite superposition of possibility results in some concrete materialization of these.

46. Heisenberg's uncertainty principle sheds light on the always-buzzing pregnant-infinity behind all appearance.

47. Euler's equation will approximate the range of materialization influenced by the dark - or light at zero speed, to light - or light at infinite speed, spectrum and can be modeled to oscillate or exist between these two bounds.

48. Imaginary numbers in Schrodinger's equation or Euler's equation interrelate antecedent layers of light with the material layer.

49. The Big Bang is the phenomenon that occurs when Light slows down to c. This slowing down materializes a vast amount of information that creates the Big Bang.

50. The process of light slowing down to c from its native state of traveling at infinity is modeled as a light-based-singularity superstructure. The term superstructure implies that different kinds of light-based-singularities can occur within it.

51. When the Big Bang occurs the materialization of information creates space, time, energy, and gravity.
52. The Big Bang is the first envisioned partial-singularity that occurs after the formation of the seed-singularity.
53. The Big Bang is a watershed event and signifies the start of a progressive, functionally rich evolution.
54. In the absence of the Big Bang evolution would have proceeded based on light existing at speed zero as the starting point.
55. Fragmented-singularities, such as any AI-based singularity, that are essentially separated from any light-based singularity, can be thought of as proceeding in a universe typified by light existing at zero speed only.
56. Partial-singularities are always the result of the persistent quantum-computation that arbitrates material reality through the interaction of the many dynamics belonging to the multiple realities set up by different speeds of light, including that of c.
57. Cosmological subtle-DNA is comprised of a downward-strand and an upward-strand.
58. The downward-strand is caused by light slowing down in quantized-decelerations, as it were, progressively concretizing more of the information in light.
59. The upward-strand is envisioned as prescribing a time-variable sequence wholly determined by the nature of the layers in the downward-strand.
60. Partial-singularities remain "partial" until all the conditions specified by the logic in the upward strand are fulfilled at which point the Second Singularity comes into being.
61. All partial-singularities and the Second Singularity itself, while occurring in the

downward strand, are the result of the interaction of the downward and upward strand.

62. Space can be thought of as containing all the seeds of uniqueness formed by antecedent layers of Light.

63. Time is a way to think of the process of maturity by which the potentiality in seeds will fructify.

64. Energy is a way to think of the essential accumulation of possibility that will allow materialization of seeds.

65. Gravity is an arrangement that binds seed to a set of seeds that will be required in the journey to its own fulfillment.

66. Space can be thought of as filled with multiple levels of superposed entanglements in static form.

67. Time appears to be the dynamic working out of the possibilities implicit in superposed entanglements materialized as seeds in Space.

68. Space, Time, Energy, and Gravity are emergent from Light and mirror or reflect Light's implicit properties of Knowledge, Power, Presence, and Harmony, respectively.

69. The process of materialization of space, time, energy, and gravity takes place in two steps.

70. Step one takes place as the speed of light approaches c from above. Logic specifying this essential four-foldness of space-time-energy-gravity is created in a layer at the quantum-levels. This layer is referred to as Four-Base-Logic-Encoding-Ecosystem (FBLEE).

71. Logic systems in FBLEE can exercise themselves through FBLEE-entanglement (FBLEEE).

72. Step two is a further precipitation that occurs as light travels at c. Space-time-energy-gravity FBLEE precipitates into a layer that houses this kind of information, referred to as the material-fabric (MF).

73. The MF can be thought of as housing genetic-type information that is pre-genetic, and yet exercises a binding logic on all constructs arising in matter.

74. The material-fabric is imbued with reality when the fourfold space-time-energy-gravity quantization is itself generated as pre-genetic code.

75. Hence the logic of space, time, energy, gravity are binding on all constructs arising in the material universe specified by light moving at speed c.

76. Space-Time-Energy-Gravity logic embedded in the material-fabric is what allows quantization and therefore materialization to take place.

77. Space-time-energy-gravity has two modes of operations. The macro-mode is ingrained in the material-fabric and puts in place the familiar large-scale universal parameters of Space, Time, Energy, and Gravity.

78. The micro-mode occurs as a composite space-time-energy-gravity quantum whose action is required for the encoding of any logic-ecosystem in FBLEE and subsequently its materialization in the MF or appropriate information-housing structure.

79. Such micro-mode space-time-energy-gravity quantization legitimizes the logic of an ecosystem into "law" by encoding it in the material-fabric.

80. Quadrumvirate space-time-energy-gravity quantization seeking to express the reality of its inherent oneness will cause matter to containerize regardless of scale.

81. The information-housing structure can be the pre-genetic material-fabric, the genetic DNA, or some post-genetic hybrid structure that may surface as the Second Singularity Biography advances sufficiently.

82. With each partial-singularity, specific logic-ecosystems are first encoded in FBLEE through a

successful activation process involving space-time-energy-gravity quantization.

83. The outcome of successful activation involving space-time-energy-gravity quantization will culminate in a state of quantum-certainty.

84. FBLEE action that alters the quantum-level interface between the material and antecedent layers of light in effect changes the basis by which matter materializes.

85. All successful FBLEE action will alter natural or human history.

86. Genetic mutation is in its essence, an attempt to change FBLEE.

87. Genetic-type mutation, the process by which repositories of instruction change, allows the wholeness implicit in the seed-singularity to materialize as the many partial-singularities with greater and greater spheres of influence, culminating in the Second Singularity.

88. Constructive mutation is the only type that can successfully change FBLEE.

89. Constructive mutation implies an integration of U with light-layers traveling faster than c. It is hence an attempt to integrate the nature of a materialized construct with more of its antecedent potentiality.

90. Destructive mutation will only affect local genetic structure precisely because it is unable to penetrate FBLEE.

91. Destructive mutation implies an integration of U with aspects of the layer of light existing at zero speed.

92. In the Second Singularity Biography the electromagnetic spectrum partial-singularity follows the Big Bang partial-singularity.

93. The primary architecture of the electromagnetic spectrum partial-singularity, and all subsequent partial-singularity architectures are determined by the light-space-time emergence equation.

94. The light-matrix in the light-space-time emergence equation delineates the essential layers of Light and the transformations that allow uniqueness-based diversity to materialize at U.

95. The essential electro (power), magnetic (harmony), wave-archetype (knowledge), mass-potential (presence) architecture, is specified by Light in its native state (the first line) in the light-matrix. Subsequent lines further detail this architecture.

96. The electromagnetic spectrum can likely be more completely described as electro-magnetic-wavearchetype-masspotential spectrum.

97. The infinite variation in wave-archetype creating infinite number of seeds in the electromagnetic spectrum, causes it to be spread out in spite of the impulse of union of the essential four-foldness in its architecture.

98. Mass containerizes in an attempt to realize its oneness when the action of space-time-energy-gravity quantization takes place.

99. Just as the space-time-energy-gravity quadrumvirate has an impact on every material emergence, so too will the action of the electromagnetic spectrum have an action on every emergence of life.

100. Segments of electromagnetic spectrum related code due to variation in wavelength and frequency creates diverse functional possibility in the intent, energetics, and mass-possibility of built-up life.

101. Cross-over of fourfold consistency from one type or partial-singularity to another, as in the case from the electromagnetic partial-singularity, to matter-based partial-singularities, to life-based partial-singularities is the hallmark of light-based singularities indicating that all that emerges in such a manner is of one light-based edifice.

102. The partial-singularity generation equation is holographic. That is, the same equation in its wholeness can be applied in the creation of the smallest emergences from Light.

103. The partial-singularity generation equation is fractal. That is, the same mathematical equation will generate material reality regardless of scale.

104. Light can be thought of as a creative medium and the partial-singularity generation equation as modeling the specification of what is to be created.

105. Mathematical biographies for quantum particle, boson, atom, and cell partial-singularities follow the electromagnetic spectrum partial-singularity and are also expressed or generated through the light-space-time emergence equation.

106. The notion of U as the bases in the formation of established-matter, comprising of all partial-singularities prior to the cell partial-singularity, simply implies that this is a baseline and is subject to all the adaptability yet to come about through the play of dynamics such as urge and will which increases with the generation of subsequent partial-singularities.

107. Common DNA existing in every cell at the material layer can be thought of as influenced by up to four antecedent-entanglements: ∞ - entanglement, K-entanglement, N-entanglement, and FBLEE-entanglement.

108. The light-space-time emergence or partial-singularity generation equation outputs genetic-type information summarized as a series of equations. Solutions to these generated equations detail the genetic-type information for that partial-singularity.

109. Life is a field that allows pre-existent complexity in Light to emerge materially.

110. The light-space-time emergence or partial-singularity generation equation is iterative and

will add to the biography of a previous partial-singularity to output a subsequent partial-singularity biography.

111. The time-matrix of the partial-singularity generation equation approximates when different layers of light are active, and importantly when the automaticity of meta-layer action yields more to dynamics such as urge and will.

112. Urge and will, will tend to become more active with the emergence of human-related partial-singularities.

113. As capacities of self such as urge, will, desire, emotion, and thought emerge, the causal effect of these on the body, is founded on genetic code that is created with the emergence of these capacities.

114. In pre-human partial-singularities the actions of meta-layers is automatically more insistent as compared with human and post-human partial-singularities.

115. Even human and post-human partial-singularities are light-based singularities with deeper light-based dynamics always active regardless of their being perceived. That after all, is the point of a light-based singularity.

116. As partial-singularities approach the human, strong wills and urges create an activation-state that can potentially open a quantum-window.

117. The space-matrix of the partial-singularity generation equation details the conditions that will allow meta-layers to become active at U.

118. A quantum-window provides access to FBLEE, and based on the levels of light active, can initiate space-time-energy-gravity quantization to change or add to the code in FBLEE.

119. All code in FBLEE is accessible by materialized constructs and can precipitate into genetic or even post-genetic structure.

120. Stable-mega-organization and sustainable-global-civilization partial-singularities are milestones toward the Second Singularity.

121. All close-to-human partial-singularities have an impact on genetic structure. This is so because the four-fold functionality that all genetic-type information essentially expresses is enhanced with subsequent partial-singularities.

122. Each partial-singularity biography essentially elucidates how fourfold functionality increases. All fourfold-functionality increasingly expresses potentiality in Light.

123. Super-matter-based partial-singularities begin to express themselves when primary activity is meta-layer based. This will cause fourfold-functionality to increase dramatically and will necessitate a different structuring of matter: super-matter.

124. Super-matter is a foundation based on will or cohesive want and sets the stage for a potentially unending willed development in which functional-richness existing in Light in its native state traveling at infinite speed, can manifest in this material universe.

125. Creation of super-matter is likely linked to the expansion of the universe. This is because increase in super-matter is linked to increase in fourfold-functionality, which will require more seeds to express themselves. More seeds expressing themselves means more Space is created. More Space means acceleration of an expanding universe continues.

126. Dark energy is hence related to an increase in cardinality of the combined set containing elements from all four manifested sets.

127. Manifested sets are a record of specific fourfold functionality elements that have materialized.

128. A persistent quantum-level computation resulting in a constant stream of genetic-type

information alters the code and the very light-based computational machinery emergent as progressively more sophisticated pre-matter, matter, and post or super-matter based light-based singularities.

129. Matter is a dynamic state linked to the very Life of the universe.

130. In situations such as Black Holes or a Cosmic Bounce the large number of seeds that are normalized into a smaller number of seeds implies that space will be compressed and time will slow down. This appears to be consistent with Einstein's General Theory of Relativity.

131. In a Big-Planet composite quantum, time, due to increased number of seeds acting independently, will be elongated or experienced faster.

132. The human is a master emergence, an unusually promising partial-singularity, in which by definition the micro and the macro become extraordinary reflections of each other, in a universe of Light where all is Light and all is of Light. This is so because 30 orders of magnitude lie on both sides of the human, into the micro and into the macro respectively.

133. If we, or any species that followed the human-species were to reach a point where we are unable to increase fourfold-functionality, it would mean that the materialization of potentiality in Light would discontinue and that the current complexification of four-foldness in this universe would likely need to be reconfigured in some fundamental way. In this case the universe would go through a collapse and possible re-ignition through a subsequent Cosmic Bounce.

134. It is also for this reason that any AI-based singularity will be unable to sustain itself: essentially because being based on mind-based processes only it will reach a limit to the four-foldness that can be expressed.

135. The Second Singularity implies fully integrated and conscious four-foldness across multiple levels. Reality at U becomes a powerful means to consciously and rapidly express four-foldness resident in Light's native state.

136. The Second Singularity implies a stable and dynamic equilibrium where potentiality in Light is able to express itself in more profound, complete, exhilarating, and astounding formations.

137. Light in material form turning on itself can alter the very nature of what it is by what it is. Such is the future tale of Ouroboros changing from swallowing its tail to entwining itself through a Cosmos in a never-ending spiral toward unforeseen flowerings.

Illustrator's Note

This note elaborates the illustrators' rationale for each of the 164 figures, by section.

Author's Introductory Note Illustrations

Fig 1: This illustration is inspired by the lines - 'the emergence of these naturally occurring partial-singularities, leading up to the possible human-founded singularity.' Here smooth curves from a darkness below spirally emerge with bright colors in symbolizing partial singularities leading to the human-founded singularity.

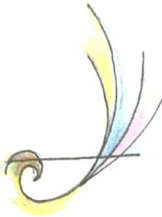

Fig 2 : Based on the lines -'Life arises, as did matter before it, and mind after it, because these were already contained in potentiality in infinite Light. .. not random or arbitrary, as physicists may suggest, but the result of a persistent quantum-level computation..' A small shrub here is rising up from seeds, and a beautiful curve points to its inspiration that is the Light - the Sun. Note that in his book Life 3.0, the MIT physicist, Max Tegmark, distinguishes between Life 1.0 in which all hardware and software are evolved, Life 2.0 in which software is designed and hardware is evolved, and Life 3.0 is which hardware and software is designed. This book will in fact suggest that all of matter and life is already and has always been Life 3.0.'

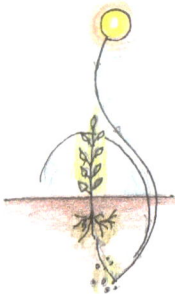

Section 1 Illustrations

Fig 3: The upward and downward strands in the process of this emergence are depicted here by two serpents crossing each other as they transcend two barriers / levels.

Fig 4 : Light travelling at potentially different speeds results in different realities. Quanta is he bridge between one layer of light and another slower moving layer of light. Here we see several waves of different wavelengths and amplitudes interacting with each other. These are shown in play of four colors (four aspects, Quadrality) This four color code is used in previous books as well and continues throughout this book s well, as light will have properties of omnipresence, omnipotence, omniscience, and omninurturance in such a space.

Fig 5: This is an artistic depiction of the words - 'Quanta have to be understood as the bridge or device that ties different layers of light together.' There are strong lines like several streams of a moving mass or fluid which are seen here as passing through a bridge which also helps them to be brought together and channelized.

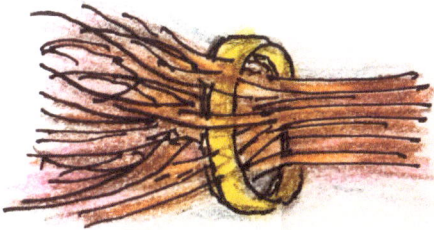

Fig 6: '..the slower speed of light implies that light will need to travel some unit distance in order to be expressed. This light is slower because it is materializing something.' This is shown as a rapid fall of a water stream through a passage which slows down by friction. As it comes to a horizontal surface it gets slower and spreads far and wide.

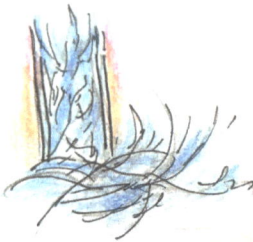

Fig 7: '..If light slows down from an infinite speed to some other lesser speed whether a multiple or a fraction of c, then the material reality will have to alter, perhaps in a similar way as the material reality between vacuum and water is different.' The concept in Fig 6 is further expanded here and connected to the mystic tale of the descent of the divine river Ganga from cosmic heights. Her force was slowed down and made calmer by Shiva, who took her in his hairlocks and gently channelized it to descend down for the earth. '..As an analogy, this can also be thought of in terms of an incredibly rapidly moving stream of water. If the water is traveling so fast, then no boundary will be able to contain it and the energy will be continuous over the length of the stream. If it is traveling slower though, then the water will be able to be held by boundaries along the length of the stream.'

Fig 8 : '..different universes created as a result of light

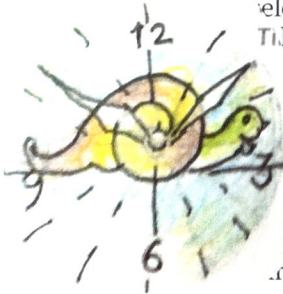

selectively slowing down to different levels. The slowing down to a different level will create a particular kind of universe.'..' so that it practically moves at a snail's pace. This slowing down has to be put into context.' These lines prompted the drawing of this figure with time and a snail .mposed on each other.

Fig 9: 'Note that there have been experiments to slow down light. Even if light were to slow down to 1 mile per second, say, from 186,000 miles per second, keep in mind that it is possible that this change in speed is likely only a miniscule fraction of the change in speed from light

traveling infinitely fast to light traveling at c, that is, 186,000 miles per second.'--this is shown here by the figure above.

ight-based singularities, hence, is t from the AI-based singularity.' This line has inspired this illustration wherein several small linear stepped AI based singularity curves are contrasted with a smoothly ascending blooming lotus depicting a light based singularity.

Fig 11: 'Matter itself being a container in which space and time can allow deeper properties of Light ..' This idea is shown here pictorially.

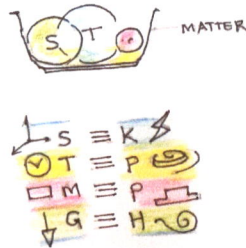

Fig 12: Infinite number of superposed states are shown here in a stack of layers with four color mix.

Fig 13: 'In digital computing, an electrical-current through a bit will determine if it takes on a value of 0 or 1.' ...these pairs are shown here in an illustration and also they are connected as complementary to each other and completed whole.

Fig 14: The Quantum-Level FBLEE Stratum: 'It is a different substratum - four base-logic-encoding-ecosystems.' (FBLEE)

502

Fig 15: 'Subtle-Libraries of Pre-Genetic Information: The progressive materialization of light can be modeled by a series of mathematical transformations.' This is shown here with the space matrix or upward strand.

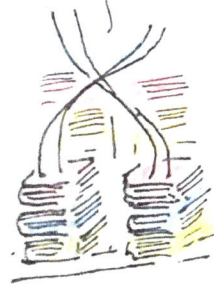

Fig 16: The Essential Structure of Subtle-DNA: The essential structure of a light-based singularity comprises of a largely pre-existent Light-Matrix or downward-strand and a resulting Space-Matrix or upward-strand.

Fig 17: This book explores four types of entanglement.

Three are due to envisioning light traveling at three additional constant speeds besides c. ∞ - entanglement, K-entanglement, and light traveling at a speed 'n', between k and c, engenders N-entanglement. This is shown here in a simple diagram.

503

Fig 18: Constructive-Mutation, Destructive-Mutation, Mutational Sequence and Post-Genetic Code are discussed in this section. 'In general there is a mutational sequence dictated as it were by the levels in the downward-strand.' Here it is envisioned as closely knit multiple upward and downward strands with four colors intermingling to create one body.

Fig 19: 'In his book Life 3.0, Tegmark suggests that memory, computation, learning, and intelligence have an abstract feel to them, and take on a life of their own that

does not reflect the details of their underlying material substance – they are substrate independent (Tegmark, 2017).' This is shown here by using letters L I F E and number 3 . and 0 to create a symbol or logo for Life 3.0.

Fig 20: 'Solving such problems can be imagined as trying to find the lowest point on a surface covered in hills and valleys……but all too often that causes one to get stuck in a "local minimum" -- a valley that isn't the very lowest point on the surface.' This is shown here in series of hills and valleys.

Fig 21: This is inspired by following wonderful insight - 'We may erect a science of how changes in the curling of the lip, creasing of the forehead, compression of the eyes, and so on, can be linked to all manner of thought. But this is a chimera. Without considering the complexity in levels and structures and processes of thought itself, it is unlikely that facially-based surface phenomena will explain thought.': Chimera - a fictitious being is a combination of several real beings - here shown with four feet as four aspects of reality- has part of all and yet it is none of them and unreal.

Section 2 Illustrations

Fig 22:   Multiple layers of seed singularity is symbolically shown here. 'The seed-singularity can be thought of as the singular light-based structure resulting from the integration of layers of light traveling faster than c. The seed-singularity therefore occurs in the upper layers of the downward strand of subtle-DNA.'

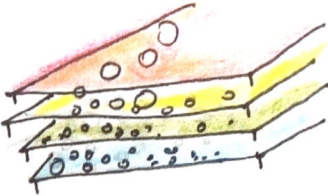

Fig 23: The Shaman shown here is recollected from an actual experience of mine in tribal areas and then by this insight: 'It is interesting to note that Stanford anthropologist, Jeremy Narby, in his book 'The Cosmic Serpent: DNA and the Origins of Knowledge' (Narby 1998) points first to a shift in vision experienced by shamans around the world after the ingestion of ayuhuasca, and second to the commonality of the ensuing vision marked primarily by two entwining serpents. His interpretation is that these shamans enter into the molecular realm and actually perceive DNA and accompanying structures at the cellular levels.'

506

Fig 24: This is a depiction of two serpents intertwining and together moving up as they face each other. This forms a double helix just like DNA helix. The symbol is common in ancient mystic traditions and also in healing practices. There are gods and goddesses which are serpents or have serpents as their weapons.

Fig 25: '..(while past is physicality,) the 'present' is 'the tremendous play of forces of all kinds to express themselves here and now. There is "vitality" that is

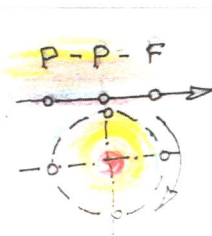

present in this play...the 'future' is 'the inevitability of what will manifest...we can summarize as an essential "mentality." These 'time' concepts are seen here as turning from linearity to circularity as creation and dissolution of matter.

Fig 26: Quantization function ..resulting in material diversity is shown here as pouring down of multiple particles from a mass of matter.

507

Fig 27: Quadrality is shown here in a novel way as the cube has all the diagonals also connected creating a wire

mesh structure which results in several triangles and rhombuses. These are again filled with four colors intermingling in several shades.

Fig 28: Superposition, entanglement, quantum computation, genetic mutation, and singularities are discussed in the section following equation 2.1.11. This is integrated here in a symbol wherein emergence of a blooming flower is seen via a DNA spiral from a reductionist and binary base.

Fig 29: System presence, system power, system knowledge, and system nurturing are put together here in a four plus four petal wheel.

Fig 30: '..the process by which a singularity materializes possibility contained in the seed-singularity to morph into partial-singularities, and then into the Second Singularity.' Here it is visualized as a process of an ancient sculpture being sculpted from stones. The cartesian boundaries of rectangular stones are surpassed to create a whole and nonlinear image of a living being.

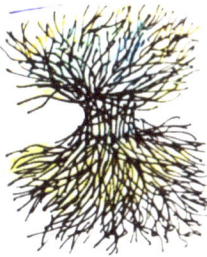

Fig 31: 'System-presence, system-power, system-knowledge, and system-nurturing that define the nature of every point in our system, become more tangible as a broader set of architectural forces that emanate from each of them.' It is seen here as a tree of ancient mystic nature with innumerable roots and innumerable branches emanating from a seed signifying in this context a broader set of architectural forces emanating from each point.

Fig 32: The illustration shows the evolutionary journey from prehuman species up to homo sapiens and then to AI and Robots.

Fig 33: 'In other words space contains seeds…and the seeds can be thought of as a result of superposition of ever present - entanglement,…' this is shown here pictorially.

509

Fig 34: 'The condition of overcoming any fixed and limiting patterns is the prerequisite for the emergence of 'Force' or for entering into the force-level.' This is shown here as some elements overcoming fixed linear patterns - like a square or rhombus - to emerge out as force.

Fig 35: 'Further, how this deep fount of creativity is present everywhere, and how light-based sets with their attendant 'K-entanglement' dynamics that further materialize the range of creative forces has also been considered. These architectural forces elaborate the possibility inherent in any system. Leveraging these sets of forces by virtue of 'N-entanglement' an equation for the uniqueness of an organization, regardless of scale, was also arrived at.' The K and N entanglement are depicted here in an artistic way - the smooth extending curves encircling basic K and N and their entangling is notable.

510

Fig 36: 'The rate of the transformation can be better envisioned when considering action of the Transformation Circle, or TC.' This is shown here in an illustration of a circle and a more ethnic design of a clock - symbolizing Space, Time, pendulum of cyclic or repetitive wave function and so on.

Section 3 Illustrations

Fig 37: This figure shows 'partial-singularity dynamics in which more comprehensive emergences manifest materially to also begin to increase partial-singularity "spheres of influence". '

Fig 38 : The drawing of T and U is inspired by the passage - 'where the subscript T implies relatively-transformed, now becomes the input, x_U, for the next iteration of the equation, where U implies relatively-untransformed. Hence through time there is greater and greater transformation that pushes experienced reality to greater and greater levels of functional-richness.' Here a caterpillar in the U-untransformed is butterfly in T-transformed.

Fig 39: 'A vast and actionable information-base therefore animates any light-based emergence. Note that this is in contrast to any AI-based emergence that is in this view a fragmented development that is not based on the persistent quantum computation and generation of genetic-type

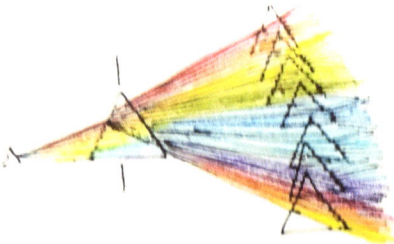

information that animates every light-based emergence.' This is visualized here as a prism generating seven colors to several prisms which in turn do the same.

Fig 40: Schrodinger's Equation and his thought experiment is shown here as a cat in his head, the cage is also a cage due to the mind and its thinking processes. *'Man eva Manushyanam karan bandhan mokshayoho'*: whih means mind is the reason of your bondage, and mind is the way to your freedom' as says an ancient saying.

Fig 41: Equation 3.4.3, 3.4.4, 3.4.5 are shown here in a diagrammatic way. It is related also to 'inter-relation between U and the collective-set of light-layers behind or above U.'-- here each subsequent layer below the top layer, describing the reality set up by an infinite speed, is emergent.

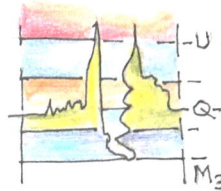

Fig 42: The Light-Space-Time Emergence equation (3.1.3) derived in Chapter 3.1, and the Quadrality of Space, Time, Energy, Gravity are symbolically drawn here - e.g. the three axis graph for Space, wheel of time or clock for Time, etc. 'There are several fundamental quantizations possible that may inter-relate a specific light-layer with other layers of light."

Fig 43: this is based on the line – "Such quantization requires patterns to be overcome, and so long as these are, this will result in the phenomenon of quantum-certainty.' Here several patterns are intermingling and are required to be overcome. The quantization of Space, holds the seeds of knowledge and that is the reason behind all that will emerge.

Fig 44: Light-Space-Time based computation is elaborated in Section 4 along with an overview of the process by which pre-genetic and genetic information is created. This is depicted here with a light, life and cell symbol.

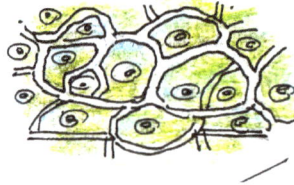

Fig 45: It is related to Eq. 3.6.4 - Constructive-Mutation and the subsequent section . 'The extent to which the antecedent, and therefore the more permanent four-base logic-encoding ecosystem (FBLEE) is altered'…'For in the model based on Light it is this category of library that is important in heredity or the longer-term scheme of things. Destructive or random mutation may alter local copies of libraries, such as exist in DNA at the cellular-level, but it is only constructive mutation that can alter the deeper level subtle libraries.' The stack of books

514

on he table and in he background, and the strands emerging depict this.

Fig 46: Is a ray diagram drawn for K > C and other variables discussed in this section.

Fig 47: The universe within is shown here as a person deeply introspective with folded hands,

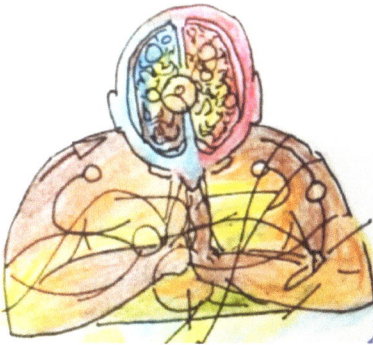

and his brain / mind details are also seen with several planets, stars, and galaxies revolving around. 'As explained by Neil Turok in his book The Universe Within' (Turok, 2012)…its sinusoidal oscillation between narrow bounds as x increases.

515

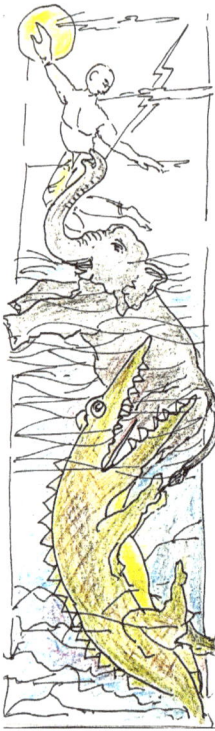

Fig 48: 'Equation 3.6.8, Wavefunction for Random-Mutation, suggests the mixed bases for organizations, as specified by dynamics of both the untransformed (U) and the meta-levels (M_x), which therefore results in random mutation.'...'Equation 3.6.10, Wavefunction for Constructive-Mutation, suggests some transformed bases for an organization, as specified by dynamics of the meta-levels (M_x), which therefore can result in a constructive mutation.' These paragraphs have inspired this hierarchy of emergence of unique entities / identities. From water species to crocodile to elephant to human and then to the sun in the sky. Each wants to liberate and ascend higher, escaping from its lower form and aspiring for the higher form and truth. This is also a modern take on the ancient mystic tale of liberation of an elephant named Gajendra subject to the clutches of a crocodile from below and the grace of divine intervention from above.

Fig 49: Application of Qualified Determinism to Genetic Mutation -
The possible impact of different layers or levels of light on genetic mutation is studied in the previous section. 'The genetic mutation results in altering or creating new genetic-type information, which is positioned as being the biographical language of any partial-singularity.' Here it is shown as alphabets Q and D shaped to be mapping on the double helix.

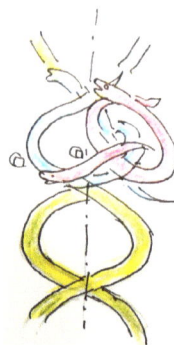

Fig 50: Illustrated here are Light matrix equations - '…the light-space-time matrices, central to the mathematics in this treatise, and additionally in terms of a wavefunction form, central to the probabilistic foundation in quantum theory…this has led to a non-probabilistic approach by application of a process of qualified determinism to genetic mutation. In such a process structural forces influencing mutation are disaggregated into vertical and horizontal components. … - Dynamic Interaction (DI) that has a 'vertical' and a 'horizontal' component. Hence, this mathematical model is suggesting that any situation, including micro and nano-level situations, rather than having a random outcome, has a 'qualified deterministic' outcome. In the introduction to his book "Where is Science Going?" (Planck, 1933), James Murphy points why Planck spent so much of his time giving lectures on causation.

Fig 51: Here in Eq 3.7.5 - Establishing the Nature of the Change - is explored. It further says that 'If T, implying that the action of one of the meta-levels has caused the transformation or mutation, then application of one of the following integrals will determine which level is the likely source for change….in the vicinity of the change…is greater than some threshold value $Threshold_{Signature}$, then the signature dynamics are likely the source of change, where 'a' can be thought of as some kind of gene-footprint. This is summarized by Equation 3.7.6 - Signature Dynamics as the Source of Change. This is depicted here as a seamless combination of several ascending curves and shapes which also resemble an ascending style signature.

Fig 52: The Eq 3.7.7 - Architectural Forces as the Source of Change , and Eq 3.7.8 - System Properties as the Source

of Change is visualized here as a flow /current of elements just a objects float and move with the current of a river stream. The objects floating here are from pure abstract mathematical operators on the left side to progressively becoming a mesh and matrices and patterns, which in turn become a cell, and atoms, and then

seeds to finally emerge as beautiful lotus flowers on the
extreme right.

Section 4 Illustrations

Fig 53: If 'the pre-genetic and genetic code that imbues everything becomes the result of continuous, real-time Light-Space-Time Matrix based computations implicitly involving quantum-levels'...'there is likely code within all living cells that has to do with space, time, energy, gravity, the functioning of the electromagnetic spectrum, and all the stages of material elaboration that precede the emergence of a living cell.' This is visualized here as a matrix of shrublike elements in each intersection of rows and columns and the central portion is given a whirling impetus to locally shift the overall pattern. This can be a seed / beginning of further transformation. The generated pre-genetic code belongs to the pre-material-fabric genre since it will remain in a subtle domain exercising its influence through ∞ - entanglement where possible. Further, the generated pre-genetic code also narrates the biography for a composite singularity in that it is whole and complete, even though it is describing only a part of the seed-singularity. Hence, strictly speaking (4.1.1) is describing a partial-seed-singularity.

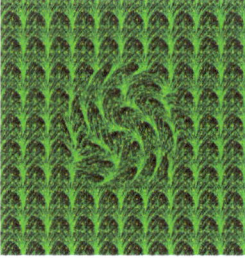

Figs 54,55,56,57,58,59,60: This series delves into several concepts including Partial-Seed-Singularity Pre-Big Bang Pre-Precipitation Zero-Speed Pre-Material-Fabric Pre-Genetic Code, the Light-Based-Singularity Superstructure,

520

Partial-Seed Singularity, Partial-Seed-Singularity Pre-Big Bang K-Entanglement Pre-Material-Fabric Pre-Genetic Code, The Seed-Singularity Architectural Forces K-Entanglement Pre-Genetic Library & Dynamics, Partial-Seed-Singularity Pre-Big Bang K-entanglement Pre-

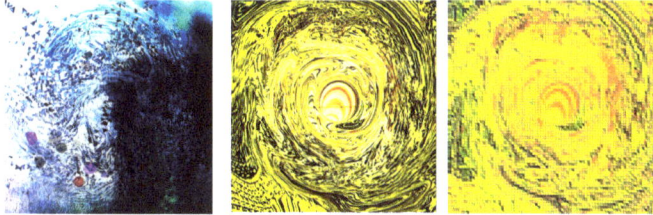

Material-Fabric Pre-Genetic Code, Organizational Uniqueness N-Entanglement Pre-Genetic Library & Dynamics, Big Bang Partial-Singularity Pre-Material-Fabric Pre-Genetic Code. Generated FBLEE-segments exist at the quantum-levels and can be thought of as exhibiting entanglement, referred to as FBLEE-entanglement (FBLEEE).

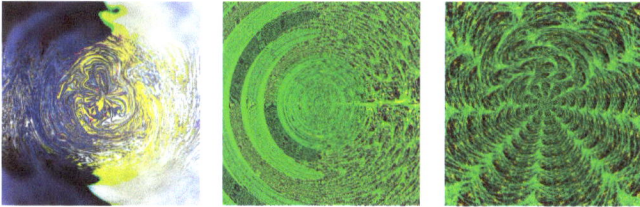

This is visualized here by a series of digital drawings with unique variation and an abstract image…'it is a bridge between form and formless, expressible and inexpressible, potential and manifest.'

Fig 61: The 'light layers implies the presence of libraries of pre-genetic pre-material-fabric information, and a much quicker and potentially richer evolution than would be possible if only c_0 existed.' ... 'that the emergence through matter, life, and mega-organization can proceed.' This is shown here as emergence of multiple entities resembling a star shape ascending on a smooth curve of evolution and resulting in the flame of Light.

Fig 62: 'An overview of various fourfold code projections that accelerate material evolution....' This is shown here as a rising wave of he sea but differently coloured and ascending into the sky.

522

Fig 63, 64, 65, 66: Quantum Particle Partial-Singularity Material-Fabric Pre-Genetic Code Iteration. In Equation 4.2.1, the starting point, x_U, is 'EM Spectrum', and the

ending point, x_T, is 'Quantum Particle' Here the naming of subscripts implies that the code generated is actually housed in the material-fabric. In Equation 4.2.4, Cells Partial-Singularity Genetic Code, is discussed. This whole section is visualized here as a novel

adaptation of the dance of Shiva as Nataraja - however here it is not in one pose but in several poses of hands and legs superimposed on each other, and there are four quadrants in his four sides : wherein subatomic, atomic, cellular

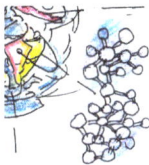

and particle structures are seen in four corners differently coloured and yet related and connected to each other.

Fig 67: LSTE signifies iterations(s) of the Light-Space-

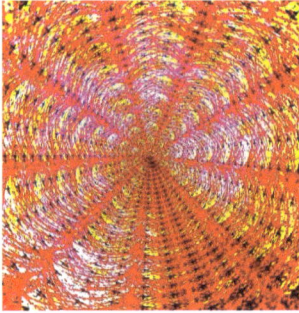

Time Emergence equation....through the device of entanglement, or perhaps even by the material-fabric code segments being embedded in cellular genetic code. This again is depicted in an abstract design where growth outward from the center is visible, and symmetric al in polar axes.

Fig 68: Humans Partial-Singularity Genetic Code Iteration. 'In Equation 4.2.5, Humans Partial-Singularity Genetic Code, the starting point, x_U, is 'Cells', and the ending point, x_T , is 'Humans'. Clearly there will be a vast number of iterations before humans emerge which also covers vast tracts of the plant and animal kingdoms that will cause an appropriate divergence in genetic code.' Taking the concept of Fig. 63 forward the multiple

layered multiple mudra and posed dance is also connected to a square with four quadrants for four colors and quadrality. It is also made to resemble the Archimedes homeo-figure...this is also then linked to Eq 4.2.6: Truer Individuality Partial-Singularity & Genetic Code.

Fig 69: Sustainable Global Civilization FBLEE-Based Partial-Singularity Genetic Code Iteration as in equation 4.2.7. It is shown here as Mother Earth caressing the face of a child which can be that of humanity and of all living and nonliving beings who are her children.

Section 5 Illustrations

Fig 70: This illustration is inspired by the following passage - 'In Equation 4.2.8, Super-Matter Partial-Singularity Post-Genetic Code, the starting point, x_U, is Molecules, and the ending point, x_T, is Super-Matter. All the preceding Light-Space-Time Emergence equation iterations progressively change the reality of matter through subtle quantization that allows space, time, energy, and gravity to operate

differently...The emergence of super matter is also discussed in the book Super-Matter (Malik, 2018a), with (4.2.8) included here to suggest computational realities yet to emerge.' In the illustration an origin point yields a number of wave patterns / sinusoidal curves that vary in amplitude and pitch. also at times coinciding with other waves. This results in a beautiful pattern as shown here.

Fig 71 : Generation of Electromagnetic Spectrum Partial-Singularity Material-Fabric, Pre-Genetic Code is discussed in this section wherein 'the Light-Space-Time Emergence equation (3.1.3) can be used to model emergence as it proceeds from simpler four-fold to more complex four-fold manifestation. In essence (3.1.3) can also be thought of as the equation by which the

biographies for any light-based-singularity, including all partial-singularities such as the electromagnetic spectrum partial-singularity are generated.' Here the attempt is to extend the previous figure into a next phase. The result is a complete object resembling a multicolored conch shell. (This is very satisfying as this symbol of the conch shell was arrived at through a different intuition, and used in previous books as well)

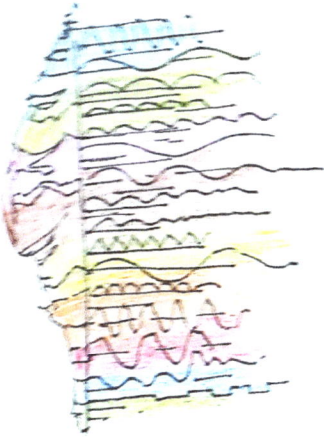

Fig 72: The same concept is expanded further to have a half section of the conch shell. It's various layers are then connected to several waves of different natures.

Fig 73: Generation of Electromagnetic Spectrum Partial Singularity 'Wave Archetype' Material-Fabric Pre-Genetic Code is depicted here. A wave archetype is symbolized as a small curling surface which transforms a two dimensional drawing into a three dimensional drawing.

Fig 74: Eq 5.1.3, Structure of EM Spectrum, is shown. Here is an X and Y axis corner with a continuously enlarging spiral which is also connected with a number of arcs with multi-color shades in it. What this implies is that the significance or intent of the different types of waves that exist can also be expressed by this general equation where the X and Y elements will vary. Here, Generation of Electromagnetic Spectrum Partial-Singularity 'Mass Potential' Material-Fabric Pre-Genetic Code is visualized.

Fig 75: Summary of Electromagnetic Spectrum Partial-Singularity Material-Fabric Pre-Genetic Code is

visualized here as a peacock dancing in the rain. The peacock feathers are in full bloom. The peacock feather is a beautiful harmonious perfect and bewitching expression with multifarious colors, and the bloom is associated with joy.

528

Section 6 Illustrations

Fig 76: Generation of Partial Singularities in the surfacing of matter (Section 6) is visualized here as a golden globe emerging out of the earth, and out of the earth consciousness. Of course one will be immediately reminded of the magnificent and such a significant Matrimandir at Auroville, near Puducherry.

Fig 77: Containerization of Matter in partial singularities is illustrated here as series of curves intersecting each other and resulting in a container like shape. Universality Versus Localization of Space, Time, Energy, and Gravity is depicted by the formation of some small and big solid objects along with fluid kept inside. The layers from the bottom to top show the several layers in the Light matrix and other equations including 6.1.1, 6.1.2.,6.1.3 and 6.1.4, wherein the relation of seeds to space, time, gravity and energy/matter is explored.

Fig 78 : Material localization at multiple scale emerges from the passage as follows - 'Since quanta hold within themselves the essence of what must be projected from

properties or function in higher-velocity light to lower-velocity more-material light,..' here the lines emerging like rays from a single point at the bottom curve up, and gradually resemble a globe of the earth filled with Light.

Fig 79: This composite-quantum in an expanding

universe is seen in equation 6.1.7 and the adjoining figures also explore composite-quantum at the black hole level which illustrates a relatively different quadrilateral. This is seen here as a whirl of lines expanding out from a black hole center.

Fig 80: '…from the perspective of the seed-based analyses foundational to a model based on multiple layers of light.'…'The electromagnetic spectrum or perhaps more accurately, the electro-magnetic-wavearchetype-masspotential spectrum, describes one of the first layers or translation of light into pre-genetic material manifestation.' This is visualized here as a intricate mandala with multilayered multidimensional

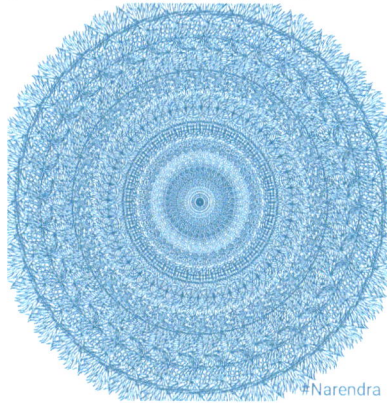

#Narendra

expansion from a seed in the center.

Fig 81: 'Generation of Quantum Particle Partial-Singularity 'Quark' Material-Fabric Pre-Genetic Code' is visualized here. This is shown in a mandala but with a variation of many arcs emerging out of it. A few oval shapes are rotated around the center to resemble atomic structure as well as petals. The use of faint subtle color shades help the effect.

531

Fig 82: Here a very novel multi axis perspective sketch is made to give a three dimensional feel so that the farther

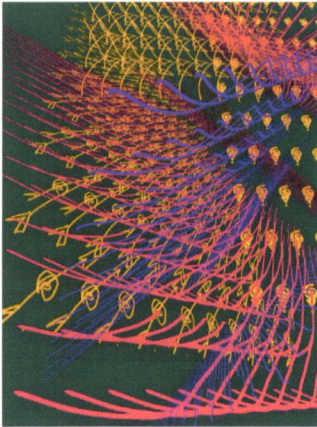

things look smaller. The design resembles several small elements and patterns which are self repeating over a space and time on a different axis of symmetry. Dark bottle green background and bright yellow pink and blue colors set linearly and in matrices, plus a spiral design, resembles a peacock feather bloom as well as the symbol for Saraswati, Goddess of Knowledge and Perfection. 'It can be seen first that the nucleus of an atom is made up of a combination of quarks. Specifically, a proton is composed of two "up" quarks and one "down" quark. Quarks have unusual names – up, down, charm, strange, top, bottom, with each subsequent pair belonging to a different "generation" of quarks.'

Fig 83: 'Atomic number in turn uniquely identifies the element from the periodic table.

Note that (6.2.3) already implies that Lines 1 – 5 in the Light Matrix (6.2.1) have been activated, and that the logic of the lepton-ecosystem will automatically precipitate into the material-fabric through the action of Line 6-7 of (6.2.1). This logic is none other than the pre-genetic code as specified by (6.2.3).' Here Quadrality is symbolized by making a symmetrical design about two perpendicular axis wherein several spherical or round shapes for atomic particles are used. The insertion of a lotus and pairs of eyes in all intersections add to the deeper significance and meaning. The use of faint, rich shades makes it even more attractive.

Fig 84: Generation of Quantum Particle Partial-Singularity 'Boson' Material-Fabric Pre-Genetic Code. 'The bosons on the other hand are thought of as force-carriers. They are what allow all known matter particles to interact. The three fundamental bosons in this category are the photon, the W and Z bosons, and the gluon. The carrier particle of the electromagnetic force or spectrum is the photon. The carrier particle of the strong nuclear force that holds quarks together is the gluon. The carrier particle for the weak interactions, responsible for the decay of massive quarks and leptons into lighter quarks and leptons, are the W and Z bosons.' Here an intricate mandala is drawn, and interrelations and interactions are more sukshma – subtle. The colors and the light spots used is an attempt to add to the effect. 'Bosons can be thought of as the precipitation of what creates relationship and harmony at the quantum level. Hence, they can be thought of as the precipitation or emergence of Light's property of Harmony at the quantum level and this is further expressed in Eq 6.2.6: Generalized Signature of Particle. So, we see that again the underlying properties of light - Knowledge, Power, Harmony, and Presence - emerge as quarks, leptons, bosons, and Higgs-bosons respectively.'

Fig 85: Summary of Quantum Particle Partial-Singularity Material-Fabric Pre-Genetic Code follows the quantum particle stage iterations of the Light-Space-Time Emergence equation. The fundamental architecture of these aspects is an emergence of the properties of Light from ∞. Here the circular pattern is multilayered and is a superimposition of curves and shapes which resemble the roots of a tree symbolizing consciousness on one side, and the curvature reminiscent of the Sun rays on the other side.

Fig 86: 'Line 3 in the Light-Matrix, C_K: $[S_{Pr}, S_{Po}, S_K, S_H]$, elaborates the sets for Presence, Power, Knowledge, and Harmony, each containing multiple elements.' Here various elements derived from the four sets not just define the behavior of photons but probably are also functions such as 'pervasiveness', 'multiple wave handler', etc. ..as a way of their being. This is shown here as a matrix pattern of several small linear and square entities (means logical) entities which are being progressively transformed by a flames of fire from one side and soon will be giving light. 'A holistic "ecosystem" with its own "boson logic" as it were. The wholeness has now precipitated into the material-fabric and is available to be consciously and unconsciously tapped into.

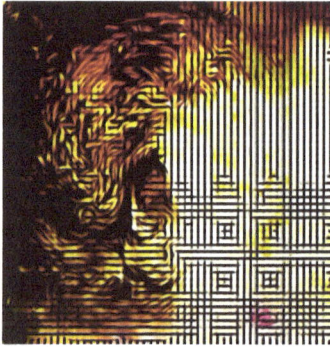

534

Fig 87: The Generation of Boson Partial-Singularity 'Photon' Material-Fabric Pre-Genetic Code. Here The photon is the carrier particle of the electromagnetic force and the electromagnetic force pervades everything and is foundational to the reality we are experiencing. This is shown here as a rising flame of light which has a small bluish pink beginning. The darkness at the bottom is slayed by the goddess of light whose image can be seen emerging in the flame.

Fig 88: The Potential Effect of Levels of Light are explored here in the Genetic-Type Information equation. There are

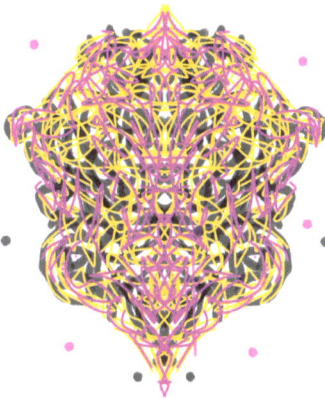

specific types of mutation and code-segments that are being generated as seen in sections 3.1.3 to 3.6.5 and more. The creation of the code-segments that defines the logic of the atom ecosystem are generating the atom partial-singularity. This is visualized here as a mirror symmetry of several intricate curves of different thickness and different color lines with a few atoms dropping out, resulting in a single shape eventually.

Fig 89: Generation of Atom Partial-Singularity 's-Group' Material-Fabric Pre-Genetic Code is explored here, and the s-Group consists primarily of what are known as alkali metals and alkali earth metals...here when they lose

535

electrons, energy is gained, but when the electrons are taken up by other atoms in proximity there is a lot of energy released. And so when stars shine it is because

they are transmuting vast amount of hydrogen into helium, both of which are s-Group elements. Thus, they are an emergence of light's property of power: here two s-shape curves are in a mirror image and also are shaped to resemble two back to back peacocks. The full blown peacock feathers shine with stars embedded in them. Stars and suns are also known to be the crucibles where all the different kinds of atoms are created. So these furnaces of power by virtue of their heat and high pressure are able to force electrons and protons and neutrons to come together to create all the different types of atoms known in the universe.

Fig 90: Equation '(6.4.6) already implies that Lines 1 – 5 in the Light Matrix (6.4.1) have been activated, and that the logic of the p-Group-ecosystem will automatically precipitate into the material-fabric through the action of Line 6-7 of (6.4.1).' This illustration is a mirrored image on bold white lined design, precipitating on the substratum of earlier designs. The central pink and yellow design is adapted from the earlier figure, 88.

536

Fig 91: Generation of Atom Partial-Singularity 'f-Group'

Material-Fabric Pre-Genetic Code is traced here with this observation - 'The f-Group comprises of the Lanthanides and Actinides.

Philosophically, elements in the f-Group consist of 6 lobes around the nucleus within which an electron may be found. 6 lobes will exist in multiple planes around the nucleus and suggests the notion of extended relationship and collectivity...the attempt to build larger and larger bonds within a small space. Considering this it is likely that the f-Group is an emergence of Light's property of Harmony.' This harmony element is visualized here as a mandala with closely knit thin threads of many colors which create a circle and omkar in the middle - the primordial sound which is associated with the creation of the material universe.

537

Fig 92: Summary of Atom Partial-Singularity Material-Fabric Pre-Genetic Code follows the atoms stage iterations of the Light-Space-Time Emergence equation, through equation 6.4.20, Active Atom Partial-Singularity Material-Fabric Pre-Genetic Code Segments. It seems that the generation of partial-singularities result in the surfacing of life. Here in a very bright mirror symmetry with colors and several patterns interacting. With a pair of prominent involutes at the top and an involutionary curved womb-like recess at the bottom of figure. The figure is symbolizing emerging life.

Section 7 Illustrations

Fig 93: 'The quantum-level computation and the generation of genetic code is associated with the emergence of life, and it happens through the cell,

complex human attributes such as thoughts and emotions / feelings, and most importantly in the uniqueness of individuality. As suggested genetic code articulates a biography, and in this case, the emerging biography depicting an increase in light-based singularities' sphere of influence through the emergence of additional light-generated "laws".' This passage has inspired this digital mandala with a tiny brown seed with an orange surface in the center, which blossoms in an intricate design with polar symmetry. There are small expanding spirals drawn in individual sectors throughout the illustration, as also on the outermost periphery.

Figs 94,95, 96,97,98,99: This group of sketches are inspired by discussion and equations in section 7 -

equations 7.1.3, 7.1.4, 7.1.5 and

539

till the end of this section. 'There is likely a build up of mass in such a manner that facilitates the creation of these systems. An assumption can be made that there is a 'red-mass' x-element and perhaps other y-elements that yield

the different systems when built up. Equation 7.1.3, Reproductive System – Red Mass Possibility' .. similarly 7.1.5 discusses orange intent,'… 'In its association with life the energy-profile of 'yellow' is suggested to engender qualities to do with 'digestive stimulant', and 'lymphatic stimulant' amongst possible others. Assuming that the primary or x-element has to do with 'stimulant', then various y-elements in combination can create the other engendered qualities. These relationships may be depicted in Equation 7.1.7. The mass-potential or body-part most linked to lemon is suggested to be 'joints', 'gallbladder', 'diaphragm' and 'vision', amongst possible others.' ..'An assumption can be made that there is a 'lemon-mass' x-element and perhaps other y-elements that yield the different systems when built up. Equation 7.1.12, Joints – Lemon Mass Possibility, suggests how this can be depicted: ..Note that (7.1.13 – 15) are proposed to be part of the material-fabric. As such they are of a pre-genetic type but also inform or even help form the cell-based genetic code...Cyan is associated with the wavelength 490 to 520 nanometers. In its association with life the energy-profile of 'cyan' is suggested to engender qualities to do with 'cleansing', ..Indigo is associated with the wavelength 430 to 460 nanometers. In its association with life the energy-profile of 'indigo' is suggested to engender qualities to do with 'sleep-inducing' and 'muscle relaxant', amongst possible others. Eq 7.1.17 shows Cyan Intent…The mass-potential or body-part most linked to cyan is suggested to be the 'arms and hands', 'parathyroid glands', and 'skin'. ..Blue is associated with the wavelength 460 to 490 nanometers. In its association with life the energy-profile of 'blue' is suggested to engender qualities to do with 'calming', 'anti-inflammation', and 'antipyretic', amongst possible others. (Eq 7.1.19: Blue Energy) ..The mass-potential or body-part more linked to violet is suggested to be the 'cerebral cortex', 'pineal gland', and 'memory, amongst possible others. So there is likely a build up of mass in such a manner that facilitates the creation of these systems. (Eq 7.1.27: Cerebral Cortex - Violet Mass Possibility)...They are of a pre-genetic type but also inform or even help form the cell-based genetic code. :

The effect of the electromagnetic spectrum on life is a clear example of how the sphere of influence of emerging partial-singularities increases...cohesive complexification of four-

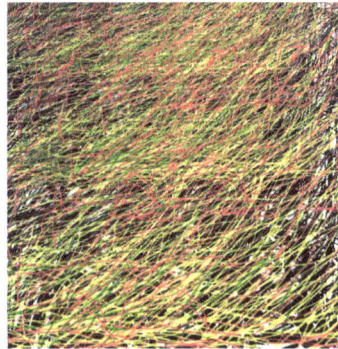

foldness in the pre-genetic material-fabric, with the comprehensive logic of the space-time-energy-gravity quadrumvirate with the emergence of the Big Bang partial-singularity. 'Variation in wavelength in the electromagnetic spectrum is the mechanism by which antecedent functional possibility encoded in Light moving at speeds greater than c materializes in an initial pre-material medium at light traveling at c. '

In this set of drawings instead of using mono colors like orange, cyan in each drawing which will be the obvious way of explaining the color discussed, deliberately all colors in the electromagnetic wavelength bands are splashed in each sketch on a dark background, and every time a few areas are later excluded or erased (except sketch for orange where white background is retained). This instinctive erasing has resulted in some significant shapes including those like a tree, conch shell, cloth fabric, butterfly etc.

Fig 100: Generation of Cellular-Level Partial-Singularity

'Proteins' Genetic Code - 'Proteins are the cells work-horses. Look anywhere in a cell and one will see proteins at work. Proteins are built in thousands of shapes and sizes, each performing a different function…some are built simply to adopt a defined shape, assembling into rods, nets, hollow spheres, and tubes. Some are molecular motors, using energy to rotate, or flex, or crawl…Proteins hence, exist for service, to bring about perfection at the level of the cell, are characterized by extreme diligence and perseverance, and so on. ..Proteins can therefore be thought of as a precipitation of system-presence at the cellular level.' This is symbolized in a sketch here with a peculiar curve resembling a rod plus a knot with a snail / spiral shape in the center. This knot ascends in a reverse curve which looks like a neck of an animal. Together this also resembles the front part of a horse, signifying the energy aspect of proteins.

Fig 101: The theme adapted for the earlier set till figure 99 is picked up again here to depict a deeply embedded relationship founded on genetic code that is created with the emergence of these capacities. Thus 'the quantum-level computation that suggests how light emerges as these capacities, how these precipitate into the post material-fabric genetic housing structure, and what the specific equation-segment function-based code manifests as' are illustrated here as a more intricate and multilayered representation of mirror symmetry with all the colors pf the EM wavelengths splashed in and later erased. Instinctively the portions radially emerge out.

544

Fig 102: Light's Emergence as Fundamental Capacities of Self : 'We become present to Presence through the device of sensation.' Eq 7.3.2 focuses on Sensation and leads to equations for hearing, seeing, tasting, touching, and smelling. But there is also a deeper experience of sensation and that is beyond and behind the sense and objects in fullness of light ..'with all its potentiality and possibility, in the smallest thing we look at? Do we see that the whole universe and more is present in all its fullness in the least thing that we easily ignore, or belittle, or loathe? When we touch things is it the seeming concreteness of the play of the particles or atoms or chains of molecules that we touch? Or is it the Love and Light and the vastness of all that IS that allows itself to be as a small corner that we touch so as to make infinity be felt by something so finite?' This beautiful thought is depicted here as a tiny branch of tree with little leaves which when touched, seemingly smile and shy away to go within themselves.

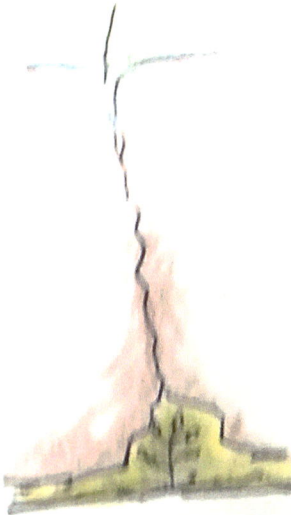

Fig 103: While discussing the generation of Fundamental Capacities of Self Partial-Singularity 'Urges, Desires & Wills' Genetic Code it is said that urges and desires and wills are in fact a play of the emergence of Light's property of Power. 'In the mystery of focus, the vastness of Light has projected itself in us into an apparent smallness that is in reality everything that is.' This is depicted here as the

545

small shrub below the surface of the earth which due to a strong urge and will to rise above, finds its way through narrow cracks in the ground to emerge up to the sky and the light.

Fig 104: 'And this smallness is trying through urge and desire and will to connect viscerally or even intentionally to other smallnesses that similarly are nothing other than the fullness of Light projected into a small smorgasbord of selected function. So the urge or desire for food, or companionship, or of possession, or of climbing a peak, is nothing other than Light's compressed property of Power, trying to reach more of the fullness that it is through a fulfillment of the urge or desire or will that it masquerades as. Hence urges can be represented as Equation 7.3.4.' This is visualized here as a person climbing a hill. At a moment he is tired and his head is touching the inclined surface, but soon he gathers himself, raises his head, looks at the top of the mountain and starts climbing again. 'Elements may be of the type of 'grasp', 'possess', 'deeply connect', amongst others.'

Fig 105: Generation of Fundamental Capacities of Self Partial-Singularity 'Feelings & Emotion' Genetic Code is discussed here. What can be more expressive of feelings and emotions than human eyes: here the eye of a human is drawn with several subtle lines and shades. The attempt is to feel the complexity of nature and the multiplicity of emotions.

Fig 106: 'Light's property of Harmony or Nurturing. Its instrument is the Heart, and it generates an array of emotions that are an indication or active radar of whether we are moving toward or away from a reality of harmony, whether based on our small self or some larger Self of Light. Gradually, by navigating with these emotions and feelings we can get to a state where we always feel positive emotions which basically means we have more truly entered into relationship with some larger continent of Light. An equation for feelings is as represented by Equation 7.3.5.' This is drawn here with a large pink heart radiant and bright which is embraced by a human being much smaller than the heart: as if the human heart can become as large as light or ocean or sky surpassing the human body and sense limits. This is shown in Eq 7.3.5: Feelings.

Fig 107: Generation of Fundamental Capacities of Self Partial-Singularity 'Thought' Genetic Code are depicted here as 'Thoughts are a play of the emergence of Light's property of Knowledge. Through the thought we can become greater or conceptualize things greater or begun to enter into relationship with some things other than our small self.' Eq 7.3.6 explores Thoughts and it is shown here as a human face with a skull with spirals and wheels of thought. Also included are some spikes and ascents, with a lightening idea having a rare presence.

547

Fig 108, 110, 111 and later Fig 125: These have a common thought thread which is used to visualize the ' Summary of Fundamental Capacities of Self Partial-Singularity Genetic Code' Here a robotic hand, and especially the palm and fingers in various grips or mudras is drawn. In contemporary popular depiction of robot arms it is generally shown precisely with machine parts made up of mechanical elements and materials. It is often seen with metallic shining pieces joined and put together with knuckle or spherical joints to resemble finger joints, wrist joints, palm, elbow, shoulder, etc. Such an AI symbolic robot palm is depicted here with a twist to suit the central idea of Life 3.0 where the singularity is going to be far more integral and real than any AI singularity. Here therefore it is seen to be holding an oil lamp for worship, wearing a silk thread tied at the wrist, or gently holding a lotus in its fingers in an open palm supporting DNA and other living organizations evolution, or in Yogic mudra like chin, chinmay, or abhay mudra…'Summarizing, after the fundamental capacities of self iterations of the Light-Space-Time Emergence equation are complete, the result is Eq. 7.3.7, Active Fundamental Capacities of Self Partial-Singularity Pre-Genetic and Genetic Code.'

Fig 109: Focuses on the section dealing with the 'final code segment following 'LSTE <...>' contains the code for all possible $(\sum x)$ sensations, urges, feelings, and thought, respectively, and depicts the growing biography by which the singular light-based edifice expresses its

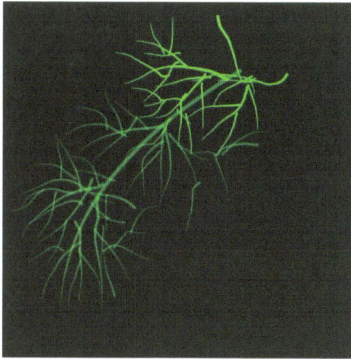

materialization, and at this stage, through the fundamental capacities of self'...Seen here as a leafless branch of a tree which also resembles lightening in the sky or a neural network--- which is spreading out, branching further and piercing the darkness of limitations and ignorance.

Fig 110: Generation of Truer Individuality Partial-Singularity refers to the series theme explained earlier in figure 108. It is taken further here...'At the core individuals can be thought of as projections of seeds formed in a vast continent of Light. So that core must always be there but it is often covered by surface dynamics of the physical, the vital, and the mental.' Seen here as a

palm holding lotus and in yoga mudra with thumb and first finger tips joining each other.

Fig 111: The hand is holding an oil lamp with luminous Jyoti or flames as if in a process of worship. In (7.4.1), 'the first line from the top, $C_\infty: [Pr, Po, K, H]$, specifies the architecture of truer individuality as will be further discussed in this chapter. Hence, Knowledge-type individuals are an emergence of Light's property of Knowledge, Power-type individuals are an emergence of

Light's property of Power, Service-type individuals are an emergence of Light's property of Presence, and Harmony-type individuals are an emergence of Light's property of Harmony. The fundamental architecture of these aspects hence, is an emergence of the properties of Light at ∞.'

Fig 112: Explores the matrix structure as described by -
'Line 3 in the Light-Matrix, $C_K: [S_{Pr}, S_{Po}, S_K, S_H]$, elaborates the sets for Presence, Power, Knowledge, and Harmony, each containing multiple elements...various elements derived from the four sets would define the architecture of Harmony-type individuals and could be functions such as 'driven to connect people together', amongst others, hence collectively describing Harmony-type individuals...Line 6, ($\downarrow R_{C_U} = f(R_{C_N})$) , specifies quantization between the layer where the seeds are formed, and the physical layer, and as explored in Chapter 3.5 and 3.6, will result in Line 7, $C_U: [P, V, M, C]$, hence changing the post material-fabric housing structure...The possibilities represented by Lines 1 through 5 hence concretize through the quantization represented by Line 6 to become or further enhance the

unique individuality with its subtle physical (related to Presence), vital (related to Power), mental (related to Knowledge), and connection (related to Harmony) aspects now existing in material reality typified by Light moving at c.' This passage has inspired the illustration wherein the vegetable kingdom, to amoeba, to a fish, to tortoise, to crocodile, to a bird, and then a deer, to a Homo Sapiens - human being is seen in a hierarchy. In one of the earlier figures an elephant was seen rescued by a divine force and this figure at first glance can remind one of that figure. But here there is a very striking difference. Instead of being a prey to a lower being and seeking help from a higher being to escape from the lower as was depicted in that previous sketch, here all are seen to be connected with a bond of mutual cooperation and symbiosis. The love and care with which the human is holding the deer in one hand while trying to reach the sun above is the zenith of this matrix and can be seen in this regard.

Fig 113: Further Service-Type, Power-Type, and Knowledge-Type individuals as well, with infinite variation in the precise nature of the core seed, as captured in Equations 7.4.3-5: '... (7.4.2 - 5) already implies that Lines 1 – 5 in the Light Matrix (7.4.1) have been activated, and that the logic of the truer-

individuality ecosystem will precipitate into the post material-fabric cell-based genetic structure through the action of Line 6-7 of (7.4.1). Fig 113 shows this as four leaf branch of a tree and each leaf has a symbol of Power, Knowledge, Service and Harmony.

Section 8 Illustrations

Fig 114: 'As generation of partial-singularities in the surfacing of complex organization...Having traced the computations resulting in the partial-singularity

emergences of space-time-energy-gravity, the electromagnetic spectrum, matter, life, and even human becoming as manifest in sensation, urges, feelings, and thoughts, and truer individuality, we now turn our attention to study the computation resulting in the emergence of partial-singularities related to complex organizations.' Here the four quadrants are seen with four symbols of Quadrality.

Fig 115: This figure is about Chapter 8.1 which examines

the generation of the stable mega-organization partial-singularity and about Chapter 8.2 which examines the generation of the sustainable global civilization partial-singularity. The nature of the partial-singularity is FBLEE-based, as opposed to being both FBLEE and MF-based as both these are post-human structures and will require some post-cell genetic-type structure. Section

553

9 on super-matter based partial-singularities explores the materialization of such functional richness further. Here an abstract design is rotated in symmetry using polar axes and a pentagonal structure. Number five is significant as alphabets in FBLEE, as five mahabhutas or cosmic powers, five senses, five fingers, five prana in yoga or five yajnas etc….it is also significant as five transcends four in a way which is symbolic of post human structures, the next future sustainable civilization and so on.

Fig 116: in Chapter 8.1 i.e. Generation of Stable Mega-Organization FBLEE-Based Partial-Singularity the emergence of the four properties of Light through the primary fourfold emergence of space-time-energy-gravity, the electromagnetic spectrum, through matter, through life, and even through human becoming as manifest in sensation, urges, feelings, and thoughts, and truer individuality is traced. At each of these levels it is a balanced combination of all four of the properties that creates stability and sustainability. This is shown here by a multidimensional, multilayered subtle but stable designed mandala with four basic colors (used as a theme through out this series). These are again seen to be present as distinct elements and yet merging seamlessly in the others to make the whole look integral and one.

Fig 117: This is clearly an artistic way to show Figure 8.1.1 Sustainability of CAS 'An equation for the sustainability of systems, $Sustainability_{Systems}$, where the interaction between the four families of forces is instrumental can be created.'

555

Fig 118: Specifies the architecture of mega-organization. Academic organizations are an emergence of Light's property of Knowledge, Political organizations are an emergence of Light's property of Power, Social organizations are an emergence of Light's property of Presence, and Commercial organizations can be thought of as an emergence of Light's property of Harmony. Line 3 in the Light-Matrix,

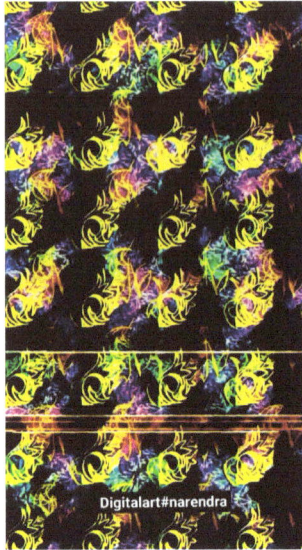

Digitalart#narendra

$C_K: [S_{Pr}, S_{Po}, S_K, S_H]$, elaborates the sets for Presence, Power, Knowledge, and Harmony, each containing multiple elements.

$$h \in \begin{pmatrix} [Q]: \text{Constructive zone,} \\ [Q]: \text{Constructive zone} \land \text{Constructive mutation,} \\ [F]: \text{Destructive mutation,} \\ [R]: \text{Random mutation} \end{pmatrix}$$

This is shown in a unique matrix wherein each element has an abstract design, and connects to other elements in a hierarchical basis - at first randomly, but then also destructively, and constructively.

Fig 119: This is again a very different and perhaps uniquely visualized illustration wherein the inspiration is the passage - 'Line 1 from the top in the matrix is simply a static form of (3.6.1) the Simplified Light-Space-Time Emergence equation. The static form is designated by 'STATIC', and implies that fundamental operations true of (3.6.1) are being highlighted in (3.6.5). In other words (3.6.1) already has all the operations highlighted in (3.6.5) in it, but by 'freezing' it by making it static, the essential

556

dynamics leading to possible mutations at the genetic level can more clearly be highlighted.'

Here a polygonal body with very aesthetically woven embroidered (if cloth) or etched (if metal or silicon chip) or engraved (if ancient stone or metal handicraft) design on it is progressively seen to be repeating on higher and higher levels. This creates a 3D effect at progressive layers and an impression of emergence from the bottom. This effect is added with a number of fine light rays tangent to the body and design. The golden intricate design on the body looking like a wafer or chip or an ancient ornament, is raised by gold and blue rays of light on the darker brown plus maroon background. This depiction is further clarified by the following passage - 'Line 1 is then subjected (×) to a determination of the dominant levels of light that may be active, designated by Line 2, ' $\left(\left(Y > U : Z_Q \right) \vee (Y \leq U : Z_F) \vee (Y = U : Z_R) \right)$'. Unpacking this, 'Y > U' implies meta-levels are active and as a result it is possible that Z_Q is going to take place (the subscript 'Q' implies quantum-level action). This also implies activation and potential change of FBLEE. The call from below, as it were, may invoke some function..'

.

Fig 120: Generation of Stable Mega-Organization FBLEE-Based Partial-Singularity Genetic-Type Code - as suggested by inductively derived (8.1.1), a sustainable organization is the result of two dynamics: maturity along each of the architectural force dimensions, and a robust interaction between all for sets of forces.' ..this is explained with the example of Silicon Valley. Here the two dynamics are visualized as two different symmetrical designs integrated in one. There is a circular design in white color with polar symmetry and concentric circles emerging from a center illustrating the first dynamic. There are also a series of rays crossing the first design showing a progressively enlarging motif which resembles a bird with spread wings or flowers in full blossom. The size of the motif differs as distance from the back and from the right and left changes.

Fig 121: '8.1.3 - 6 implies that Lines 1 – 5 in the Light Matrix (3.1.3) have been activated, and that the logic of the stable mega-organization-ecosystem will precipitate into FBLEE and may also impact post material-fabric cell-based genetic structure through FBLEEE. ' this is visualized here through a multilayered multipatterned design symbolizing post material fabric cell based genetic structure where bright yellow flowers are notable as a source of light at regular intervals as per seeds in rows and columns.

Fig 122: 'Summarizing, after the stable mega-organization iteration of the Light-Space-Time Emergence equation is complete, the following code-segments will have been generated as specified by Equation 8.1.7'…here an intricate mirrored mandala is digitally drawn to convey this point of 'Active Stable Mega-Organization Partial-Singularity Genetic-Type Code:'…here the Active Stable Mega-Organization FBLEE-Based Partial-Singularity Genetic-Type Code.' It is a variant of figure 92 discussed earlier with more clarity in demarcation lines after freezing or crystallizing the earlier version.

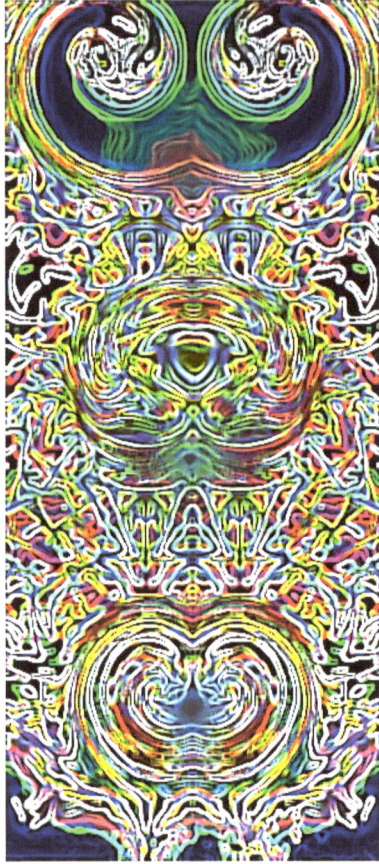

Fig 123: This illustration references 'Generation of Sustainable Global Civilization FBLEE-Based Partial-Singularity in Chapter 8.2. It is visualized with rows and columns matrix as in one of the earlier illustrations. But here one can see more illuminating motifs and brighter color scheme…this is to signify Light's Emergence as Sustainable Global Civilization…Material fabric housing structures, so long as the bases involved are driven primarily by a meta-level.

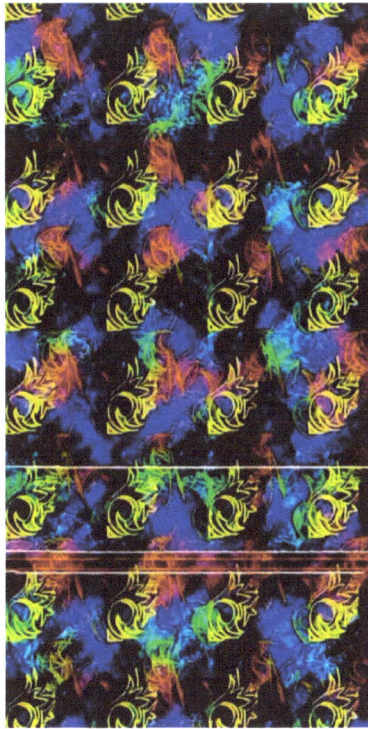

Fig 124: The passage inspiring this illustration is - "Starting with the Light-Matrix, the top left-hand matrix in (4.2.7), the first line from the top, $C_\infty: [Pr, Po, K, H]$, specifies the architecture of sustainable global civilization as will be further discussed in this chapter…This requires maturity by large organized parts of the world, whether nations or regional blocs. The maturity is such that the fourfold properties of Light are adequately expressed.' Here an intricate Mandala is drawn with multilayer, multidimensional

design and polar symmetry. The addition of a couple - male and female pair - symbolizes the genesis of Adam and Eve or Manu and his consort who were the nucleus of all the later growth of human race and human civilization.

Fig 125: 'The fundamental architecture of the combination of such emergences, which as per (8.1.1) is required for sustainability, is hence an emergence of the properties of Light at ∞. …Line 3 in the Light-Matrix, $C_K: [S_{Pr}, S_{Po}, S_K, S_H]$, elaborates the sets for Presence, Power, Knowledge, and Harmony, each containing

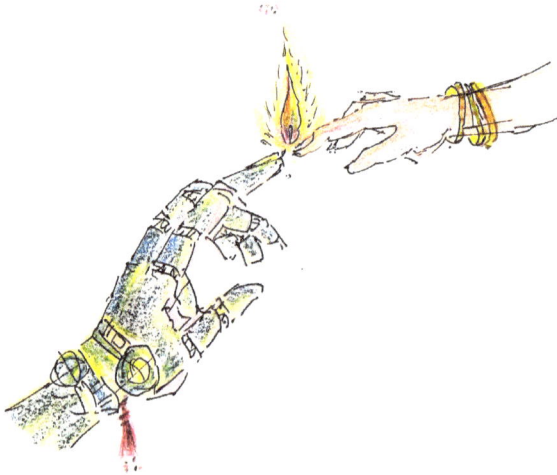

multiple elements…As a reminder FBLEE exists at that barrier where the speed of light is c, but at the quantum-level only. Note that just as Line 6 represents a process of quantization relating the layer of reality created by Light traveling at c with the antecedent layers, so too Lines 2 and 4 as previously discussed, also represent quantization of a more subtle kind that ultimately plays a critical part in allowing FBLEE to express infinite diversity. This harmony and touch of suprarational, post human with the human led AI singularity revolution is depicted here by resemblance to the famous ancient painting of a human being connecting to the divine through the touch of a finger (Michelangelo, *The Creation of Adam* as seen on the Sistine Chapel ceiling in the Vatican). The addition here is a light flame which is a symbol of a new pathbreaking illumination due to this harmony.

Fig 126 and Fig 127: The two illustrations are related as the later is blossoming from the seed of the former. The passages which prompted this is - 'Typically it is the process as captured by the Space-Matrix that will determine if Line 6 is activated. Specifically patterns at the untransformed layer, U, will need to be overcome, as specified by the second-line from the bottom of the Space-Matrix: $(\uparrow > P_x)$.'

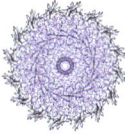

Fig 127: Here the seed is in the center and is used to expand every tangent to the seed circle, and points on the outer surface centripetally, like throwing out fluid from an impeller. The stable self driven unit symbolizes Generation of Global Sustainable Civilization FBLEE-Based Partial-Singularity Genetic-Type Code.

Fig 128: 'Thus, those civilizations that have endured typically have a balance of all four families.' (Sri Aurobindo, 1971). Civilizations that have become extinct typically have had a focus on few drivers of innovation. Resources, climate changes, diminishing support from friendly societies, hostile neighbors, and inappropriate attitudes for change (Diamond, 2005)…"societies decay

because of their over-reliance on structures that helped them solve old problems.' (Toynbee, 1961). This section has inspired this intricate mandala with a calm but

assuring oil lamp burning in its center. The layers of brown and blue are distinct and so is a small window or bridging outlet which supports the lamp and a octagonal star shape (composed of two squares at 45 degrees offset: 4 signifies Quadrality of four factors) created in it.

Fig 129: 'Summarizing, after the sustainable global civilization stage iteration the Light-Space-Time Emergence equation is complete, the following code-segments will have been generated as specified by Equation 8.2.5, Active Sustainable Global Civilization

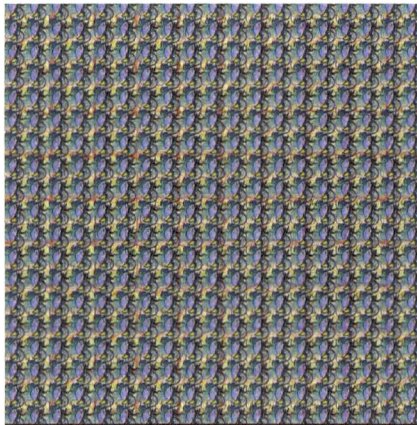

FBLEE-Based Partial-Singularity Genetic-Type Code:'…this has further led to section 9: Extrapolation to Super-matter based partial-singularities. Here the rows and columns matrix in earlier illustrations is made further fine and intricate.

564

Section 9 Illustrations

Fig 130: 'Space-time emergence equation, which with the emergence of partial-singularities is also known as the partial-singularity generation equation… Components of such biographies include the essential architecture of the singularity generated from the four principal characteristics of light…unique seeds created from combinations of elements derived from the four sets, and the essential dynamics of maturation culminating in possible quantization.' This is visualized here as a rows and columns matrix with a design that has zoomed in the figure 129.

Fig 131: This illustration is related to 'emerging living cell partial-singularity from Chapter 7.2, Generation of Living Cell Partial-Singularity.' Shown here is a luminous thin linework resembling a leafless branch of a tree on the backdrop of a lush green forest.

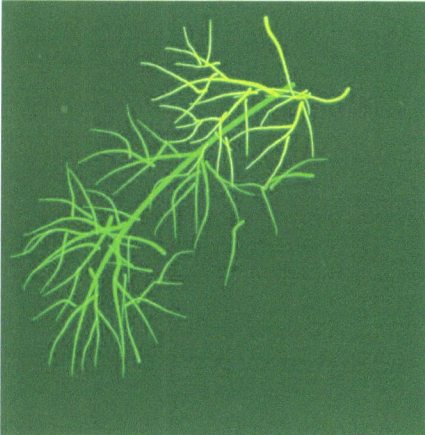

565

Fig 132: The illustration shows a fabric with intricately interwoven linear patterns which also resembles a jig saw puzzle embedded in a circuit board. The following passage triggered the illustration - 'The visceral urge, then, may stimulate interaction with a collective-intelligence or 'specific-species-intelligence meta-function' at FBLEE, which allows the generation of a new and specific 'predator-overcoming meta-function' so that the hypothetical insect in question can now go through an adaptation to survive at least some kinds of predator attacks.'

Fig 133: 'In this example, S_{System_P} relates to a set of Polysaccharides, as introduced in Chapter 7.2, and suggests that the chains of sugar molecules will adapt to become a shell to protect the insect. The primary element X, is therefore an element of the set or Polysaccharides.

$S_{System_{Pr}}$ refers to the set of Proteins.

S_{System_K} refers to the set of Nucleic Acids.

S_{System_N} refers to the set of Lipids. Y as the secondary element will invoke the action of some

566

proteins, some existing polysaccharides, some nucleic acids, and some lipids in bringing about the adaptation as specified by $Sig_{hard-protective-shell}$.' This is visualized as multilayered flower petals with beautiful interconnected and color bands separating a section of petals.

Fig 134: 'Equation 3.5.7 models gravity-quantization. Recall that gravity-quantization has to do with inter-relation between the species and surrounding objects and will change the very nature of gravity to allow a subtle new balance in the species interaction with its surrounding so that the deep urge of the species can more easily be fulfilled.' This is visualized with a perspective drawing of three rows and columns. A closely knit design with green color shades signifies life is pervading it, where the nearer they are to ground the size of design motif increases to give a three dimensional effect and to show the pull of gravitational force.

Fig 135: 'Leveraging the discussion on the broad lines of the emerging sustainable global civilization partial-singularity from Chapter 8.2, Generation of a Sustainable Global Civilization FBLEE-Based

567

Partial-Singularity, this chapter will specifically consider how aspects of this biography are determined in the case of the development of a sustainable global civilization. Such development signals the successful change in biography of the sustainable global civilization partial-singularity, … Such an urge-to-quantization process sheds insight into an increasingly important aspect of the dynamics of partial-singularities, especially when humans begin to play a part.' This is visualized with a mandala with several threads emanating from the center.

Fig 136: 'While the key aspect, or primary X-element in the nation-essence-foundation meta-function may be an element from the set of Presence, such as 'nation-essence appreciation', there will be many parts or sub-functions, that could be represented by Y-elements, such as the structure, processes, institutions, culture, cohesion that

will also need to be created to ensure such nation essence foundation. Cohesion, as a sub-function, represents an important denominator in breaking patterns at the individual and collective levels and as in Equation 9.2.2, Nation-Essence-Cohesion, is represented with the primary element belonging to the set of Harmony or Nurturing.' The cohesive design with polar symmetry here has a light emanating from the center which results in the manifestation on the physical plane.

Fig 137: 'Hence, S_{System_K} relates to the set of Thoughts required to contextualize and frame nation-essence. $S_{System_{Pr}}$ refers to the set of Sensations required to assess

and interpret the constantly arising signs of the new development – the sensory cues as it were that will allow any individual or collectivity to sense that they were on the right path. S_{System_P} refers to the set of Urges and Wills constantly required to ensure that the goal of cohesion was attained. S_{System_N} refers to the set of Feelings and Emotions that will need to be generated to attain cohesion.' This is shown in this illustration by a four quadrant symmetrical design wherein the fine lines in each sector and their pattern formation symbolizes thoughts and sensation.

Fig 138: 'The new meta-function changes the very nature of matter, which is quantized to allow the material fabric of existence to promote the new meta-function of cohesion in unimaginable ways. This quantization of energy, resulting in a subtly different matter is modeled by Equation 3.5.6, reproduced here for convenience...' This is visualized here as an atomic structure with several concentric circles. The design in it is distinctly thrown out by centripetal forces are splashed out from each blade of such an impeller. Subtle and very soothing shades of colors are used to add to the effect.

Fig 139: Reference 'Chapter 9.3: 'Macro-Conditions for Transitioning to Super-Matter Based Partial-Singularities Super-Matter is positioned as being fundamentally different from matter. While matter can be thought of as the result of Nature's automatic working, super-matter can be thought of as the result of a conscious will and cohesive wanting that causes a deliberate process of quantization by which the very fabric of matter is changed. Quantization occurs in any case even when Nature's automatic working drives change. But with conscious will this process can be accelerated, broadened,

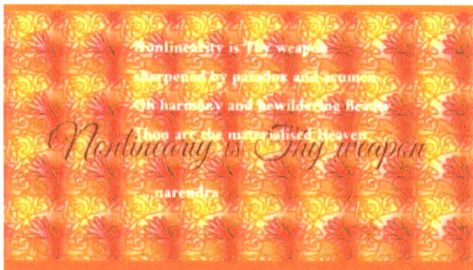

570

and heightened, so that a wider and higher possibility of function may consciously materialize.' This insight is presented here in a rectangular array wherein a each point of intersection is a budding flower cluster signifying new creation and super matter. This is done with bright color combinations of orange and red, and with a four line poem inserted as was intuitively thought by the illustrator.

"Nonlinearity is thy weapon
Sharpened by paradox and acumen
Oh, Harmony and bewildering Beauty
Thou art the materialized heaven"

'Equation 4.2.8 suggests computational realities yet to emerge. The essential action of space-time-energy-gravity quantization can cause the materialization of potentially infinite four-base logic-encoding ecosystems. It is conceivable that such a variation of space-time-energy-gravity quantization.'

Fig 140: 'Coupled with an increasing ability to easily move back and forth between the material and antecedent realms, perhaps the outcome of an enhanced function driven by an initiating will itself, makes the need to house genetic information in a form such as DNA alone, burdensome.

It is conceivable that four-base logic-encoding ecosystems, or even the ∞-entanglement, K-entanglement, and N-entanglement libraries may be able to more directly act at the material level in some composite subtle-material post-genetic form that also includes DNA. This possibility is in general referred to as post-genetic code.'...'matter becomes more dynamic and adaptable as a result of this.' This is visualized here as a cube with several smaller and hollow cubes of different colors fitted in its each sector. When they interact and connect, one can see a curious formation of a star - two triangles intersecting with one descending and the other ascending. This a great significant symbol related to Sri Aurobindo an The Mothers Darshana of The Life Divine.

Fig 141: 'As explained by Neil Turok in his book The Universe Within (Turok, 2012), Euler's formula reproduced below, can be used to model many naturally occurring phenomena because of its sinusoidal oscillation between narrow bounds as x increases. Reproducing Euler's formula:

$e^{ix} = \cos x + i \sin x$

The sum of the squares of the ordinary and complex parts, on the right side of the equation, is one. In quantum theory this ensures that the probabilities for all possible outcomes add up to one. Hence this formula is useful when summarizing the macro-level effects of (3.6.1) which necessarily has diverse drivers of phenomenon.' This is picked up and related from the earlier series of similar figures erased or etched from multi coloured background in darkness. The tree etched is symbolic of the life as well as the Universe within.

Fig 142: 'From discussion in the previous chapter it is clear that if the cumulative effect of overcoming habitual

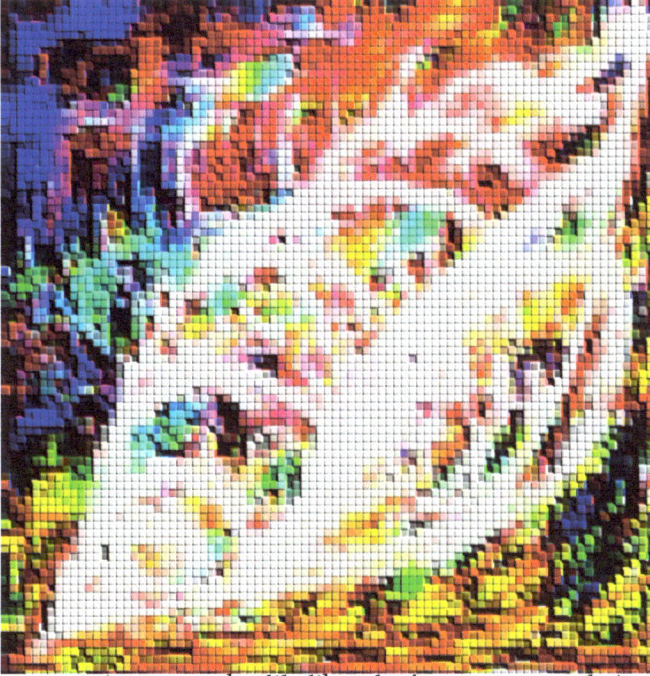

patterns increases the likelihood of super matter being formed, this implies the increase too of four-fold quantization. Space, as the seeding ground of knowledge, may therefore also require expansion to continue to allow seeding to take place.'...'Keeping in mind the complexification of matter as it journeys from the field-level through the quantum-, atomic-, and cellular-levels, and beyond, it may be possible to re-interpret the supposed expansion-contraction dynamics of cosmology in relation to the mathematical model presented in this book.' This is again picked up from an earlier section and series, and the etched or erased figure here resembles a conch shell as well as fish...both are very profound symbols.

Fig 143: 'This implies that the manifested-set, MS, is growing at a certain threshold level. Assuming this threshold level, $MS\_Growth_{Threshold}$, is a property of space related to dark energy it may be possible to restate the condition of cosmological expansion and contraction. Hence, so long as the MS is increasing at a certain rate that

exceeds....' So comes another illustration based on the theme of the series discussed earlier. Here it is more abstract and formless. It resembles the thrashing against rocks of rising waves in the ocean, which calms down temporarily only to come back with renewed vigor with the onslaught of a new series of waves.

574

Fig 144: 'Further iterations of (4.2.8) will lead to a possibility of all patterns in the Space-Matrix being broken, thus leading to the Second Singularity. This possibility will be taken up in Section 10, The Second Singularity.' Here the earlier patterns are broken in small granules with still some traces faintly visible, assuring a new assembly / creation.

Section 10 Illustrations

Fig 145: 'The Second Singularity by highlighting the difference with an AI-based singularity, suggesting a set of representative operators that will expedite its emergence, and tracing its emergence…' 'Hence: four-fold emergences of properties codified in Light have all contributed to the emergence of the human. Here again the earlier patterns are broken in small granules with still some traces faintly visible, assuring a new assembly /creation. 'Essentially the materialization of potentiality in Light may lead to The Second Singularity, the singularity whose power will far surpass any AI-based singularity precisely because of the level and sophistication of integration, and the continual real-time re-computation of reality and generation of genetic-type information driven by enlightened heart, will, and mind dynamics.'

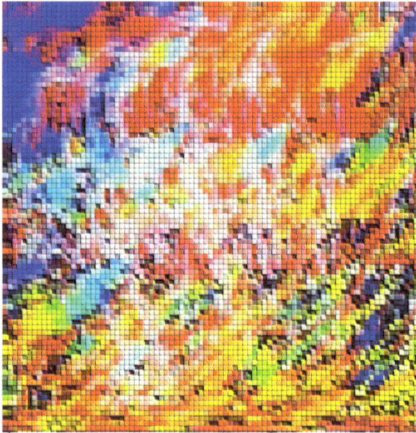

576

Fig 146: This is to visualize Chapter 10.1: Summarizing Fourfold Adherence and Implicit Wholeness in Light-Based Singularities. The preceding sections have

elaborated the architecture of light-based singularities that are all built from the four essential properties of Light. 'There is hence, "singleness" in the emergence of any singularity that binds it with all previous light-based singularity so that a subsequent partial-singularity can be thought of as having a greater sphere of influence than preceding ones by virtue of all previous emergent laws being active also in it.' This is visualized here as a dark purple background on which a fair design is symmetrically emerging. It has a light source with a mantra or a mystic formula to inspire it. This will dispel darkness and will usher new light creating a new universe.

Fig 147: Figure 10.1.1 Fourfold Adherence and Levels of Implicit Wholeness in Representative Partial-

Singularities, is artistically represented here by a comparative architectural masterpiece from Indian architecture. The hierarchy of patterns reminds us of Fractals. 'Equation 10.1.2, System-Power (Time) at Space-Time-Energy-Gravity Level, summarizes how power is related to 'maturation of seeds', since anything that gets in the way of the maturation will generally be overcome. This is shown here in a novel way with a typical ancient architecture with several layers and hierarchy of elements in it, as if it is a mathematical formulae in matrix form.

Fig 148: This figure again is based on a similar concept as in an earlier series now made more subtle. It is broken in essential building blocks and blurred to show process of dissolution and integration one after the other. '…Hence, the general relationships of the four-fold wholeness may be understood by knowing just two variables, C, and either, λ or v….The emergence of the EM Spectrum adds another fold as it were to the partial-singularity biographies that can progressively be seen as a complexification of four-foldness.'

Fig 149: 'At time, t = 13.8 x 10^9 years, there is a fifth clear expression of the same four-fold order as the bases of an even more complex organization, that of cellular life and all that is founded on it.' This too is connected with an earlier figure with four foldness emerging. Here it is modified to be more subtle and granular or having bytes of information in each element to create the whole.

579

Fig 150: 'By contrast AI-based singularities are fragmented in that there is no connection with any technology that is not mind-based. The essential elements of AI-based technologies are all derivatives from capacities created with the emergence of a "mind" partial-singularity and would include sensing, memory, and the use of logic and empiricism.' This is visualized by another abstract drawing emphasizing 'fragmentation' and 'reductionism' which are the essence of any mind based model. This is visualized with a bright color combination and fragmentation of every element as if a glass collage is broken in several further smaller elements.

Fig 151: 'Energy could have elements such as 'concentrated', 'diffuse', 'intense', 'subtle', amongst possibly infinite other energy-elements. The Set of Energy, Set_{Energy} , is summarized by Equation 10.2.7, Set_{Energy} .' This is visualized as several rings intersecting and interacting with each other. Elements are bursting out of the system formed by this interaction.

Fig 152: 'In the Second Singularity emergences are characterized by integrated light-based functioning and

would have deeper access to the founts of innovation resident in deeper libraries at the antecedent layers of light. Emergences would be the tip of a sophisticated light-based edifice where no characteristics of previous partial-singularities would be absent. The ability to effectively concretize knowledge in real-time, based on a foundation....' The illustration depicts an ancient stupa or gompa - a pyramid like structure for prayer and meditation in Buddhist faith, surrounded by many rectangle-like houses and roads in the surrounding area. On the other hand this can also resemble a hemispherical astronomical planetarium or hemispherical atomic reactor site...it may be even seen as a digital circuit with several elements, chips, drivers etc. embedded and assembled on it.

Fig 153: 'Leveraging Mathematical Operators to Expedite the Arrival of the Second Singularity...on modeling the

mathematical structure of light-based-singularities, ...These operators are the result of considering some of the dynamics of each of the levels in the seed-singularity, and integration of the multiple levels of the overall light-based-singularity mathematical

model. This is then developed in Eq. 10.3.2: Fullness: 'To see such 'fullness' in everything is to begin to see as

581

perhaps Light sees, and this will assist in closing the gap between the material layer and layers of Light antecedent to it.' This is visualized here with a beatific color combination of early dawn when the sky is a faint reddish blue shade. The sun slowly spreads light all around as it rises up and bring the surrounding trees and shrubs to life.

Fig 154: 'In general for any two points 'A' and 'B' it can be further suggested that the Fullness behind A is the

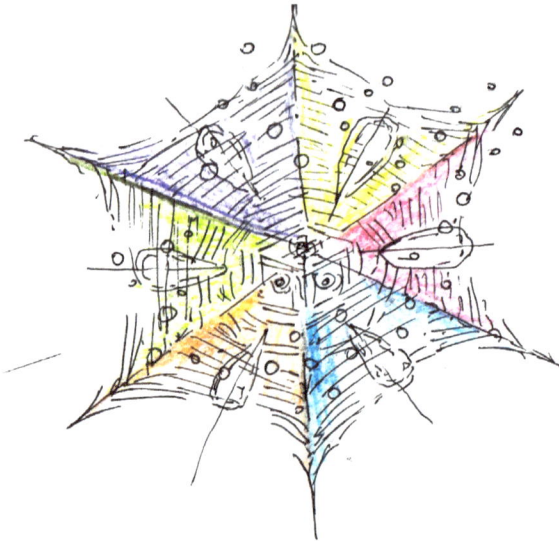

same as the Fullness behind B, as in Equation 10.3.3, Equivalence of Fullness. This suggestion may also be arrived at by considering Einstein's General Theory of Relativity (Einstein, 1995) that states: "All bodies of reference K, K1 etc. are equivalent for the description of natural phenomena (formulation of the general laws of nature), whatever may be their state of motion."
Equation 10.3.3: $Fullness_A \equiv Fullness_B$ - Equivalence of Fullness. This is visualized here as a novel figure with a cube with six faces (four sides, up and down, 0 results in a web of a hexagonal shape with a full figure (fullness) formed in each sector. All sectors are identical and fullness in one is equivalent to that in any other. A web of such shape also has deeper significance as it is used as one

of the symbols of how the universe was created and is functioning as per one of the ancient Upanishads.

Fig 155 and Fig 156 are visualizations of unique time and space coordinates and hence characteristics, as in Figure 10.3.1: Figure 10.3.1 shows 'Uniqueness in Time-Space Continuum.' But even for any organization, Vaastu shastra and Shilpa shastra i.e. science of architecture is closely connected with fractals. Eq. 10.3.8: depicts Fractal 'Organizational complexity refers to an order of magnitude change as in from a person to a team, or from a team to a business unit, and so on, for example. In Nature such fractal arrangements abound in the way the human body is constructed to the very structure of galaxies (Briggs, 1992).' This is shown in these two figures. The first one modifies the curve in an ancient gompa or shrine or temple shikhara. The second takes it further to more clearly link fractal geometry with all layers of a temple.

Fig 157: "X' can be thought of as an element from the Set of System-level architectural forces.'…'*Flowering* refers to the possibility that any time-space boundary-n, depicted as TS_n (as in Fig. 10.3.1) will have more potential or possibility associated with it than a time-space boundary-(n-1). Putting this into equation format yields Equation 10.3.12.' This is visualized with hollow circles in different rows which vary in number and

position, but have the same pitch. As one starts joining these in all permutations and combination, it becomes a beautiful fullness of a figure, like a flowering or a light bulb, chandelier, or an artistic lamp shade.

Fig 158: '*Higher* refers to the possibility that over time the direction will always tend to move to a higher meta-level. This can be summarized by using the notion of the core-matrix (2.6.7) yielding Equation 10.3.13. ' This as well as equation 10.3.14, Nurturing-Based Mathematical Operators, is visualized here as two parts of a grain or cereal. These are mirror images of each other and four involutes showing quadrality. The single pulse or cereal may look small but has potential to become a complete / full tree. The four color scheme which is consistently used

584

in this series is seen here vividly as it is seen in most of the earlier illustrations.

Fig 159: "*Remember*' has to do with remembering that there is something in each organization that existed before the existence of any organization, and further, that there is something in each organization that exists in every other organization. This captures a key dynamic that accompanies changing biography in any partial-singularity. This dynamic can be thought of as going back to a time-space moment of zero, and subsequently of expanding into the Time-Space Continuum keeping that connection in mind.' Here the earlier figure concept of grain/cereal in two mirror parts is taken further and imposed on the 'remembrance' curve. (Figure 10.3.2: Remembrance in Time-Space Continuum)

Fig 160: This is based on the beginning of Chapter 10.4: Altering the Microcosmic-Macrocosmic Balance. The link between the expansion of the universe and the changing nature of matter due to continued precipitation of fourfold functional richness which was explored in the previous section is taken further here. So, the 'inter-galactic-bangs in such a functionally rich universe precipitates cycles of development based on other speeds of light, that in turn sets

up a more conscious fusion of the multiple layers of existence so created, into one incredible Universe where an entirely new type of super-matter can form.' This is seen here in another novel creation wherein four quadrants of four colors have elements from farthest to the closest to the center (i.e. from macrocosm to microcosm), and have four trishul or trident shapes in quadrants. There are threads or ribbons or rings linking the elements of this trident across quadrants. All point to one center which has a rhombus (four sides for four elements and quadrality) in the center of the entire figure.

Fig 161 & Fig 162: 'The final code-segment in Equation 10.4.2, Future Code, is a generalization (Σ) of all the development that can take place along known

dimensions of physicality, vitality, mentality, and integrality. This equation implies a constant stream of materialized functional-richness, an evolving super-matter, and an evolving post-genetic structure for encapsulating such future code.' The discussion about Ouroboros follows this - the ancient and mystic symbol of

586

a snake devouring its own tail - and means many things including eternity, cyclic universe and so on. Here it is drawn in two variants different from the normal

depiction of one serpent shown eating its own tale. There are two or more serpents forming a multilayered pattern shown in different colors. This variant of multiple serpents is found in some of the ancient sculptures and architectures across cultures and civilizations.

Fig 163: This is related to Chapter 10.5, The Second Singularity Biography. 'In a society driven by rapid technological advancement it is easy to come to believe

that the future of life will be synonymous with such advancement. There has been much talk of an AI-based singularity – an event or point in time in which human capability will be marginalized by a singular, global, vastly intelligent AI. But this ignores the fundamental light-based nature of Life.' Here it is visualized as a series of dots which are joined in a way similar to traditional rice flour rangoli like Alpana. The resultant shape also reminds of multiple multilayer ouroboros serpents. It's also using mirror symmetry, containerization, hierarchy of wholes...several concepts which were used in many of the previous sketches in this book. In a way this figure attempts to summarize the various sections in the book.

Fig 164: The concluding sketch is Rudra (used in one of the previous books) who is the destroyer of old

foundations and paves the way to new reality, new foundation. He is dancing with Sati on his lap: a form of Aditi and Adimaya -- her sacrifice for self-identity and honor has triggered his rupture dance and majestic stillness which led to manifestation of this existence on a new foundation, a new singularity. His third eye signifies opening to Kutasth Chaitanya, the higher consciousness. This is also associated with the Ajna chakra in Yoga and an aspiration and strong Will to transcend and transform, and create a new light-based singularity to dispel darkness.

Relevant Background and Follow-up Information

The Author's Early Books

1. The Flowering of Management
2. India's Contribution to Management

The Fractal Series

1. Connecting Inner Power with Global Change: The Fractal Ladder
2. Redesigning the Stock Market: A Fractal Approach
3. The Flower Chronicles: A Radical Approach to Systems and Organizational Development
4. The Fractal Organization: Creating Enterprises of Tomorrow

The Cosmology of Light Series

1. A Story of Light: A Simple Exploration of the Creation and Dynamics of this Universe and Others
2. Oceans of Innovation: The Mathematical Heart of Complex Systems
3. Emergence: A Mathematical Journey from the Big Bang to Sustainable Global Civilization
4. Quantum Certainty: A Mathematics of Natural and Sustainable Human History
5. Super-Matter: Functional Richness in an Expanding Universe

6. Cosmology of Light: A Mathematical Integration of Matter, Life, History & Civilization, Universe, and Self

The Application of Cosmology of Light Series

1. The Emperor's Quantum Computer: An Alternative Light-Centered Interpretation of Quanta, Superposition, Entanglement and the Computing that Arises from it
2. The Origins and Possibilities of Genetics: A Mathematical Exploration in a Cosmology of Light
3. The Second Singularity: A Mathematical Exploration of AI-Based and Other Singularities in a Cosmology of Light
4. Triumph of Love: A Mathematical Exploration of Being, Becoming, Life, and Transhumanism in a Cosmology of Light

The Artistic Interpretation of Cosmology of Light Series

1. The Mandala Illustrated Story of Light
2. Musings on Light: A Meditative, Non-Mathematical Summary of a Cosmology of Light
3. The Illustrated Oceans of Innovation: The Mathematical Heart of Complex Systems Depicted in Indian Arts
4. Emergence Illustrated: A Mathematical Journey from the Big Bang to Sustainable Global Civilization Depicted with Indian Mythological Arts
5. The Dawn of Flame-Beings: Mythological Musings Based on a Cosmology of Light

6. Quantum Certainty Illustrated: A Mathematical Journey Through Natural and Sustainable Human History Depicted with Art
7. Super-Matter Illustrated: Functional Richness in an Expanding Universe Depicted with Drawings
8. Cosmology of Light Illustrated: A Mathematical Integration of Matter, Life, History Depicted with Art
9. The Emperor's Quantum Computer Illustrated: An Alternative Light-Centered Interpretation of Quanta, Superposition, and Entanglement and the Computing that Arises from it Illustrated with Art
10. The Origin and Possibilities of Genetics Illustrated: A Mathematical Exploration in a Cosmology of Light
11. The Second Singularity Illustrated: A Mathematical Exploration of AI-Based and Other Singularities in a Cosmology of Light Enhanced with Art

Note on Genesis of Books

In the earlier stage I wrote 'The Flowering of Management' and 'India's Contribution to Management'. The impetus for both these books was similar in that they were reactions to the environment that I was placed in at the time. When I first began working in the corporate world the reality of the environment struck me as decidedly anachronistic. I had a different sense of what life could offer and wrote 'The Flowering of Management' to capture aspects of a vision I thought corporations and money existed for. Similarly, when I wrote 'India's Contribution to Management' it was the result of the dissatisfaction I experienced when confronted with the prevalent interpretation of India. This was precipitated by my working with a US-based company, A.T. Kearney, in India. I sought to reverse that interpretation with 'India's Contribution to Management' which aimed to capture my understanding of the essence and deeper capacity of synthesis of India.

The next phase was marked by a strong interest in fractals that primarily stemmed from my beginning to see similar patterns in seemingly distinct areas of life. I wrote 'Connecting Inner Power with Global Change: The Fractal Ladder' as a theoretical framework of fractals. The fractals that I envisioned included the added complexities of emotional and thought components. This was followed by 'Redesigning the Stock Market: A Fractal Approach' which was an application of the theoretical fractal framework to the then recent global financial crises of 2008. 'The Flower Chronicles' sought to make the gist of the ubiquitous fractal I had described in the previous two books easily graspable at the visceral level primarily through many practical examples spanning diverse walks of life. 'The Fractal Organization: Creating Enterprises of Tomorrow' was a comprehensive summary of the fractal framework that included the basic theory, the applications, and a practical field guide that had been developed while I was working at the Organizational Development group at Stanford University Medical Center.

The most recent phase has focused on creating a mathematical framework to take the previously developed fractal framework further. The development of such a mathematical framework that seeks to frame innovation in complex adaptive systems was also the focus of my doctoral work. This gave birth to an exciting period and will result in multiple series of books.

The first series, comprising of six books, extended my

inquiry into mathematics and complex adaptive systems to an interesting limit culminating in the nature of Light and the Cosmos. The fractal mathematics I propose is at the heart of this series: Cosmology of Light.

The first book, 'A Story of Light: A Simple Exploration of the Creation and Dynamics of this Universe and Others' contains the main ideas in the mathematics, in non-mathematical terms, that are further explored mathematically in the remaining books in this series. The second book 'Oceans of Innovation: The Mathematical Heart of Complex Systems' describes my interpretation of the mathematical foundation of complex systems. The third book, 'Emergence: A Mathematical Journey from the Big Bang to Sustainable Global Civilization' applies the mathematics to several layers of matter and life. The fourth book, 'Quantum Certainty: A Mathematics of Natural and Sustainable Human History' describes a process culminating in space, time, energy, and gravity quantization by which history is made. The fifth book, 'Super-Matter: Functional Richness in an Expanding Universe' describes a process for the creation of super-matter-based on a need for continued functional-richness. A link is made between the resulting quantization of space and an expanding universe. The final book, 'Cosmology of Light: A Mathematical Integration of Matter, Life, History & Civilization, Universe, and Self' proposes an integrated mathematical framework that flows through all things, hence unifying matter, light, civilization, history, universe, and self.

The second series further explores the implications of "one mathematics flowing through all things". The first book in this series 'The Emperor's Quantum Computer: An Alternative Light-Centered Interpretation of Quanta, Superposition, Entanglement and the Computing that Arises from it' describes an alternative narrative for quantum computing to the one commonly expressed today. The second book in the series, 'The Origin and Possibilities of Genetics: A Mathematical Exploration in a Cosmology of Light' explores pre-genetic, genetic, and post-genetic possibilities in a cosmology of light. This book, 'The Second Singularity: A Mathematical Exploration of AI-Based and Other Singularities in a Cosmology of Light' explores the limits of AI-based singularities with respect to light-based singularities. This book, the final in this series explores transhumanism in a cosmology of light.

The third series, Artistic Interpretation of Cosmology of Light, is intended to make Cosmology of Light more accessible by interpreting it artistically. The first book in the series is 'The Mandala Illustrated Story of Light'. The objective is to lead the reader through the story of light using mandalas as an aid in the journey. The second book 'Musings on Light' is a meditative book set to graphical illustrations. The illustrations focus on 50 key concepts derived from the ten-book joint Cosmology of Light series. The third book 'The Illustrated Oceans of Innovation: The Mathematical Heart of Complex Systems Depicted in Indian Arts' uses Indian Arts to illustrate the mathematical heart of complex systems. The fourth book 'Emergence Illustrated: A Mathematical Journey from the Big Bang to Sustainable Global Civilization Depicted with Indian Mythological Arts' uses Indian mythological arts to express the concepts of Emergence. The fifth book 'The Dawn of Flame-Beings: Mythological Musings Based on a Cosmology of Light' uses illustrations to help focus on the mythological aspects of a cosmology of light. The sixth book is 'Quantum Certainty Illustrated: A Mathematical Journey of Natural and Sustainable Human History Depicted with Art' that leverages pencil drawings and other art to shed insight into the mathematical development of history. The seventh book, 'Super-Matter Illustrated: Functional Richness in an Expanding Universe Depicted with Drawings,' suggests the destiny of matter and its implicit relationship with an expanding universe. The pencil drawings provide visceral insight into the concepts and mathematics in the book. The eighth book is 'Cosmology of Light Illustrated: A Mathematical Integration of Matter, Life, History Depicted with Art'. The ninth book is 'The Emperor's Quantum Computer Illustrated: An Alternative Light-Centered Interpretation of Quanta, Superposition, and Entanglement and the Computing that Arises from it Illustrated with Art'. The tenth is 'The Origins and Possibilities of Genetics Illustrated: A Mathematical Exploration in a Cosmology of Light. The eleventh and current book is 'The Second Singularity Illustrated: A Mathematical Exploration of AI-Based and Other Singularities in a Cosmology of Light Enhanced with Art.'

About the Author

Dr. Pravir Malik is the founder and chief technologist at QIQuantum and the Forbes Technology Council group leader for Quantum Computing.

He is a systems thinker and approaches computation at the quantum level by seeing the layers of matter and life as a single system. This perception has resulted in the articulation of a fractal model connecting patterns at the cellular, atomic, and quantum particle levels to root patterns at the quantum level and has been elaborated through a ten-volume mathematical treatise on a cosmology of light, and numerous IEEE technical articles.

Dr. Malik has a patent pending on a fourfold fractal-based quantum computer. Dr. Malik's quantum computational theory derives from a unified theory and mathematics of organization developed over a couple of decades. In all, he has penned over 25 books related to this. In his doctoral dissertation on the mathematics of innovation in

595

complex adaptive systems, he created a top-down in contrast to the already existing bottom-up mathematics for complex systems. Beyond its application in quantum systems, this mathematics has been practically applied to the development of pricing systems, emotional intelligence systems, and AI, and theoretically applied to the areas of genetics, pharmaceutical technology, nanotechnology, space sciences, and computational theory.

Dr. Malik has funded activities at QIQuantum through modeling behavioral, organizational, and other complex systems to help different stakeholders practically navigate and understand possible futures. In his capacity as the Forbes Technology Council group leader for Quantum Computing, he has led numerous quantum computing events focused on alternative trajectories for quantum computation. The group that he helped found earlier in 2023 has now grown to over 140 executives belonging to Forbes Council member companies.

During the pandemic, he led a multi-part Organizational Sciences Certification program for Forbes, which was attended by executives from 250 companies. This certification was based on his views of complex systems, cosmology of light, and the importance of maturing teams through emotional intelligence. Dr. Malik believes that each individual is endowed with a sophisticated EQ radar-system that can give profound insight into how complex systems are changing. He developed a practical EQ-based App that has been used by the Indian Armed Forces, Zappos, Stanford University Medical Center, and Nucor, amongst other organizations, to help teams reach higher levels of self-directed operational maturity.

Previously, Dr. Malik has held a number of diverse leadership positions. These include being Head of Organizational Sciences and Head of Pricing Operations at Zappos.com - an eCommerce company, Managing Director of BSR - a global sustainability and environmental consulting company, and a member of the Founding Team of A.T. Kearney India - a global operations-focused management consulting company.

He has a Ph.D. in Technology Management with a focus on Mathematics of Innovation in Complex Adaptive Systems from the University of Pretoria, an MBA from Northwestern University's J.L. Kellogg Graduate School of Management, an MS in Computer Science from the University of Florida with a focus on AI, and a BSE in Computer Engineering from the Case Western Reserve University.

Pravir is a global citizen who has worked, lived, and been educated in many parts of the world.

About the Illustrator

Dr. Narendra has three decades of extensive experience in academics, industry, projects, grassroot work on social and cultural issues, editing magazines and journal & his recent passion has been in multidisciplinary research and its artistic expressions.

He has completed his graduation and post-graduation in Production engineering from University of Mumbai and ranked first in his Master's program. He completed several technology, management and entrepreneurship certifications from reputed institutes like IIT Bombay, IIM Ahmedabad, ISB Hyderabad, NEN-STVP and IIM B. He completed his Ph.D. from Hindu University of America. His thesis was on an unusual subject: 'A study of philosophy and Futurology of Artificial Intelligence in the light of Sri Aurobindo's integral thought.'

Inspired by the message of Swami Vivekananda and Sri Aurobindo for future of India and humanity and intrigued by the shift in foundations of Physics and the future shock, he switched over from his industrial career and worked as full-time dedicated worker for a leading NGO, Vivekananda Kendra, Kanyakumari and was posted in Northeast India in various remote areas of Assam and Arunachal Pradesh after extensive training in Kanyakumari for one year. Perplexed by the complex socioeconomic developmental especially cultural and human problems there, he started studying works of Sri Aurobindo more deeply. He was given a research project which was later published as a book named 'Ashwattha' by Vivekananda Kendra. This book has hundreds of illustrations by him and since then he has written many papers for conferences, journals, and magazines on diverse topics from technology, whole brain thinking, creativity and innovation, management, Indian arts, consciousness studies, history, evolution of consciousness, appropriate technology and sustainability, science and technology in Ancient India, Indian ethos in management, Indian philosophy, psychology and Sri Aurobindo studies. His paper on Indian ethos in management won the award from BMA for best paper for the year. He has served in industry and in academics holding various key positions including principal, project head, academic consultant, etc. He was benefitted in this journey by association and guidance from many experts.

Dr. Narendra has worked as Principal for eight years and faculty for nine years in engineering fields and he is proactive in assessing training & development needs and effectively aligning programs / interventions with business / social objectives. He is deft in designing innovative programs for industry as well as academics while catering to emerging industry needs. He has a distinguished inclination to social responsibility

598

exemplified in interdisciplinary research projects, field work and development of innovative projects dedicated to integral and futuristic studies and also for discovering the underlying currents of culture and psychology in it. At present he is working as Project Director for Vivekananda Prabodhini, Mumbai. He also provides Academic consultancy especially for distant, digital and open university models wherein students from underprivileged and remote areas have urge to learn further.

However more than all these things, he is an artist by heart and has developed passion for sketching, painting in traditional as well as digital art over the years. His sketches are now part of several magazines, blogs, books and forums. He has found that illustrating a text helps in deeper thought expression and provides further insights especially for path breaking and transdisciplinary subjects such as this book.

Selected Author Online Presence

- Amazon Author Page: https://www.amazon.com/Pravir-Malik/e/B002JVAEZE
- LinkedIn Profile: https://www.linkedin.com/in/pravirmalik/
- Forbes Profile: https://councils.forbes.com/profile/Pravir-Malik-Chief-Technologist-QIQuantum/2258eb58-6350-495f-8d26-16581994fcc4
- Google Scholar Page: https://scholar.google.com/citations?user=7DWWWZ8AAAAJ&hl=en
- IEEE Profile: https://ieeexplore.ieee.org/author/37086022058

- YouTube Page: https://www.youtube.com/user/Aurosoorya
- Twitter: https://twitter.com/PravirMalik
- Research Gate Profile: https://www.researchgate.net/profile/Pravir_Malik
- Eventbrite: https://www.eventbrite.com/o/pravir-malik-30159112262
- Medium: https://medium.com/@PravirMalik
- Company website: http://www.deepordertechnologies.com/

REFERENCES

1. Amato, I. 1991. "Speculating in Precious Computronium". Science. https://web.archive.org/web/20040622210059/http://leitl.org/amato.pdf

2. Arabatzis, T. 2006. Representing Electrons: A Biological Approach to Theoretical Entities. University of Chicago. Chicago

3. Aylward. K. 2014. http://www.kevinaylward.co.uk/qm/ensemble_interpretation.html

4. Beinhocker, E. 2006. The Origin of Wealth. Boston: Harvard Business School Press.

5. Carse, J. 1986. Finite and Infinite Games. Free Press: New York.

6. Chown, M. 1990. Can Photons Travel Faster Than Light? New Scientist 126(1711)

7. Cottingham, W & Greenwood, D. 2007. An Introduction to the Standard Model of Particle Physics. Cambridge University Press. Cambridge

8. Deppe, A. 2013. Therapy with Light, a Practitioner's Guide. Strategic Book Publishing.

9. Diamond, J. 2005. Collapse: How Societies Choose to Fail or Succeed. Viking Books: New York

10. Einstein, A. 1995. Relativity: The Special and General Theory. New York: Broadway Books.

11. Faye, J. 2019. "Copenhagen Interpretation of Quantum Mechanics". The Stanford Encyclopedia of Philosophy. Spring 2019.

12. Feynman, RP. 1985. QED The Strange Theory of Light and Matter. New Jersey: Princeton University Press

13. Fuller, B. 1982. Synergetics: Explorations in the Geometry of Thinking. MacMillan Publishing Co.: New York

14. Goldstein, S. 2017. "Bohmian Mechanics", *The Stanford Encyclopedia of Philosophy* (Summer 2017 Edition), Edward N. Zalta (ed.), URL = <https://plato.stanford.edu/archives/sum2017/entries/qm-bohm/>.

15. Goodsell, David. 2010. The Machinery of Life. New York: Springer

16. Gottlieb, M. 2013. The Feynman Lectures on Physics: III. California Institute of Technology. http://www.feynmanlectures.caltech.edu/III_16.html

17. Gray, T. 2009. The Elements: A Visual Exploration of Every Known Atom in the Universe. Black Dog & Levental Publishers. New York.

18. Gubser, S. 2010. The Little Book of String Theory. Princeton University Press

19. Harvard-Edu. 2014. http://news.harvard.edu/gazette/story/2014/04/harvard-to-sign-on-to-united-nations-supported-principles-for-responsible-investment/

20. Harvard-Smithsonian Center for Astrophysics. 2004. https://www.cfa.harvard.edu/seuforum/de_whatmight.htm#

21. Hawking, Stephen. 1988. A Brief History of Time. New York: Bantam Books

22. Heiserman, D. 1991. Exploring Chemical Elements and their Compounds. McGraw-Hill. New York.

23. Holland, P. 1995. The Quantum Theory of Motion: An Account of the de Broglie-Bohm Causal Interpretation of Quantum Mechanics. Cambridge: Cambridge University Press.

24. Hope, Chris; Fowler, Stephen J. (2007). "A Critical Review of Sustainable Business Indices and Their Impact". Journal of Business Ethics. Vol. 76. S. 243–252. Springer: New York

25. Hyperphysics. 2016. Department of Physics and Astronomy. Georgia State University. http://hyperphysics.phy-astr.gsu.edu/hbase/mod3.html#c1

26. Isaacson, W. 2008. Einstein: His Life and Universe. Simon and Schuster. New York.

27. Jeans, J. 1932. The Mysterious Universe. Cambridge University Press.

28. Kaufmann, S. 1995. At Home in the Universe. New York: Oxford University Press.

29. Kaufmann, S. 2003. The Adjacent Possible, Edge. https://www.edge.org/conversation/the-adjacent-possible

30. Lloyd, S. 2007. Programming the Universe: A Quantum Computer Scientist Takes On the Cosmos. New York: Vintage

31. Logue, A. 2008. Socially Responsible Investing for Dummies. Wiley Publishing: New Jersey

32. Lorentz, H.A. 1925. The Science of Nature. Vol. 25, p 1008. Springer

33. Malik, P. 2009. Connecting Inner Power with Global Change: The Fractal Ladder. New Delhi: Sage Publications

34. Malik, P. 2015. The Fractal Organization. New Delhi: Sage

35. Malik, P. 2017a. Doctoral Thesis. University of Pretoria Graduate School of Technology Management. https://repository.up.ac.za/handle/2263/62779?show=full

36. Malik, P. 2017b. A Story of Light. Amazon Kindle.

37. Malik, P. 2017c. Oceans of Innovation. Amazon Kindle.

38. Malik, P. 2017d. Emergence. Amazon Kindle.

39. Malik, P. 2017e. Quantum Certainty. Amazon Kindle.

40. Malik, P, Pretorius, L, Winzker, D. 2017.
Qualified Determinism in Emergent-Technology
Complex Adaptive Systems. IEEE TEMSCON.
41. Malik, P. 2018a. Super-Matter. Amazon
Kindle.
42. Malik, P. 2018b. Cosmology of Light. Amazon
Kindle.
43. Malik, P. 2018c. The Emperor's Quantum
Computer. Amazon Kindle.
44. Malik, P. 2019a. The Origins and Possibilities of
Genetics. Amazon Kindle.
45. Malik, P, Pretorius, L. 2018. Symmetries of
Light and Emergence of Matter. Indian Journal
of Science & Technology.
http://www.indjst.org/index.php/indjst/articl
e/view/110789.
46. Martel, A. 2018. Light Therapies: A Complete
Guide to the Healing Power of Light. Rochester,
Vermont: Healing Arts Press
47. Mora, C, Tittensor, D, Adl, S, Simpson, A,
Worm, B. 2011. How Many Species Are There on
Earth and in the Ocean? Plos.org.
http://journals.plos.org/plosbiology/article?id
=10.1371/journal.pbio.1001127#abstract0
48. Narby, J. 1998. The Cosmic Serpent: DNA and
the Origins of Knowledge. New York: Penguin
Putnam.
49. NASA-darkmatter.   2016.
http://science.nasa.gov/astrophysics/focus-
areas/what-is-dark-energy/
50. NASA-supernova. 2001. News Release Number:
STScI-2001-09: "Blast from the Past: Farthest
Supernova Ever Seen Sheds Light on Dark
Universe".
http://hubblesite.org/newscenter/archive/rele
ases/2001/09/
51. NASA-WMAP,   2014.
https://map.gsfc.nasa.gov/media/121238/inde
x.html

52. Neven, H. 2013. Launching the Quantum Artificial Intelligence Lab. Google AI Blog. https://ai.googleblog.com/2013/05/launching-quantum-artificial.html

53. Nowak, M., Highfield, R. 2011. Super Cooperators: Altruism, Evolution, and Why We Need Each Other to Succeed. New York: Free Press.

54. Olive, K.A et al. 2014. Particle Data Group. Chin. Phys. C, 38, 090001.

55. Openshaw, J. 2015. http://www.marketwatch.com/story/socially-responsible-investing-has-beaten-the-sp-500-for-decades-2015-05-21

56. Panek,R. 2010. Dark Energy: The Biggest Mystery in the Universe. Smithsonian Magazine. http://www.smithsonianmag.com/science-nature/dark-energy-the-biggest-mystery-in-the-universe-9482130/?no-ist

57. Parker, A. 2003. In the Blink of an Eye: How Vision Sparked the Big Bang of Evolution. New York: Basic Books.

58. Particle Data Group. 2015. Lawrence Berkeley National Laboratory. http://www.cpepphysics.org/main_universe/universe.html

59. Pauli, W. 1964. Nobel Lectures, Physics 1942 – 1962. Elsevier Publishing Company. Amsterdam.

60. Perkowitz, S. 2011. Slow Light. London: Imperial College Press

61. Planck, M. 1933. Where is Science Going? Ox Bow Press. Connecticut.

62. Portugali, J., 2012. *Self-organization and the city*. New York: Springer Science & Business Media.

63. Prigogine, I. 1977. Time, Structure, and Fluctuations. *Nobelprize.org.* Nobel Media AB 2014. Web. 5 Mar 2016.

<http://www.nobelprize.org/nobel_prizes/che mistry/laureates/1977/prigogine-lecture.html>

64. Ridley, M. 1999. Genome: The Autobiography of a Species in 23 Chapters. New York: Harper Perennial.

65. Rovelli, C. Reality Is Not What It Seems. New York: Riverhead Books

66. Snyder, M. 2010. Stanford Medicine. http://med.stanford.edu/news/all-news/2010/03/what-makes-you-unique-not-genes-so-much-as-surrounding-sequences-says-stanford-study.html#.html

67. Spitler, H. 2011. The Syntonic Principle: In Relation to Health and Ocular Problems. Eugene, Or: Resource Publications.

68. Sri Aurobindo. 1971. Social and Political Thought. Sri Aurobindo Ashram Press: Pondicherry

69. Stewart, Ian. 2012. In Pursuit of the Unknown. Basic Books. New York.

70. Smith, J, Szathmary, E. 1995. The Major Transitions in Evolution. Oxford: Oxford University Press

71. Tegmark, M. 2017. Life 3.0: Being Human in the Age of Artificial Intelligence. Vintage Books: New York.

72. Toynbee, A. 1961. A Study of History, Volumes I – XII. Oxford University Press: Oxford

73. Turok, N. 2012. The Universe Within. Anasi Press: Ontario

74. Tweed, M. 2003. Essential Elements: Atoms, Quarks, and the Periodic Table. Walker & Copmany. New York.

75. Van Obbergen, P. 2014. Traite de Couleur Therapie Pratique. Paris: Guy Tredaniel Editeur.

76. Wheeler, J. Ford, K. 2000. Geons, Black Holes, and Quantum Foam: A Life in Physics. New York: W. W. Norton & Co.

77. Whitaker, A._2006.   Einstein, Bohr and the Quantum Dilemma: From Quantum Theory to Quantum Information. Cambridge: Cambridge University Press

78. Wilczek, F. 2016. A Beautiful Question: Finding Nature's Deep Design. New York: Penguin Books

79. Willis, A. 2003. The Role of the Global Reporting Initiative's Sustainability Reporting Guidelines in the Social Screening of Investment. Journal of Business Ethics. Volume 43. Springer: New York

80. Wimmel, H. 1992. *Quantum Physics & Observed Reality: A Critical Interpretation of Quantum Mechanics*. World Scientific. p. 2. Bibcode:1992qpor.book.....W. ISBN 978-981-02-1010-6.

81. Wolchover, N. 2013. A Jewel at the Heart of Quantum Physics. Quanta Magazine . https://www.quantamagazine.org/20130917-a-jewel-at-the-heart-of-quantum-physics/

82. Wright, R. 2009. Evolution of Compassion. https://www.ted.com/talks/robert_wright_the_evolution_of_compassion/transcript?language=en. TED 2009.

83. Yates, F.E. 2012. *Self-organizing systems: The emergence of order*. New York: Springer Science & Business Media.

84. Young, E. 2016.   I Contain Multitudes: The Microbes Within Us and a Grander View of Life. New York: HarperCollins.

www.ingramcontent.com/pod-product-compliance
Lightning Source LLC
Chambersburg PA
CBHW041932220326
41598CB00055BA/5